The Psychological War for Vietnam, 1960–1968

The Psychological War for Vietnam, 1960–1968

Mervyn Edwin Roberts III

University Press of Kansas

Published by the University Press of Kansas (Lawrence, Kansas 66045), which was organized by the Kansas Board of Regents and is operated and funded by Emporia State University, Fort Hays State University, Kansas State University, Pittsburg State University, the University of Kansas, and Wichita State University.

Library of Congress Cataloging-in-Publication Data

Names: Roberts, Mervyn Edwin, III, author.

Title: The psychological war for Vietnam, 1960–1968 / Mervyn Edwin Roberts III.

Description: Lawrence, KS : University Press of Kansas, [2018]

| Includes bibliographical references and index.

Identifiers: LCCN 2017052562| ISBN 9780700625833 (cloth : alk. paper)

| ISBN 9780700625840 (ebook)

Subjects: LCSH: Vietnam War, 1961–1975—Psychological aspects.

| Vietnam War, 1961–1975—Campaigns. | Psychological warfare—United States—History—20th century. | Propaganda, American—Vietnam—History—20th century.

Classification: LCC DS559.8.P7 R | DDC 959.704/34—dc23.

LC record available at https://lccn.loc.gov/2017052562.

British Library Cataloguing-in-Publication Data is available.

Printed in the United States of America

10 9 8 7 6 5 4 3 2 1

The paper used in this publication is recycled and contains 30 percent postconsumer waste. It is acid free and meets the minimum requirements of the American National Standard for Permanence of Paper for Printed Library Materials Z39.48–1992.

Contents

Illustrations

Maps

Figures

Photographs

Acronyms and Abbreviations

AAIL	American Anti-Imperialist League
ACS	Air Commando Squadron
ACTIV	Army Concept Team in Vietnam
AFVN	Armed Forces Vietnam Network
agit-prop	agitation-propaganda
AP	Associated Press
APT	Armed Propaganda Team
ARCS	Air Resupply and Communications Service
ARVN	Army of the Republic of Vietnam (South Vietnam)
BJU-1	US Navy Beach Jumper Unit 1
CA	Civic Action or Civil Affairs
CANLF	Committee to Aid the NLF
Chieu Hoi	GVN's open arms program to induce defections
CIA	Central Intelligence Agency
CIAA	Coordinator of Inter-American Affairs
CIDG	Civilian Irregular Defense Group (South Vietnam)
CINCPAC	Commander in Chief in the Pacific
COI	Coordinator of Information
Comintern	[Soviet] Communist International
CORDS	Civil Operations and Revolutionary Development Support pacification program
COSVN	Central Office for South Vietnam (NLF headquarters)
CPI	Committee on Public Information
CTZ	Corps Tactical Zone
DARPA	Defense Advanced Research Projects Agency
DOD	US Department of Defense
DRV	Democratic Republic of Vietnam (North Vietnam)
FBIS	Foreign Broadcast Information Service
GPWD	General Political Warfare Department (South Vietnam)
GVN	Government of [South] Vietnam, 1964–1966
HES	Hamlet Evaluation System

ICP	Indochinese Communist Party (Dong Cong San Dong Duong)
ICS	Information and Censorship Section
JCS	Joint Chiefs of Staff
JUSPAO	Joint US Public Affairs Office
KMT	Chinese Kuomintang
Krestintern	[Soviet] Peasant International
LBS	Liberation Broadcast Service
LLDB	Luc Luong Dac Biet—South Vietnamese Special Forces
LPA	Liberation Press Agency
MAAGV	Military Assistance and Advisory Group, Vietnam
MACPD	MACV Psychological Operations Directorate
MACV	Military Assistance Command, Vietnam
MACV-SOG	Military Assistance Command, Vietnam–Studies and Observations Group
MAF	Marine Amphibious Force
MEDCAP	Medical Civic Action Program
MI2	Military Intelligence 2 Section
MO	Morale Operations Branch (OSS)
MTT	Mobile Training Team
NCOs	noncommissioned officers
NLF	National Liberation Front (National Front for the Liberation of South Vietnam)
NSAM	National Security Action Memorandum
NSC	National Security Council
OPC	Office of Policy Coordination
OSS	Office of Strategic Services
OWI	[US] Office of War Information
PATs	People's Action Teams
PAVN	People's Army of Vietnam (North Vietnam)
PBR	Patrol Boat, River
PDC	Propaganda Development Center
PLAF	People's Liberation Armed Forces (the VC military force)
POET	Psychological Operations Exploitation Team
POLWAR	political warfare (South Vietnam)
PORT	Psychological Operation Reaction Team
PRG	Provisional Revolutionary Government of the Republic of South Vietnam
PSYOPs	psychological operations
PSYWAR	psychological warfare
PWB	[US] Psychological Warfare Branch

PWD/SHAEF	Psychological Warfare Division/Supreme Headquarters Allied Expeditionary Force
RB&L	Radio Broadcasting and Leaflet
RVN	Republic of [South] Vietnam (1954–1963, 1967–1975)
RVNAF	Republic of Vietnam Armed Forces
SACSA	Special Assistant for Counterinsurgency and Special Activities
SORO	Special Operations Research Office
SOW	Special Operations Wing
SSPL	Sacred Sword of the Patriots League
TIRS	Terrorist Incident Reporting System
TO&E	table of organization and equipment
UPI	United Press International
USAB&VAPAC	US Army Broadcast and Visual Activity Pacific
USAID	US Agency for International Development
USIA	US Information Agency
USIS	US Information Service
USOM	US Operations Mission (American embassy in Saigon)
VC	Vietcong
VCII	Vietcong Initiated Incidents reporting system
Vietminh	League for the Independence of Vietnam (Viet Nam Doc Lap Dong Minh Hoi)
VIS	Vietnamese Information Service (South Vietnam)

Acknowledgments

Most importantly, I wish to thank my wife, Gail, for her support, patience, and encouragement throughout the long process of research and writing. I am grateful as well to Robert Citino for providing useful advice at key points along the way. Geoffrey Wawro and Michael Leggiere made research trips possible through the assistance of the UNT Military History Institute. Nancy Stockdale provided constant encouragement to me on this and other projects, as did Richard McCaslin. This project would never have come about without the support of Adrian Lewis. Amy Mondt at the Texas Tech Vietnam Archives, and the staffs of the Lyndon B. Johnson Library, Richard M. Nixon Library, the US Army Special Operations Command Archives, and the US Army Heritage and Education Center were all very helpful in finding materials. Dong Ha and Terry Ferguson were both patient sounding boards for ideas.

Throughout my military career our company, battalion, and later group commander Jeffrey Scott taught me much of what I know about PSYOPs and provided advice on this project. Blake Whitaker, Frank Johnson, Michael Barger, and Andrew Whiskeyman each provided useful comments and support. Herb Friedman at the *Psywarrior* website assisted throughout this and other projects. And finally: to all my fellow PSYOP soldiers I owe much credit.

The Psychological War for Vietnam, 1960–1968

Introduction

It is early morning on 8 December 1966, the bright dawn of a relatively cool, dry day in South Vietnam. A battalion from the United States 1st Infantry Division surrounds the village of Chanh Luu, 30 kilometers north of Saigon. The troops are part of Operation Fairfax, General William Westmoreland's push to clear Communist guerrillas of the National Front for the Liberation of South Vietnam (National Liberation Front, NLF, the Front) from the area surrounding the capital. Soldiers painstakingly comb the red dirt road in search of mines as they approach the village. At 0800 hours, mobile field teams of the 246th Psychological Operations Company carrying loudspeakers, food, and psychological operations (PSYOPs) posters and leaflets enter the village, which until recently had been controlled by Vietcong. The children of the village eagerly swarm around as the Americans distribute candy.

Meanwhile, work begins to remove all traces of enemy propaganda images from the village. In their place are fresh, colorful posters. Troops paint a progovernment slogan on the wall of the only stone building in Chanh Luu. A wizened grandmother crowds the team to get a specially designed calendar as soldiers make the rounds of the village shops. In connection with the mission, male villagers of fighting age are assembled for intelligence interviews. There are many suspected Vietcong in this village.

Suddenly . . . *crack!* A few villagers open fire when a combined team of South Vietnamese and American soldiers attempt to round them up. The situation is quickly brought under control, leaving nine enemy dead and several wounded. American casualties are light.[1]

This small firefight—lasting all of five minutes—was a minor action in a very long war. Yet it marked a shift to what became the most sustained, intensive use of psychological operations in American history. The United States Army launched numerous PSYOP programs and distributed as many as 50 billion leaflets in an area the size of California during the course of the war. And yet it remains a mostly unexplored area of American history.

Propaganda and the War

What brought these men to the remote village of Chanh Luu on that bright Tuesday morning? I seek to answer this by analyzing psychological operations in Vietnam. Focusing on the period 1960–1968, this book analyzes the development of PSYOP capabilities, the introduction of forces, and the decisions that created the Joint US Public Affairs Office (JUSPAO) and the 4th PSYOP Group to manage American PSYOP activities in the war. Rather than pursuing an isolated examination, I place these operations within the wider context of Vietnam and the Cold War propaganda battle that the United States fought simultaneously. We will examine such operations across the entire theater, by all involved US agencies. This book addresses the complex interplay of these activities, as well as the development of PSYOP doctrine and training in the period prior to the introduction of ground combat forces in 1965. Finally, it will show how the course of the war itself forced changes to US doctrine.

Although this is primarily a look at the American effort, PSYOPs never occur in a vacuum. The enemy and friendly forces, as well as civilians, all impact the information environment. This forces a constant reevaluation of programs and a shift in focus. Without setting these operations within the context of the wider war, American PSYOPs make little sense. As such, this study also discusses the development of alternative sources of information, especially from the governments of North and South Vietnam and, to a lesser extent, those of Australia, South Korea, and the Philippines.

Due to the ebb and flow of a long war like Vietnam, many historians have made the mistake of concentrating on this or that short period, often centered on 1968. Or they err in declaring the war lost by 1965 when clearly it was not. In the end, war unfolds in the realm of chance. Understanding a long war like Vietnam requires historians to take a long view. Perseverance on both sides is vital in a desperate struggle such as the Vietnam War. The last side standing, no matter how bloodied and battered, no matter how many mistakes were previously made, is the winner. This truism is particularly true of psychological war, which unfolded over many years with numerous twists and turns.[2]

The debate rages to this day over what to even call the various combatants in this war, such was the degree to which propaganda infused all aspects of the conflict. Participants assumed names for themselves with an eye toward influencing audiences. Similarly, they applied pejorative labels to the enemy, such as "puppet" or "bandit," to diminish support. While the South Vietnamese government styled itself as a "democracy," it was no such thing during the period in question. It might have evolved into one, perhaps similar to South

Korea. However, events halted its evolution. As such, this study will avoid using the terms *free* or *democratic government* to describe the Republic of South Vietnam.

A similar debate rages over the antigovernment insurgency. The NLF was a creation of the Communist government in North Vietnam. It used the so-called united front tactic developed by V. I. Lenin and refined by Ho Chi Minh to create mass organizations that could magnify the power of small, dedicated Marxist groups seeking power. Inherent in this tactic was keeping the Communist core hidden behind a shroud of seemingly legitimate mass organizations. The NLF was a propaganda weapon for the North in this war.

In recent years, it has become common in Western academic circles to shun the term *Vietcong* as a presumed pejorative. The debate over such labels, like much in the study of the Vietnam War, arose out of the propaganda efforts by both sides. The Vietminh, the original resistance movement, is a contraction of Viet Nam Doc Lap Dong Minh—the League for the Independence of Vietnam. This was the front created by Ho Chi Minh in 1941 to resist the Japanese occupation of Indochina, and the term remained in fashion after the 1954 Geneva Accords on Vietnam's temporary division and France's withdrawal to the South. The Vietnamese Communist Party (Viet Nam Cong-san Dang) was formed in October 1930. The term *Viet Cong* (rendered as *Vietcong* in modern usage) appeared in Saigon newspapers beginning in 1956 as a contraction of Viet Nam Cong-san. By using this, South Vietnamese and American officials attempted to delegitimize the original movement by identifying it clearly as a Communist front. The *Cong* is the same as in Vietnam Cong Hoa, or the Republic of Vietnam, and holds no inherent negative undertone.[3] The negative aspect arises when it is used to describe Communists who are trying to hide their connection to the NLF. Thus, the negative perception is associated with an attempt to keep the truth hidden. To be precise, the term has no negative connotation, unless one has a negative view of Communists.

In contrast to the NLF, the term *Vietcong* is more descriptive of the organization. In 1975, the correctness of the term became manifest when the North Vietnamese won the war and immediately changed the country's name to the Socialist Republic of Vietnam. Without delay they disbanded the NLF, arresting some of the members who had been duped into joining. Members of the NLF sacrificed their lives, liberty, and honor for an organization that acted simply as camouflage. As Truong Nhu Tang, a founding member of the NLF and minister of justice for the Provisional Revolutionary Government of South Vietnam (PRG), later wrote, his betrayed comrades "believed they were sacrificing themselves for the humane liberation of their people."[4] Truong himself fled Vietnam when the truth became clear. Thus, despite its propaganda origins, the term *Vietcong* was more descriptive of what the force was. It is also the term for the enemy more commonly understood by Americans. As such, this study primarily uses *Vietcong* to describe the organization, although it does use the NLF title as well.

This is a study for English speakers. For ease of reading Vietnamese words, no diacritics or special characters are used. Most place-names are written as known to Americans. For instance, this study uses *Vietnam*, rather than the more correct *Viet Nam*. Some quoted material does not follow this rule. When in quotes, all terms are as originally written, so both styles of Anglicization will appear in this study. Similarly, most period documents used the Wade-Giles system for Chinese words, so this study mostly uses that system.

Evolution of Psychological Operations

Throughout the years, terminology related to psychological operations has shifted meaning and connotations. Buzzwords for each generation further confuse the issue. Thus, prior to an investigation into the historical use of PSYOPs by the United States, we need a statement of definitions and an introduction to the arcane terminology of psychological operations as it was then understood.

The term *propaganda* is derived from the Congregatio de Propaganda Fide (Congregation for the Propagation of the Faith) founded by Pope Gregory XV in 1622. He formed this group to spread Christianity to other nations. Until World War I, the term maintained a benign connotation of spreading information. This began to modify due to the use of discredited atrocity propaganda during that war. By the end of World War II, the term had fallen into general disrepute in America, but it still retained its benign meaning in military doctrine.[5]

The phrase *psychological warfare* (PSYWAR) came into vogue after 1940 to describe wartime propaganda. However, in the Cold War era, the less hostile-sounding *psychological operations* gained popularity in America. As then described in US Army doctrine, PSYOPs encompassed "those political, military, economic, and ideological actions planned and conducted" to influence foreign citizens. This included PSYWAR, propaganda, and psychological actions, referring to military activities performed for their planned psychological value. By 1962, PSYWAR was narrowly defined as PSYOPs directed against hostile nations or armies.[6]

This is not, strictly speaking, a study in communications theory. It is an attempt to provide a detailed historical framework for understanding the psychological war in Vietnam. From a communications perspective, future researchers may want to dig into the material using such theories as reasoned action and social learning to better analyze what programs were effective and why. For this study, I primarily work through the application of cultural narratives.

All societies have fundamental narratives or myths that enable members to communicate. Humans often understand the world through these stories. A key tactic in propaganda is to frame an argument within preexisting national narratives understood by the audience. As long as possible, people will hold on to that narrative as the propagandist inserts contrary messages. As the narrative reaches a tipping point, the old dogmas are undone and not likely to be responded to. Related to this is the concept of the filter bubble. People seek messages that agree with their worldview, creating a challenge for those involved in PSYOPs to overcome. The North Vietnamese often used the direct method, attacking the narrative head-on. As noted below, this was not successful over the long term. As Benjamin Franklin found, subtle propaganda that leads the target away from his narrative to that preferred by the propagandist is often more effective. Much of propaganda attempts to create cognitive dissonance in the target audience. Deeply held beliefs are difficult to change, especially once emotions become involved. The propagandist needs to attack these beliefs indirectly or the target will build a defensive wall. PSYOPs focus on changing behaviors in the short term, but some beliefs must change to accomplish long-term shifts. By carefully presenting facts that force the target to call into question his current understanding, the target becomes susceptible to an alternative narrative.[7]

Along with this, message reinforcement and confirmation through multiple media and high-status individuals help drive home the internal questioning. One form of this process is laundering information through third parties. The North Vietnamese were skilled at this process, making an announcement, having it repeated by friendly foreign parties, then replaying that story as fresh news or proof of the facts.[8]

Another propaganda tactic at which the North Vietnamese excelled was agitation-propaganda (agit-prop). This system consisted of two mutually reinforcing actions. Propaganda in this case refers to the detailed explanation of party views on a given topic. These are provided to party leaders so that they can use this line in agitating the population. Agitation typically consisted of slogans geared at raising anger among the target audience, thereby short-circuiting rational thought. Closely aligned with this is critical theory, a concept developed by German Marxists in the 1920s. Critical theory argues that the way to foment successful revolution is to criticize all aspects of the society. Such agitation will eventually strike at a target's individual grievance, opening avenues for agitation and agit-prop. The goal is to identify a grievance and follow this seven-step chain:

> Dissatisfaction → agitation → resentment → slogans → mobilization → creation of conflict → rise from ashes with a new society

Repressive tolerance is a tool within agitation. As Herbert Marcuse described it, *tolerance* is simply a code word for those ideas the propagandist approves of. All other ideas must be repressed through ridicule or force. Agitation aims to short-circuit thinking by the target by encouraging members to actively take part in and internalize the repression of counter ideas. This can be very effective, especially in functioning civil societies at peace. However, societies at war, such as Vietnam in the 1960s, seem to be less susceptible. Perhaps because war, as opposed to political conflict, works at a baser level of life and death, agitation of minor grievances is less effective.[9] In a study of the radical organizer Saul Alinsky of Chicago, Hillary Rodham (Clinton) described the use of agitation in America. She noted that Alinsky saw conflict as "the route to power" because agitation polarized the masses, opening them to exploitation.[10] Rodham quoted Alinsky as arguing that the role of the propagandist/organizer was to serve "as an abrasive agent to rub raw the resentments of the people of the community." This is representative of the agit-prop system used by both Ho Chi Minh and the American antiwar protest movement during this period.[11]

Contrary to popular myth, the psychological war was not simply about winning hearts and minds. What matters is behavior—not whether an audience likes you. Behavior that supports your military and political objectives is the goal. Attitudinal change may be nice and, over a multiyear period, may enhance cognitive dissonance that supports behavioral change. However, in wartime the priority is behavior change. As was found in Afghanistan during the US war there, focus on marketing models for propaganda may lead to situations in which the target audience agrees with your message but acts in contrast to it due to other influences taking precedence. Market-share increase or positive branding may work in a peaceful setting. However, there is not a precise analogy between choosing a cola and choosing to risk one's life to call a tipline. PSYOPs share more in common with the rough-and-tumble of an election campaign or a police investigation. Clouded understanding of PSYOPs often leads operators to focus on the wrong messages.[12]

Another myth about propaganda is that it primarily uses lies. Certainly, it is about subjective rather than objective truth, and lies may be as effective as the truth in the short run. However, in the long run, the truth will surface. The corrosive effect of lies on credibility must always remain in the forefront of the operator's mind. As shown below, the Vietcong and North Vietnamese often had nimble propaganda programs, largely based on subjective truth. But the lies and inability of words to match deeds diminished that program over the years. The same can be said of the American public relations operation under President Lyndon B. Johnson. Credibility, once lost during wartime, is nearly impossible to regain.

The mutual goal of tactical-, operational-, and strategic-level PSYOPs is to find weaknesses or vulnerabilities in the target audiences that make them susceptible to messages and thereby influence their behaviors. For instance, a group of soldiers cut off and low on food might be highly susceptible to a theme that offers lenient treatment and food for surrendering. The key task for PSYOP intelligence analysts is to identify a target audience that is reachable and susceptible to messaging. The audience selected must be accessible to the messages: it must be able to receive the radio broadcast, to retrieve leaflets, or to meet face-to-face. The target must also be susceptible to the messages. In other words, the target must be willing to listen to the specific argument, theme, or emotional appeal made and that can cause the desired change in behavior. Finally, the target audience must be able to carry out the change in behavior sought.[13]

To conduct effective psychological operations, it is necessary to determine clear psychological objectives, asking: What is the change desired? For each supporting objective, a specific target audience must be selected so that effective themes can be devised to appeal to specific circumstances. A high level of empathy is necessary to gauge how products may be viewed through the eyes of often incongruous target audiences. Inherent in such empathy is cultural understanding. This requires detailed intelligence and analysis of the so-called human terrain—that is to say, the culture and ethnic factors that affect the battlefield during an insurgency. A person from outside a group may have difficulty navigating the nuances of a culture (yet another lesson of the Vietnam War).

Desirable behaviors in the context of Vietnam might include increased support for the South Vietnamese government or decreased support for the Vietcong. The objective must be a measurable change in behavior by the target audience. From this basis, the psychological operator can derive discrete psychological objectives that will help achieve the larger objective. If the larger objective is to decrease support for the Vietcong, a supporting objective could be that the target audience provides tips on VC locations or that Vietcong personnel rally to the government side. Both are specific changes in behavior that support the goal of decreasing support for the VC and are measurable. There is no way to measure what is in a person's heart, but one can measure the number of calls they make to a tipline.

An example is targeting South Vietnamese families with members in the VC, using themes that good treatment awaits those who rally to the Saigon government. They are accessible via multiple media, may be susceptible to themes that will allow family members to return, and they can urge family members to rally. For the psychological operator, the effect of targeting this audience can be measured by correlating operations along this line in a specific area with so-called rallier numbers (i.e., those who rallied to the South Vietnamese government). This will always be an imperfect correlation, but it is an indicator in the analysis of operational effectiveness.

By contrast, persuading North Vietnamese civilians living in a totalitarian society to change government behavior is not a reasonable target. They might be targetable by leaflet or radio, or they may be susceptible to a message calling for the end of the war, but in a totalitarian nation they are powerless to effect change. Thus PSYOPs might make them sad while leading to little measurable change in the short term. Such activities could, however, be part of a longer-term campaign to foment discontent leading to revolution or internal strife.

In the US military, PSYOPs represent the primary military element specifically tasked to communicate with and influence foreign populations. Although other branches frequently encounter civilians, their attention is on specific tactical missions. Infantry focuses on closing with the enemy. Artillery units focus on indirect fire. Military intelligence is typically focused on gathering order-of-battle information (identifying and locating enemy conventional forces). All branches focus on security. In this context, most personnel in the field often see foreign civilians as an obstacle whom they must keep at arm's length. However, the primary mission of PSYOP personnel is to interact with civilians.

More so than other branches of the military, PSYOP units can operate at once on all levels of war. A tactical PSYOP team with a loudspeaker may be the only method of dissemination for a strategic message. Also, strategic PSYOP messages aimed at international audiences to build support for an action can quickly spread through the international press to affect the operational and tactical levels as stories are acquired by the local press. Thus, effective PSYOPs require coordination of messages to prevent information fratricide.

The weapons of the PSYOP unit include loudspeakers, leaflets and handbills, films, face-to-face talking points, and gifts. In PSYOP parlance, these are generically referred to as "product." By working with key local leaders, developing relationships, and utilizing product, tactical PSYOP teams hope to influence the target audience. Consistency of message, themes, and symbols aimed at informing and influencing behavior are essential for PSYOP effectiveness.

Beginning in the administration of President Dwight D. Eisenhower and continuing into those of John F. Kennedy, Lyndon B. Johnson, and Richard M.

Nixon, many in key leadership positions saw PSYOPs as an essential element in maintaining influence during the Cold War. Consequently, the United States developed civilian and military organizations to implement this. However, PSYOP doctrine was formulated without understanding the prerequisites for conducting psychological operations in a counterinsurgency environment such as Vietnam. Additionally, PSYOPs alone cannot achieve strategic objectives. At best, PSYOPs can assist with tactical changes in behavior and support long-term shifts in outlook among the targeted population. Policy makers occasionally failed to understand this.

Propaganda Development in Vietnam

North and South Vietnam simultaneously developed their own propaganda programs, the latter heavily influenced by the United States, the former by Ho Chi Minh's experience. Between 1960 and 1965, the United States had to improvise and adapt current PSYOP structures and doctrine to meet the Vietnamese insurgency challenge. The process was influenced greatly by Edward G. Lansdale, a proponent of psychological operations and longtime adviser on operations in Vietnam. In this opening phase, the United States acted primarily as an adviser to South Vietnamese programs. The Americans largely succeeded in this advisory role. During the rapid expansion of forces between 1965 and 1968, American PSYOP structures were tested constantly. As they assumed greater roles, PSYOP units struggled to meet the surge in demand. Individuals often had to learn on the job while adapting operations to a wide spectrum of missions and cultures. Although imperfect, they succeeded in many campaigns. At the same time, the South Vietnamese organization for political warfare (POLWAR) made great strides.

During the Vietnam War, the United States followed a doctrine that included strategic, tactical, and consolidation PSYOPs. Strategic PSYOPs aim to influence large segments of the target nation's population, using themes that "exploit economic, military, psychological, and political vulnerabilities" and that are "usually designed to reduce the effectiveness and internal control apparatus of the target government."[14] Taking place at the strategic level, the focus was more on broad themes and global target audiences. Often, the goal is to influence a relatively small number of actors, such as leaders, to behave in a way consistent with military goals. Tactical PSYOPs are focused on the immediate tactical situation. Moving civilians off the battlefield and encouraging enemy surrender are key tasks of tactical PSYOPs. At the operational level are such activities as assisting deception operations and, during the Vietnam War, so-called consolidation PSYOPs. Consolidation focused on the civilian population "with the objective of facilitating operations and promoting maximum cooperation among the civilian population." At the strategic and operational levels, radio broadcasts and targeting of international press became principal methods of dissemination during the Vietnam War.[15]

Within all levels of war, psychological operations may be conducted using what were then known as "white," "gray," and "black" messages. White, or overt, propaganda are messages clearly attributed to the element publishing the message. In other words, the author of the message is not hidden; Voice of America radio broadcasts is the classic example. Gray messages disseminate products with no discernible author, such as leaflets or posters without attribution. Black, or covert, propaganda is attributed to a party other than the one producing it. During the Vietnam War, white and gray PSYOPs were primarily conducted by the US Army and the Joint US Public Affairs Office, while the Central Intelligence Agency (CIA) and, later, the Military Assistance Command, Vietnam–Studies and Observations Group (MACV-SOG) were principally responsible for black PSYOPs, including deception operations aimed at causing the enemy to waste resources or to react. During the Vietnam War, the US Army had primacy for American military propaganda and the US Information Agency (USIA) was responsible for the civilian program.[16]

Despite the tremendous effort put forth during the war, very little was written about the PSYOP activities afterward. Although a wealth of primary source material exists covering PSYOPs in Vietnam, few secondary sources analyze it. Within this small universe, even less has been written by someone not directly involved in operations in Vietnam. Thus, little exists that is written by someone who can remain historically dispassionate while still understanding the arcane art of tactical PSYOPs. This study seeks to fill that gap. It argues that American and South Vietnamese PSYOPs were poised for success by 1965 and that, by 1968, had achieved a number of objectives. However, PSYOPs alone were not sufficient to win the war.

Collectively, most books about the Vietnam War, and specifically PSYOPs, resemble the story of the three blind men trying to identify an elephant by touching different parts. No overarching narrative or chronology exists of the PSYWAR and the complex interactions of operations by all parties. The studies may be factually correct, but still come to the wrong conclusions. Among this small body of work, there is no consensus yet on the extent of success, or the cause of failure, for the PSYOPs program in Vietnam. This is unusual because the Vietnam War arguably saw the most intensive use of psychological warfare in history. The common assumption seems to be that, because the North won the war, its program must have been more successful.

Due to politicization and positional loyalty among authors and academics regarding the Vietnam War, it is nearly impossible to derive a consensus opinion on most of its facets. Thus, although this study cites many reputable books on the war, I believe an accurate historical framework can be derived only by cutting though the propaganda substructure supporting many of these books. This study focuses on the war through the documents rather than the postwar political filters. This method contains its own problems—purposely or inadvertently incorrect documents not least among them. However, without cutting out its propaganda roots, historians of Vietnam are doomed to feed on fruit from the poisoned tree. Here, postwar studies are used mostly to provide facts, to double-check the analysis of documents, and to ensure I do not stray too far afield. However, the analysis is solely this author's, as are the failures in this area.

This is not a history of the entire war. It makes no claims about why the North won. I do not assess the rightness of either side to go to war or how that war was fought overall. Although related to the pacification effort, this work is not a study of pacification, either. It is a history of the psychological war, or, as Radio Hanoi alluded, the dogs barking in the night.[17] Understanding the organization, objectives, reactions, and themes is impossible, however, without setting these items within the context of the war and the North's counterprogram. As this book will demonstrate, by 1968 the North's propaganda was increasingly ignored in the South.

This also is not a history of the antiwar movement in America and makes no claims about the effect of that movement on US policy. However, after analyzing the North Vietnamese and VC propaganda programs, it became clear that this topic was unavoidable. The extent of the targeting of the antiwar movement by the North and the level of message reinforcement, and the actual propaganda messaging and theme development provided by members of the antiwar movement to the North, made it impossible to ignore. This study does not pass judgment on the movement's impact on the war effort, although there are indications that there was an effect. Rather, this history focuses on the targeting of that movement by North Vietnam and its surrogates and the support provided by specific individuals within the movement. I will leave it to future historians to consider the effect on American actions.

In order to make this narrative coherent, I have chosen to highlight a few individuals whose influence on the psychological war covers many years. People such as Ho Chi Minh and William Colby are well known. Edward Lansdale is also known, but to a lesser extent. Others such as Nguyen Be, influential in the South Vietnamese pacification effort, Truong Nhu Tang, justice minister for the Provisional Revolutionary Government of South Vietnam, and Frank Scotton, a USIS field operator in South Vietnam, are virtually unknown outside of researchers of the Vietnam War. Their activities help provide a useful structure to what might otherwise be a fractured narrative.

Conclusion

The American PSYOP forces went to Vietnam with little knowledge of the history and culture of the country and little experience conducting psychological operations in a counterinsurgency. As this book will demonstrate, despite these drawbacks they had considerable success during the period 1960–1968. In advising, innovating techniques, and developing equipment they made great strides. However, they faced an experienced enemy in the psychological war. And though the North's propaganda increasingly fell on deaf ears in the South, it developed mutually supporting linkages with the American antiwar movement. This program effectively exploited President Johnson's lack of candor on the war, as well as the sound of his uncertain trumpet, to rally a broad movement. As such, 1968 proved to be the year of collapse of morale among the Vietcong and the disintegration of American political will to continue the war under Johnson's leadership.

1 Exciting a Spirit of Disaffection

The use of psychological operations by Americans predates the existence of the nation. Benjamin Franklin, an expert in political propaganda, produced the sliced snake image for the "JOIN, OR DIE" cartoon at the time of the Albany Congress in 1754. In the years leading to the American Revolution, colonial governments and newspapers used propaganda extensively to mobilize people for or against continued unity with Great Britain. A good example here is Thomas Paine's *The American Crisis (No. 1)* (1776). As expected in any war in which popular support was so necessary for success, the propaganda effort increased after hostilities began. Among the Revolutionary War psychological operations were efforts to encourage Canadians to rebel, to urge Native Americans to maintain their neutrality, to influence foreign support, and to encourage British desertions. Most of these activities involved ad hoc organizations by interested people rather than formal structures. During the war, propaganda consisted largely of printed items like pamphlets, broadsides, and newspapers.[1]

Americans printed their first leaflet of the Revolutionary War in May 1775, shortly after the battles at Lexington and Concord. They directed it toward the British forces in Boston, calling on them to join the American cause. It was in this context that the Americans developed a program to induce Hessians, the Brits' German auxiliaries, to desert. This was the most extensive official psychological operation undertaken by the Americans during the war.[2] Shortly after Britain signed the contracts for German auxiliaries in the winter of 1776, Franklin conceived a desertion program aimed at the Hessian mercenaries fighting for Great Britain.[3] This program eventually offered Hessian deserters "free exercise of their respective religions . . . rights, privileges and immunities of natives," and, perhaps most important, fifty acres of land. Posters, handbills, and German-speaking colonials spread news of the program.[4]

The results of the program were mixed. Some Hessian soldiers deserted, but their exact numbers and motivations are less clear. Of the 30,000 German auxiliary troops sent to America during the war, about 18,000 returned to Germany at its conclusion. This left approximately 12,000 who died, deserted, or were otherwise unreported.[5] Several factors may have limited the effectiveness of the American propaganda campaign. The soldiers were all foreigners, far from their families, and most did not speak English. Many Hessians held people of German ancestry in America in contempt, believing them to be vagabonds and criminals who had fled their homeland. Additionally, repeated tales of Hessian plunder and pillage reinforced the initial fear that Americans felt toward these professional soldiers. Neither side understood the other.[6] When the war ended, Americans did not attempt to continue the propaganda organization, such that it was. The nation had made a choice—with propaganda as well as the army as a whole—to rely on militiamen and individual genius rather than organization and doctrine in developing its military.

Little evidence of the American military use of psychological operations appears in the following sixty years. There may have been some improvised operations during the War of 1812, such as General William Hull's "bombastic proclamation to the people of Canada," but the next major use of propaganda by the US military was during the war with Mexico, especially during General Winfield Scott's successful pacification campaign along the route to Mexico City.[7] The Seminole and the Creek Wars, which occurred during the 1830s, had taught Scott hard lessons on unconventional war. He hoped to avoid problems he had encountered in those earlier fights through strict discipline, prompt payment of Mexicans for provisions, and protecting their property and religious rights.[8] Throughout the campaign from Veracruz to Mexico City during the spring and summer of 1847, Scott fully integrated what are now called PSYOPs and civil affairs into his military operations.

After capturing Veracruz in March 1847, Scott imposed martial law. He paroled enemy prisoners, distributed food to hungry citizens, and issued proclamations, plastered on walls, regarding his program. Scott issued strict orders for his soldiers to respect religious property, and he even attended Catholic mass. Scott went so far as to publicly hang US soldiers for killing Mexicans, to impress upon the people that he could be counted on to follow through with his proclamations. These actions helped overcome resistance to the American occupation. Scott ordered that the "army would continue to advance—presenting at once the olive branch and the sword." Guaranteeing Mexican property enhanced his credibility with the populace. Thus, guerrilla effectiveness waned, freeing his lines of communication to the coast as he advanced. Once again, the United States depended on the genius of a lone individual rather than a cohesive organization, and the results indicate that he was the right man for the effort.[9]

America again went to war in 1861 with no official PSYOP structure. It is not surprising that no doctrine or lessons learned emerged from the Mexican War, as the American propaganda effort depended on the brilliance of

General Scott alone. However, many of America's military leaders during the Civil War had served under him, including Generals Robert E. Lee and Ulysses S. Grant, and it may be possible to see traces of Scott's brilliance in their subsequent behavior.

Possibly the most important Union PSYOP activity was the promulgation of General Order No. 100 in April 1863 to provide occupying forces official guidelines for dealing with civilians in occupied areas. This order sanctioned escalation-of-force policy to maintain actions as moderate as the situation allowed. As in Mexico, the military published these requirements via newspapers and broadsides. A second example was the Emancipation Proclamation, a type of strategic PSYOP. The target audiences for this included the British government, slaves in the South, and residents of the various states. President Abraham Lincoln crafted and timed the proclamation carefully in hopes of having maximum impact on all the intended target groups.[10]

Perhaps America's most effective Civil War propagandist was Henry Hotze. He served the Confederacy in what could be the first formal American military propaganda organization, funded by the Confederate War Department. In 1856, the Zurich-born Hotze immigrated to Mobile, Alabama, where he became a journalist and developed solid relationships with important people. He worked for the politically connected John Forsyth as the associate editor of the Mobile *Register*. Josiah C. Nott, a prominent Mobile doctor and supporter of racial inequality theories, hired Hotze to translate Arthur de Gobineau's *An Essay on the Inequality of the Human Races* from French. The Confederate secretary of war, Leroy Pope Walker, later asked Hotze "to travel to Europe to communicate with Confederate agents already on the ground in Britain and France."[11] Walker ordered Hotze to "advise the CSA as to the tone of English press, gauge the current of public journals, convince the British public that the South could maintain its independence, and keep the 'Tyranny' of the Lincoln government constantly before the people."[12] He arrived in Britain in October 1861 and was "shocked at the unchallenged monopoly the Northern press had over public opinion of the British people." Until this time, the Confederacy based its diplomacy and propaganda on the power of "King Cotton" and the hope of forcing Britain to recognize the South's independence by imposing economic hardship on English mills. British textile mills depended on ready access to Southern cotton. However, Hotze devised an enhanced propaganda plan aimed at a broader defense of the South. This necessitated a more active organization in Europe.[13]

By April 1862, Hotze made the decision to establish his own newspaper in Britain on a shoestring budget. For the next three years, his paper, the *Index*, spread Confederate propaganda themes in England. Hotze was skilled at devising appeals in tune with his target audience. Additionally, he printed pamphlets, which he distributed through British religious journals, and his *Address of the Southern Clergy to Christians throughout the World* ran 250,000 copies. In the end, the effects of the war outweighed his ability to influence it. With the Confederate defeat at Gettysburg in July 1863, the hope of British recognition faded.[14]

America went to war against Spain in 1898, again without an official PSYOP capability, but with a clear need to influence the behaviors and attitudes of foreign civilians and enemy troops. In many ways, the counterinsurgency during this war foreshadowed the American experience in Vietnam. Additionally, war in the Philippines saw the first tentative steps by US Army military intelligence into the propaganda field, mainly in the area of enemy propaganda analysis.[15]

In line with President William McKinley's desires, commanders issued orders to troops and had proclamations printed for the Filipinos, stating that America came "not as despoilers and oppressors, but simply as the instruments of a strong, free government whose purposes are beneficent." On 4 January 1899, General Harrison Gray Otis published an edited version of President McKinley's proclamation in the Philippines, which downplayed the establishment of a colonial government. However, McKinley's original message leaked, and tensions flared over the question of sovereignty for the Philippines. While Otis tried to ignore the issue, the McKinley proclamation directly affirmed US dominion over the islands. The affair illustrated the effect of strategic propaganda on the operational and tactical levels of war. In fact, interplay between strategic and tactical messaging was a hallmark of this war, as the telegraph made rapid worldwide communication possible. Information and propaganda could no longer remain isolated and disconnected from the world, and Otis's attempt to spin the president's message backfired.[16]

The Americans organized native police and used Filipinos to disseminate propaganda on their behalf urging the rebels to cease fighting. This method continued as the US military expanded its control to the outer islands. Although the combination of action and explanation had initial success, it soon encountered limits. As the rebels increased their violence against collaborators, local support for the Americans often waned. The rebels reacted to various American proclamations by tearing them down as soon as they were posted and tried to prevent the dissemination of information about the proposed liberal framework for governance.[17]

Nevertheless, word spread. On 21 June 1900, General Arthur MacArthur, the US military governor in the Philippines, issued an amnesty proclamation to encourage rebel desertion. Later, in December, he instructed department commanders to apply General Order No. 100 from the Civil War to their operations in order to isolate the guerrillas from the civilian population.

MacArthur's decision allowed for a swift implementation of the rule of law and provided a basis for the successful pacification campaign. Additionally, most commanders understood the connection between how they treated the Filipinos and the goal of pacification. American commanders punished their troops for infractions, and at least sixty-seven men went before courts-martial in 1899. This was, in part, to ensure that no basis existed for enemy propaganda alleging American atrocities. Without a doubt, some Americans committed atrocities in the Philippines, but antiwar groups in America often overstated accounts of such atrocities to influence the upcoming US presidential elections.[18]

In opposition to McKinley's policy, the American Anti-Imperialist League (AAIL) and other groups opposing the war began a propaganda campaign of their own. With prominent members such as Samuel Gompers, Andrew Carnegie, Grover Cleveland, and Mark Twain, the AAIL issued letters, petitions, broadsides, and pamphlets to rally support for their cause. Often this campaign took the form of spreading unsubstantiated claims of American atrocities, occasionally based on reports provided to the AAIL by Emilio Aguinaldo's supporters. Aguinaldo headed the most important Filipino resistance group. Active AAIL support encouraged him to continue fighting and reinforced his own propaganda themes. AAIL support rallied the waning revolt in the face of a successful pacification campaign. Aguinaldo also used AAIL's support to highlight the divisions within American opinion, schisms he hoped would ensure his eventual victory. His followers based their hopes on William Jennings Bryan winning the upcoming US presidential election. With Bryan running on a Democratic Party platform opposed to the war, they believed his victory would change American policy. The platform that year called for independence for the Philippines and attacked "the war of 'criminal aggression' against the Filipinos" fought for commercial gain at the expense of liberty.[19]

Coordinated propaganda messages between the rebels and antiwar groups in the United States facilitated the spread of atrocity themes. Although some horrible events certainly occurred, they were not the norm. However, such tales formed the narrative upon which many people understood this war. In fact, much of the later scholarship on the war drew on propaganda from the AAIL to paint a picture in the popular imagination of bloodthirsty and immoral American soldiers. Only recently has a more balanced scholarly approach avoided the insidious influence of this early propaganda.[20]

The tendency of lies to take hold and become firmly entrenched in broad narratives underscores the important issue of propaganda blowback, that is to say, unintended consequences. With the advent of the telegraph, a quote often misattributed to Mark Twain—"A lie can travel halfway around the world while the truth is putting on its shoes"—became truer than ever. This is a problem often encountered in psychological operations. Beliefs are less susceptible to change than facts because beliefs contain an emotional component. Thus, popular understanding of the war remained entrenched more than a hundred years on. Similar propaganda myths from the Vietnam War have embedded themselves into American political discourse.[21]

Upon the reelection of McKinley in 1900, the resolve of the insurrectionists slackened. William Howard Taft, the governor-general in the Philippines, concurred with MacArthur on the need to increase military pressure, and both agreed to continue the propaganda campaign to pacify the countryside. The effects of MacArthur's policy bore fruit. In April 1901, Aguinaldo surrendered and issued a proclamation urging those still under arms to surrender. This was a turning point, and the major fighting ended rapidly thereafter, though with combat continuing in some remote areas. Propaganda played only a minimal part in this victory. Superior US military skills and weapons, along with mistakes by Aguinaldo, certainly had much more impact on the outcome. However, the pacification program, publicized with printed proclamations, "brought to many a village the first law and the first peace it had known in years."[22]

American Propaganda Comes of Age

In 1917, the United States was unprepared for war, let alone psychological warfare. Although the United States had practiced psychological operations informally throughout its history, it was not until World War I that the nation created a formal information organization. Unlike in previous wars, the United States now had a well-developed advertising and marketing industry that took full advantage of modern printing technology, the advent of film, and national marketing campaigns. These mass-marketing techniques led to a very effective propaganda effort and characterized American PSYOPs in the years to come.[23]

The belligerent nations often used atrocity propaganda during World War I to elicit anger toward the enemy. During the war, all sides used emotionally compelling depictions of atrocities to arouse anger; they also relied on nationalist themes to support their causes. Messages such as the poster portraying the German execution of a British nurse in Belgium—"Miss Edith Cavell murdered October 15, 1915. Remember"—played to both. The sensationalism and hyperbole of such themes worked to a degree, but later they had negative consequences. The postwar disillusionment with the war and a new cynicism over the contrast between the war's aims and its effects grew in part as a reaction to the messages used to rouse the populace during the hostilities.[24]

The Committee on Public Information (CPI) became the United States's first formal propaganda agency of the war. It focused on both domestic and international strategic PSYOPs. Former journalist George Creel ran the CPI chaotically, however, and "agencies proliferated whenever a new idea turned up."[25] Creel worked directly with President Woodrow Wilson to draft propaganda appeals, often using Wilson's war aims to create themes.[26] One of his programs deployed "four-minute men" throughout the nation to spread propaganda themes. Members received talking points on news items to present timely commentary to people they met. By November 1917, the CPI claimed that nearly 15,000 spokesmen, using talking points and themes produced by the committee itself, had given talks at theaters, in elevators, and on street corners across the country.[27]

The CPI's international effort included influencing American film distributors to compel movie theaters in Switzerland to agree to play American propaganda films or else lose access to valuable Hollywood productions.[28] Other neighboring countries were targeted as well. This was strategic-level psychological warfare, attempting to undercut German morale as messages crossed the border. The propaganda films contained divisive themes, such as the claim that America was fighting the German government rather than its people. The CPI published printed appeals in European newspapers as well, from groups such as the Union of Friends of German Democracy and signed "In the name of Americans of German Descent."[29]

Simultaneously, work began in the Military Intelligence Section, located in the US War Department's General Staff structure, to develop a military tactical propaganda capability. By December 1917 the department had created the Psychologic Subsection as part of the Military Intelligence 2 Section (MI2).[30] Heber Blankenhorn, an American journalist and the son of German immigrants, played a key role in the MI2 program. He first joined CPI and advanced the idea of targeting enemy troops with propaganda, in opposition to the CPI's strategic focus. He hit a brick wall and sought help from Major Charles H. Mason, the chief of MI2. As a result, Blankenhorn transferred to MI2 in February 1918 to work in the Psychologic Subsection. Unfortunately, lack of integration between the CPI and MI2 hurt the overall effort.[31]

Blankenhorn immediately began recruiting "men for General Staff service who had knowledge of European history, languages, and cultures." By July this section consisted of twenty-eight persons, including the journalist Walter Lippmann. They began producing surrender leaflets within hours of arriving at their French base in August 1918. The first American combat propaganda of the war consisted of 2,000 leaflets disseminated on 29 August 1918 near Saint-Mihiel. Their purpose was to undercut German claims that the Americans had a "no surrender" policy. The leaflet simply quoted the Americans' general order on the treatment of prisoners and listed the daily ration prisoners received.[32] The response reportedly exceeded expectations: "Within days reports were received that enemy troops were reading and discussing the leaflets."[33] Among the most effective American themes concerned the steady growth of US military forces in Europe, which were inexorably tipping the balance against Germany. One leaflet paired the question "WILL YOU EVER BE AS STRONG AGAIN AS YOU WERE IN JULY 1918?" and the depiction of an endless stream of fresh American troops coming over the ocean. The answer—clear to all by this point in the war—was that Germany would never be as strong.[34]

World War I boosted mass communications, printing output, and distribution to a scale theretofore unequaled in war. As the hostilities drew to a close, between August and November 1918, the US General Staff's Psychologic Subsection designed twenty-one separate leaflets. By war's end, the Americans had distributed more than five million leaflets. One report claimed that a third of German prisoners captured in the American sector possessed US-produced propaganda when they were taken.

The Psychologic Subsection innovated methods to measure propaganda effectiveness and gave birth to the study of public opinion in propaganda. It distributed leaflets to tactical units, along with English translations and questionnaires to intelligence officers to evaluate effectiveness. After distribution by patrols, aircraft, artillery, or balloons, officers used such forms to question prisoners who possessed the leaflets.[35] The subsection also used prisoner interviews to assess the effectiveness of potential future products. Before the advent of scientific public opinion surveys, the subsection devised a morale analysis chart. Hampered by subjective factors and limited data, it nonetheless attempted to track the effectiveness of the propaganda effort and to track changes in German morale.[36]

The American Expeditionary Force's psychological warfare effort existed for less than ten months and was in action only for about ten weeks. Within six months of the Armistice, the US Army had demobilized or reassigned most of the personnel. Although it remained as a skeleton unit within the US military structure, the army disbanded the Psychologic Subsection in 1925, leaving no doctrine and few people aware of its wartime activities. Its most lasting effect was perhaps also the most pernicious. Atrocity propaganda, as well as themes stressing that America was "making the world safe for democracy," fed postwar disappointment in and skepticism of propaganda in general. In America, the campaign to influence behaviors of foreign targets came into popular disrepute.[37]

Ho Chi Minh's Propaganda System

Even as World War I was wrapping up and US propaganda efforts were winding down, one of the twentieth century's great propagandists was beginning his own rise to prominence. Born Nguyen Sinh Con in French Indochina (Vietnam), and later going by Nguyen Tat Thanh, he spent decades learning the skills of the propagandist while using yet another alias, Nguyen Ai Quoc. During his life, he often changed his name several times a year and traveled under too many passports to recount here. By the time of the wars for Vietnamese independence, he had adopted his most famous name of all: Ho Chi Minh.

Ho was an extremely dedicated and intelligent man. He eventually became fluent in Chinese, French, English, and Russian and even acted as a translator and propagandist for the Soviet Peasant International (Krestintern) and Communist International (Comintern), the Chinese Kuomintang (KMT), the Chinese Communist Party, and later the US Office of War Information (OWI) during World War II. Although by the 1960s other leaders had eclipsed his power, Ho had such a profound imprint on the conduct of North Vietnamese propaganda that details of his life are critical to understanding its development and relationship to American military operations.[38]

In 1909, Ho Chi Minh's father became a district chief in Binh Dinh Province, which remained a troublesome area to the South Vietnamese government throughout the Vietnam War. Ho became a teacher at a reformist school in 1911 and later moved to Saigon after his father lost his position because of his advocating for anti-French reform. This disgrace eventually led Ho to travel abroad to further his education. Ho worked as kitchen help on a ship, visiting Thailand, the United States, France, and England. Dates for this period of his life are imprecise, but he eventually settled in France and began a career writing propaganda for French and Vietnamese journals. He also lived for a period in Great Britain, teaching himself English.[39]

After a stint at the 1919 Paris Peace Conference, arguing unsuccessfully on behalf of Vietnamese independence, Ho helped establish the French Communist Party in 1920. He published articles in French Communist papers and also wrote for Vietnamese propaganda organs. Around 1923, Ho traveled to the Soviet Union and joined the Comintern, the organization that Lenin created in 1919 to ensure Communist control of the worldwide socialist movement and to foment revolutions abroad. Some historians credit Ho's shift to Russia to the failure at the Paris peace conference to secure Vietnamese rights. Ho worked in Soviet Russia "as a translator, propagandist and specialist in political mobilization."[40]

In the 1920s, Joseph Stalin seized power after Lenin's death and vacillated on the role of nationalism versus united front tactics to spread Communism. For a time, he favored the international Communist movement over national Communist movements. Despite the internationalist nature of the Comintern, Ho "remained rooted in the traditions of Vietnamese patriotism, the family and regional ties." The argument over whether he was first a Communist or a nationalist has continued to vex historians (as it vexed analysts during the Vietnam War). The seeming contradiction between the two positions arose due to the outcome of the Soviet revolution. As the son of an oppressed colonial people, Ho remained both strongly Communist and nationalist throughout his life. However, at times this placed him in opposition to Stalin's preferred method. Eventually the Comintern fell out of favor, and Stalin ordered the arrest and execution of many of its members.[41]

By September 1924, the Comintern assigned Ho to Canton, China, on a "loosely defined" mission. While in Canton, he worked and taught at the Peasant Movement Training Institute, which included Vietnamese students who arrived in 1925. The training simulated "the organization of an army company and [students] received training in survey and propaganda work." Coincidentally, Mao Tse-tung instructed the fifth course in 1926.[42] The Fifth Comintern Congress in 1925 sought to "increase propaganda and recruitment in colonial areas and set up a Commission on international propaganda." At the session, Ho urged "a broad appeal to 'the slaves of the Colonies.'"[43] During this period, Ho also worked for Krestintern as "a translator of French and English language materials." The Krestintern aimed at fomenting Communist revolutions in agrarian societies. Early on, Ho understood the central role that propaganda would play in supporting revolution in Vietnam. For example, he printed the journal *Thanh Nien* (Youth) in Canton and between 1925 and 1930 shipped 208 issues to Vietnam. Beyond that, the Comintern archives shed little light on Ho's activities at this time.[44]

Although Ho's precise locations and names are difficult to track, Comintern archives and French intelligence reports offer some hints on his activities. In a 1926 report to Moscow, Ho claimed to have "organized a peasant Union among Vietnamese residing in Siam" and "organized a propaganda school, for students brought clandestinely to Canton from Vietnam" and Thailand. This effort gained momentum by February 1927. He also accompanied British, American, and French Communists around Canton and solicited support from the French Communist Party's Colonial Section.[45] To support his wide-ranging activities, Ho submitted a budget to Comintern headquarters to cover training for a hundred propagandists and a support staff of "ten full-time propagandists for one year." He requested additional money to cover "publications, communications, setting up small shops as fronts for liaison posts and other organizational tasks" to support his activities.[46]

During this period, the Comintern focused on united front tactics. Ho learned the value of using united front organizations to enhance the power of his small cadre force. Concealing the ultimate purposes of his movements, Ho was adept at tapping into unmet needs and latent anger to build organizations that seemed broad in scope even as he retained all power at the core. Ho's instructions were to "fuse the national-revolutionaries among the Indochinese émigrés . . . by creating a solid Communist core" among them. However, much as many members of the National Liberation Front found in 1975, this façade ended when the core no longer needed the misinformed.[47]

Although the Comintern vacillated on whether class inequality and the need for international revolution would trump nationalist goals and united front efforts to foment revolution, Ho never wavered. Much of his propaganda focused on national independence in order to attract and motivate "converts from other nationalist parties."[48] Nevertheless, Moscow insisted that the Vietnamese Communist Party change its name to the Indochinese Communist Party (ICP—Dong Cong San Dong Duong) in October 1927. This move deemphasized Vietnamese nationalism in favor of the class struggle and reflected arguments within the Soviet Union over the level of nationalism that would be allowed to steer international Communism.[49]

By 1930, Ho Chi Minh's influence in the Comintern peaked and he returned to Hong Kong. British police arrested him in July 1931 (under the alias Sung Man Cho) at the behest of French authorities, but they refused to extradite him. Skillful legal counsel provided by the International Red Aid organization helped lead to his release. He subsequently led a move to reassert the united front program in 1932 and pushed a resolution to infiltrate comrades into nationalist parties in order to destabilize them and attract their supporters to the ICP.[50]

During this period, Ho refined his ideas on agit-prop and the organization of Armed Propaganda Teams (APTs) to motivate the masses. APTs consisted of small, armed teams to spread messages in hostile areas. Agitation contained easily spread slogans and themes designed to stir anger among the masses, making them amenable to carrying out actions the party desired. The agitator actively sought out grievances and kept scratching at them. Propaganda included detailed analysis and explanation of party decisions to enable cadres to execute the agitation program. According to Ho, the party missions included "military training of the party members; agit-prop among the army; organization of workers' and peasants' guards." The goal was to make all accept "the revolutionary movement of the workers and peasants and accept the leadership of the Party." Ho studied history and Russian language at the Institute for the Study of the National and Colonial Questions in the Soviet Union during the mid-1930s. While there, he translated the *Communist Manifesto* and Lenin's *Left-Wing Communism: An Infantile Disorder* into Vietnamese and authored *The Land Revolution in Southeast Asia.*[51]

Ho remained firmly associated with the united front effort, as he would for the remainder of his life. However, Ho began to run afoul of internal Soviet politics as the Communist Party shifted back and forth over that and national versus international Communism. As a result of Japanese aggression in East Asia, Moscow returned to the united front line in 1935 to "mobilize a broad alliance of progressive and democratic forces around the world to oppose the growing danger of malignant fascism." However, Stalin's purges during 1937–1938 led to the deaths of many Comintern activists, including many involved in the squabbles over the united front strategy. The fates shifted again after the signing of the German–Soviet Nonaggression Pact in August 1939. Ho skillfully navigated the treacherous waters of Soviet politics that consumed the lives of so many others, possibly due to protection from above. Ho continued to favor the united front as the best way to advance the international Communist revolution by harnessing nationalist movements.[52]

Ho Chi Minh later claimed that he returned to China toward the end of the Stalinist purges and worked with the Chinese Communist Eighth Route Army as a "translator (listening to radio broadcasts), as secretary of the cell"; he had also "written a brochure on Japanese atrocities, on the heroism of the Chinese combatants, [and] on the anti-Trotskyite struggle."[53] At this time, Ho was more engaged in the Sino-Japanese War than Vietnam's internal struggles. He produced propaganda to support China and to increase morale during the united front efforts between the Communists and the Kuomintang.[54] Likely owing to personal interactions in Canton in the 1920s, and to his later work with the Eighth Route Army, Ho gained awareness of some of Mao's ideas on revolutionary warfare and the concept of a so-called people's war utilizing a liberated zone as a base. ICP leaders differed over strategy. Some, like future general Vo Nguyen Giap, looked to China for inspiration on military matters. Others favored the Soviet model for spreading revolution. However, Ho carefully struck a balance between Soviet and Chinese support. In part, he was attempting to avoid domination by the giant on the northern border while acknowledging the successes of the Chinese Communists.[55]

Ho later began recruiting among Vietnamese refugee groups in China. In Jiangxi, a onetime Communist stronghold, Ho began to train villagers from among Giap's so-called border work team. As with his previous experiences, Ho organized cadre courses to indoctrinate students in Marxism–Leninism as well as military and propaganda skills. Ho taught four days of lectures, followed by a practical exercise in propaganda during which he observed and critiqued the students.[56]

The French destroyed the ICP infrastructure in Vietnam by 1941 and executed many of its leaders. As a result, a host of nationalist groups vied for leadership of the anticolonial movement. In May 1941, Ho formed the League for the Independence of Vietnam—the Vietminh. This united front group enabled the remaining ICP to mobilize opposition to Vichy French rule. In the South especially, the Vietminh encountered active opposition from other nationalist groups that formed a third force between the Vietminh and the French. This multidimensional struggle represented the beginnings of the First Indochina War that flared up after the end of World War II. Ho and the Communist core functioned successfully, using the united front structure while rebuilding the ICP. However, the united front's only purpose, as far as Ho was concerned, was to gain power for the Communist Party.[57]

A new ICP Central Committee organized in Tonkin shifted the party's center of gravity north to China because of French attacks. From there Ho Chi Minh and other party members connected with the KMT-sponsored Vietnamese liberation movement. The ties with China brought Ho and his Canton training cadre the legitimacy to lead the movement. On 6 June 1941, Ho issued an appeal explicitly calling for "Vietnamese patriotism" and exhorted the "people to follow the examples of their anti-French heroes: Phan Dinh Phung, Hoang Hoa Tham and Luong Ngoc Quyen."[58]

World War II Convergence

As Ho was solidifying his hold over the struggle for Vietnamese liberation, the United States entered World War II with no psychological operations capability, no doctrine, and only one active-duty officer who had any experience from the World War I program. About this time, the Military Intelligence Division created a military psychological warfare office called the Special Studies Office. This later became the Psychological Warfare Branch (PWB). In May 1941, Heber Blankenhorn emerged from retirement to help the PWB, using his experience from World War I. Much to his chagrin, he found "that the Army was reinventing the wheel." Apparently, the lessons learned in the previous war had been forgotten.[59]

Thus began the proliferation of related, often unconnected agencies that performed well enough but had little overall coordination. On 11 July 1941, President Franklin D. Roosevelt appointed Colonel William Donovan to be Coordinator of Information (COI). In June 1942, the COI split into the Office of Strategic Services (OSS), responsible for covert psychological warfare, and the Office of War Information (OWI), to conduct overt and gray PSYOPs. The OWI had control over "all domestic propaganda, and over white propaganda abroad except for the Western Hemisphere," which remained under the Coordinator of Inter-American Affairs (CIAA), run by Nelson Rockefeller.[60]

Brigadier General Robert A. McClure received orders in December 1942 to organize the Information and Censorship Section (ICS) of the Allied Force Headquarters in Europe. He was to have a long career in the PSYOP field, eventually being tasked to form the US Army's unconventional force after the war. The ICS coordinated propaganda activities between OWI, OSS, the British Political Warfare Executive, and US Army Combat Propaganda Companies. Later, the ICS became the Psychological Warfare Division/Supreme Headquarters Allied Expeditionary Force (PWD/SHAEF). This organization conducted wide-ranging tactical and strategic psychological operations in Europe, especially in support of the Normandy invasion.[61]

Thus, by 1944 the United States had created a series of improvised units and arrangements. No specific element oversaw all information operations. The OSS's Morale Operations (MO) branch had responsibility for covert propaganda. The army's theater-level PWD elements in Europe and the South Pacific conducted tactical and consolidation propaganda. The OWI undertook domestic propaganda and white/gray strategic propaganda everywhere excluding South America, the CIAA's area of responsibility. This profusion of agencies continued throughout the war with frequent changes in organizational names, responsibilities, and leaders.

Each theater commander defined the exact scope of responsibilities for his assigned propaganda units. At the theater level, especially after the invasion of Normandy, PSYOP staff ran AM radio stations, conducted leaflet drops and loudspeaker broadcasts, and took responsibility for consolidation PSYOPs in liberated areas as well as for command information programs. A major lesson drawn from after-action reports and PWD histories makes it clear that effective PSYOPs were a function of command influence. Initially, commanders around the world were skeptical of the usefulness of psychological operations, but as PSYOP units became more effective, they gained respect from tactical commanders. By the war's end, "even Generals like George Patton were asking for front line support because 'it was definitely recognized that the loudspeakers helped to persuade the enemy to come over with arms in the air.'"[62]

While the United States was beginning to embrace PSYOPs in 1942, the KMT arrested Ho as a French spy. Ho had been embroiled in the shifting politics between the Communists and KMT and their united front against Japan. This incident eventually led to a brief coordination between the American and Vietnamese propaganda systems. After the KMT released him in September 1943, Ho operated in southern China and northern Vietnam. On one of his trips into northern Vietnam, Ho met the American pilot Randolph Shaw, who was downed over the region in late 1944. Ho greeted Shaw in English: "How do you do, pilot! Where are you from?" He personally escorted Shaw to China, using his newfound friend as a tool to gain the confidence of US officers at the OWI branch in Liuzhou.[63] He often went to the OWI building in Kunming to "read *Time* magazine and any other new literature they happened to have" and to chat with Americans. The OWI personnel, impressed by Ho's English and his interest in the Allied war effort, sought to procure a visa so he could go to San Francisco. The officers saw Ho as a potentially influential regional leader, and they planned to use him to broadcast news to Vietnam. They dropped the plan due to French objections.[64]

In the final months of World War II, the OWI produced a monthly review of psychological operations in Asia called the *Leaflet Newsletter*, a useful source for tracking PSYOP activities in the Pacific theater. According to the *Newsletter*, Americans began dropping leaflets over Hanoi in summer 1944. These materials used a carrot-and-stick approach to encourage humane treatment for downed Allied airmen, which apparently resulted in "the good reception of the crew of a Liberator which came down north of Hanoi, on January 1, 1945." Leaflets were printed in Vietnamese on one side and French on the other and were reported to be very popular. Despite low literacy rates, the report stated that "those who are unable to read take [the leaflets] to someone who can. In fact, leaflets were bought and sold in the bazaar, the common price being one piaster."[65]

The OWI staff in Kunming produced most of these leaflets. By that winter, OWI was producing 29,000 leaflets and leaflet newspapers per week. By the end of the war, the Kunming station printed close to a million leaflets.[66] In response to leaflets warning of the upcoming bombing of the Gia Lam railway workshops north of Hanoi in December 1944, the Japanese closed the facility and transferred its workers to other centers. One source noted, however, that "the bombs reached the earth first," with eighty civilians killed. Timing problems occurred frequently. Failure to follow leaflet drops with bombings also hurt the program's credibility.[67]

The Vietminh issued its own leaflets encouraging aid to American pilots downed over Vietnam and conducted a propaganda campaign to serve its own agenda. Shortly before leaving Kunming to return to Vietnam, Ho asked for and was given an autographed photograph of Major General Claire Chennault, commander of the 1st American Volunteer Group, better known as the "Flying Tigers." Ho also made a seemingly insignificant request for several Colt .45-caliber pistols. The staff promptly provided several freshly unpacked ones to Ho. Per one OSS agent, Ho, once back in Vietnam, made use of these props at a meeting of leaders from the various resistance groups. According to a witness, Ho "sent for the automatic pistols and gave one to each of the leaders as a present. The leaders considered that Chennault had sent these presents personally. After this conference, there was never any more talk about who was the top leader." Ho used these props, particularly the photo of Chennault signed "Yours Sincerely," to help consolidate his authority. His reappearance in Vietnam, with obvious access to American agents and equipment, helped increase his standing among resistance leaders.[68]

The OWI drastically intensified leaflet drops in Southeast Asia during the last months of the war. From the start of the war until May 1945, it dropped only four million leaflets into Indochina. In the last three months of the war, the Allies dropped nearly three times that number. The dominant theme urged noncooperation with the Japanese. Meanwhile, the Kunming staff prepared leaflets supporting a notional American landing on the South China coast under the Pastel deception plan. This simulated invasion planning was a ruse to draw Japanese troops away from planned invasion sites. (However, it is unclear if anyone used the leaflets in light of the Japanese surrender in August 1945, which obviated the need for the deception operation.)[69]

Meanwhile, OSS agents sought to expand their role in propaganda aimed at Vietnam. Eventually, OSS headquarters in Kunming encouraged installing a covert radio station aimed at the region. Interest continued to focus on the need to protect downed airmen and the safety of Allied prisoners held in the region. However, rivalries among the French, Chinese, Vietnamese, and conventional US military forces hampered the OSS's ability to implement operations in Vietnam.[70]

The Vichy French retained nominal control over Vietnam until Japan created a puppet imperial state, the Empire of Vietnam, in March 1945. Throughout the war, the Vietminh continued its resistance and propaganda. Ho and Giap formed the first armed propaganda brigades on 22 December 1944, eventually growing them into the Vietnamese Liberation Army. The units used terror to persuade government officials and regime sympathizers. These armed propaganda units laid the groundwork to mobilize villagers and the urban population to help seize power when the moment seemed right.[71]

Major Archimedes Patti, an OSS officer, arrived in Kunming in spring 1945 to organize an intelligence network in Vietnam. At a conference held in April 1945, shortly after his arrival, a representative from the Air Ground Aid Service informed Patti that "an 'old man' known as Ho Chi Minh" had agreed to help establish the network. They recognized that Ho had done "psychological warfare work with the Chinese Propaganda office in cooperation with OWI." Due to the pressing needs of the war, most OSS agents overlooked Ho's Marxist background. Patti was interested in using Ho, but the Japanese surrender that August altered his plans. The OSS immediately implemented its Mercy Team plan to send teams to secure Allied POWs held by the Japanese in Indochina. The OSS tasked the French-speaking Patti with overseeing POW repatriations from Vietnam.[72]

At the same time, villagers around Hanoi mobilized by Vietminh Armed Propaganda Teams began streaming into the city, seizing it from the Japanese. Immediately upon Patti's landing in Vietnam, the Vietminh sought to meet his team to add legitimacy to the Vietminh's seizure of power in Hanoi. Patti noted, "The Vietnamese came to be seen with the Allies and acquire status in the eyes of their adversaries." At a military parade on 26 August, the crowd waved American flags while Patti and his team saluted the US, Soviet, British, and Chinese flags and, more important symbolically, the Vietnamese flag. On 2 September, during the proclamation of a provisional government, Chairman Ho Chi Minh read a declaration quoting from the opening lines of the American Declaration of Independence and the French Declaration of the Rights of Man and of the Citizen, delineating the abuses of power by the French. By happenstance, two American P-38 Lightning twin-engined fighters "swooped down low over the crowd," further cementing the idea that Ho had the Americans' support. This is not to say that America put Ho in power but rather that Ho, a trained agit-prop operator, skillfully used his access to the Americans to consolidate his power at the expense of other nationalist leaders.[73]

Divergent views in the United States over the shape of the postwar world hampered the issuance of formal American recognition of his Vietminh Front as the legitimate representative of the Vietnamese people, as well as the provision of assistance. The death of President Roosevelt on 12 April 1945 had changed US policy toward colonial areas, particularly Vietnam. At a 25 April 1945 meeting, the OSS deputy chief in China, Colonel Willis Bird, noted clear directives prohibiting "equipment or arms . . . given [to] FIC [French Indochina] under any circumstances. . . . OSS may do as much as they can in FIC for intelligence purposes only." Under President Harry Truman, American policy shifted away from decolonization and toward anti-Communism.[74]

Ho last used the alias Nguyen Ai Quoc in August 1945 because his Comintern name had become a liability. The world now knew him as Ho Chi Minh, the president of the newly independent State of Vietnam. After declaring the Provisional Government of the Democratic Republic of Vietnam, Ho and other leaders affected a restrained political face in a calculated move "to win the support of a broad cross-section of the people in order to focus on the key problem of containing the threat of foreign imperialism." To support this subterfuge, the Vietnamese Communist Party formally dissolved itself and re-formed as the Marxist Study Association. The strategy did not always work, and some Vietminh operators continued to engage in class warfare and assassinations. Despite rumors that Ho was the moderate face of the party movement, it is more accurate to say that he adroitly used his adversaries' fear of Communism to pose as a moderate to thereby gain concessions.[75]

Ho styled himself as the kindly, khaki-clad "Uncle Ho," winning admiration among the Vietnamese people and from OSS officers, including Patti. According to the historian William Duiker, "There was, of course, an element of calculation in Ho Chi Minh's pose, and to some intimates he occasionally let slip that he was well aware of the artifice." This successful tactic supported domestic and international propaganda themes throughout the Vietnam War.[76] In the wake of this victory, the Vietnamese Communists began to use Ho himself as a propaganda prop to "create a party history of unity and impeccable decision-making." Ironically, as Ho Chi Minh's real power waned in his homeland, his reputation on the world stage waxed.[77]

Ho Chi Minh's "ascendancy in 1945 as the symbol of the Vietnamese independence movement" was not inevitable. Although he deserves much of the responsibility for bringing Marxism–Leninism to Indochina, he was not the only source. Additionally, despite his skillful navigation through the shoals of Soviet politics, he lost influence in that relationship. Thus, even though he remained a dedicated believer in Marxism, he allowed little Soviet authority in Vietnam. In 1947, Stalin treated Ho with open contempt during Ho's visit to Moscow. Using the same ruse he had performed on Chennault, Ho sought Stalin's support, or at least the implication of support, by asking him "to autograph a copy of the magazine *The USSR in Construction*." Unlike Chennault, however, Stalin had his guards secretly repossess the prop. After the Communists' victory in China in 1949, the Vietminh and Ho consistently sought to leverage support from the Soviet Union as well as China into valuable aid from those two competing powers.[78]

Ho's rise paralleled that of another key figure in Vietnam War propaganda who surfaced at this time. Far from the image that North Vietnamese propaganda later created, Ngo Dinh Diem was an intelligent, active, and independent nationalist leader—and a fierce anti-Communist. Diem, the son of a Royal Court official in Hue, graduated first in his class in 1921 from the prestigious School of Administration and became a junior official in Thua Thien Province near Hue. He quickly rose through the ranks of the colonial bureaucracy, becoming province chief of Ninh Thuan by age thirty. Diem was minister of the interior in Emperor Bao Dai's cabinet before World War II, but he resigned in protest over France's failure to grant the emperor sufficient authority. Bao Dai selected Diem, in part, "because of his intransigence and his fanaticism" that "could be counted on to resist Communism."[79] Diem detested Communists; he led counterinsurgency sweeps against them as the province chief. His attitude likely intensified when the Vietminh executed one of his brothers during the 1945 revolution that proclaimed independence. The Vietminh even imprisoned Diem during the chaos of 1945, but Ho ordered his release in an attempt to rally nationalists to the cause. During this chaotic period, Diem charted a neutral path between the French and the Vietminh, all the while awaiting a suitable time to seek power as an alternative force. Eventually, he left the country.[80]

The United States employed perhaps 2,000 personnel in PSYOPs at any given time during World War II. It developed a formal bureaucracy and published field manuals. According to Dwight Eisenhower, "the expenditure of men and money in wielding the spoken and written word was an important contributing factor" in the Allies' victory. The Allies dropped an estimated eight billion leaflets in the European theater alone. However, in the rush to demobilize, the Americans dissolved their military PSYOP capability as well. Despite urging from several quarters, the US Army conducted no official postwar historical analysis of PSYOPs' effectiveness and operations.[81]

Paul M. A. Linebarger, a science fiction writer and political science professor, wrote an unofficial study of the wartime operations, titled *Psychological Warfare* (originally published in 1948). (Linebarger had served in the OWI in China during World War II.) Despite its unofficial status, Linebarger's history was a bible for a new generation of psychological operators. And it became the textbook that US personnel in Vietnam used for PSYOPs. As the chapters ahead will demonstrate, the United States created a multilayer bureaucracy to conduct psychological operations during the Cold War pursuant to its commitment to fight Communism through psychological warfare. This structure played to American strengths in mass marketing and tried to replicate the successes and lessons from World War II.[82]

Conclusion

As post-1945 Indochina lurched into civil war, participants developed PSYOP techniques uniquely adapted to the Vietnamese context. Three distinct approaches developed during World War II affected operations in the Vietnam War. The first was the American approach of multilayered bureaucracy described above, which remained the lens through which the US military tended to view PSYOPs. In part, this was due to the influence of Linebarger's book.

The second approach, developed by Ho, was fully integrated PSYOPs using agit-prop techniques, Armed Propaganda Teams, ruthless (if selective) use of violence, and a united front strategy. Although Ho was later eclipsed in the leadership role, by training the cadres and creating the structures, his imprint is indelible. Ho developed this structure during his time in the Comintern and fine-tuned it during his experience with the Chinese. Agit-prop was effective in mobilizing information resources and exploiting opportunities as perceived by the Communist leadership. However, tactical success did not necessarily lead to strategic success.

Finally, the Vietnamese nationalists, represented by Ngo Dinh Diem, constructed their own system, one based on APTs and agit-prop that focused on nationalist themes. However, they had difficulty creating a coherent, simply understood ideology around which to rally the people. In contrast to the united front strategy, which could mislead people into joining a group whose leadership they might not agree with, the Diem forces had to create a national identity. This task was made more difficult due to Diem's connections to French colonial power prior to 1954.

2 Cold War Developments

Because of the tremendous devastation and loss of life caused by World War II and the fear engendered by the advent of nuclear weapons, the United States placed a greater emphasis on alternatives to direct military confrontation in the post-1945 era. The Truman Doctrine stressed containment of Communist expansion. This was a policy to which psychological operations were critical. Truman's decision eventually evolved into a reliance on the doctrine of massive retaliation and a willingness to maintain at least a minimal PSYOP capability. President Dwight D. Eisenhower expanded on this, stressing the importance of covert action. President John F. Kennedy continued Eisenhower's emphasis, but he added an expanded Special Forces capability and a focus on counterinsurgency and shifted to a flexible nuclear strategy. Meanwhile, in Vietnam, both the Vietminh and the French-controlled State of Vietnam (1949–1955, part of the French Union) refined their own propaganda capabilities. After the 1954 Geneva Accords, President Ngo Dinh Diem continued this trend in the South, with the added burden of creating a national identity while separating himself and his nation from France. It was in this context that the US Cold War PSYOP capability developed.

In the immediate postwar era, the American PSYOP force demobilized rapidly, along with the rest of the military. Likewise, the nation deactivated the OSS and its Morale Operations propaganda branch in September 1945. The newly formed Central Intelligence Group and the US Army maintained elements of these forces, but for all practical purposes it was a period of "virtual psychological disarmament" for the nation. Although military leaders acknowledged the wartime value of PSYOPs, the rush to disarm left few elements of the military unscathed. Consequently, many operators returned to civilian pursuits. Meanwhile the army undertook no comprehensive historical study of wartime psychological operations.[1]

Rebuilding PSYWAR Capability

In 1947, General Eisenhower, the US Army chief of staff at the time, testified on behalf of the US Information and Educational Exchange Act (the

Smith-Mundt Act). The act became law in 1948 and authorized US international and cultural programs. Eisenhower also used his influence on the Joint Chiefs of Staff (JCS) to advocate careful study of the World War II experience in psychological warfare to, as he put it, "keep alive the arts of psychological warfare."[2] Eisenhower exhibited a keen appreciation of the possibilities of PSYOPs during the war. He had, in fact, become an enthusiast much earlier, while serving under General Douglas MacArthur in the Philippines in the 1930s. There he engaged in nation building and developed an understanding of the importance of psychological operations. It was no surprise, then, that Eisenhower became a chief supporter of expanding PSYOP capability during the Cold War.[3]

After initially gutting the PSYOP force, President Harry Truman rethought this position as the Cold War rivalry deepened. National Security Council Directive 10/2 expanded the charter of the nascent Central Intelligence Agency in June 1948 to include covert PSYOPs under the Office of Policy Coordination (OPC). This office developed and supervised covert activities, including covert propaganda, as well as direct action, unconventional warfare, "and support of indigenous anti-Communist elements in threatened countries of the Free World."[4]

The PSYOP structure that matured during the early Cold War consisted of three components. First, the US military, primarily the army but also the newly established United States Air Force, focused on combat propaganda. Second, overt, strategic PSYOPs came under the control of the new US Information Agency, working in conjunction with its overseas division, the US Information Service. The US Agency for International Development (USAID) provided additional technical aid to build communications systems in friendly nations. Third, the CIA conducted covert propaganda. All three elements occasionally practiced gray propaganda.

In 1949, General Douglas MacArthur, commander of the Eighth US Army in Japan, formed a special staff section under his intelligence staff "to plan Psywar measures to counter aggression in Asia." This section consisted of five men by the time the Korean War broke out in 1950. Although the section was not sufficient for the situation, MacArthur's foresight at least provided some capability for an immediate PSYOPs response to the invasion. Thus, at the start of the Korean War, unlike past conflicts, the United States had a functional, albeit tiny, PSYOP structure in place.[5]

A woeful shortage of PSYOP capacity plagued the entire military, however. The US Army had one PSYOP unit, the Tactical Information Detachment at Fort Riley, Kansas, when the North Koreans invaded South Korea in June 1950. This twenty-four-man detachment, part of the US Army Aggressor Center, immediately reorganized as a Loudspeaker and Leaflet (L&L)

company and deployed to Korea in September. The company consisted of a headquarters and publication, operations, and loudspeaker platoons. Authorized at 107 men, the unit never attained full strength due to personnel turnovers.[6]

With the departure of the 1st L&L Company from Fort Riley, the army organized the provisional 5021st Psychological Warfare Detachment, to act as a holding unit while the Psychological Warfare Board developed future PSYOP unit organizations and screened military records for personnel with the esoteric skills required for such units. The 5021st PWD's responsibilities also included running the PSYWAR staff officers course and PSYWAR classes at Fort Riley for officers and noncommissioned officers (NCOs).[7]

Due to problems in acquiring personnel and equipment, the 1st L&L Company did not become operational in Korea until January 1951. By this time, United Nations forces were under heavy Chinese attack and were struggling to form a cohesive defensive line south of Seoul. Loudspeakers mounted on vehicles and aircraft became a primary means of conducting tactical PSYOPs during this phase of the Korean War. Additionally, PSYOP units operated against Communist partisans in the South and had responsibility for consolidation PSYOPs in rear areas, dealing with the flood of refugees fleeing the Communist advance.[8]

It was not until later in 1951 that the US Army formed the 1st Radio Broadcasting and Leaflet (RB&L) Group and deployed it to conduct strategic level PSYOPs. Together, the 1st RB&L Group and the 1st L&L Company often produced more than 23 million leaflets per week, most of them disseminated by aircraft and howitzers. The 1st RB&L Group operated the Voice of the United Nations Command radio station and provided PSYOP programming to broadcasters in Japan and Korea.[9]

That same year, Brigadier General Robert A. McClure organized the Office of the Chief of Psychological Warfare to further develop psychological warfare units and training standards. He was instrumental in developing both the PSYWAR and unconventional warfare capabilities of the US Army and restoring neglected special operations skills training. Unconventional warfare focused on using diplomatic, military, economic, and political means. The establishment of a psychological warfare department at the Army General School in Fort Riley, Kansas, helped standardize PSYOP doctrine, which ultimately developed the military PSYOP organization sent to Vietnam. Meanwhile, the Operations Research Office at Johns Hopkins University wrote manuals, performed analysis of World War II and Korean War activities, and conducted basic regional research. Such activities were among the first analysis of the US military's PSYOP experiences and helped disseminate lessons learned and ideas for future operations. However, due to their brevity and topical specialization, these studies did not represent a comprehensive analysis.[10]

As the war dragged on in Korea, the 6th RB&L Battalion formed in April 1952 from elements of the 5021st PWD at Fort Riley and moved to Fort Bragg, North Carolina, the following month. The battalion's mission was to provide PSYWAR support at the theater of operations level, between the RB&L Group and L&L Company level. According to doctrine at the time, the battalion was the basic unit for conducting wartime PSYOPs. Unlike a standard US Army unit in which the table of organization and equipment (TO&E) prescribed a specific number and organization of platoons or teams, PSYOP units were designed to be flexible enough to manage an assortment of teams and specialties. Rather than an integrated company structure, each team was self-sufficient and detachable to combat units for direct support. It was a semicellular unit, with each cell being flexible enough to form task organizations to fit any eventuality. However, the doctrine placed cells under the standard command group staff sections, forming its semicellular structure.[11]

Studies conducted to assess Korean War PSYOPs found several problems. An Operations Research Office study in January 1951 claimed that a lack of trained PSYOP personnel plagued operations. At this time, the army's primary training consisted of PSYWAR extension courses and reading field manuals. According to the report, such minimal training hindered operation quality at all levels. The Eighth US Army in Asia had few PSYWAR personnel with any training or experience. At the division level and below, virtually all PSYWAR officers were untrained. The study also found a lack of command emphasis, which is critical to effective PSYOPs. Units below the division relied on dual-slotted personnel to run the PSYOP effort in Korea, meaning that they had two jobs. This left little time to focus on PSYOPs in the heat of battle.[12] To overcome these shortcomings, a pamphlet explaining PSYOP capabilities and how to request support was disseminated among forward units. However, without personal contact between combat and PSYWAR units, any effect was limited. "These officers might use leaflets if they know what leaflets exist and how readily available they are," noted the author, but amid combat operations they did not take the initiative to consider a program that might monopolize their time and was of dubious effectiveness.[13]

On the positive side, a combat report by the 14th Infantry Regiment from March 1952 stated that "psychological warfare, in our present situation, is worthwhile."[14] Another report from the 158th Field Artillery in August 1952 noted that "enemy morale does not seem as high as in previous periods [and] the number of deserters surrendering to friendly forces has increased during the period," crediting both these developments to PSYOPs.[15]

Drawing on experience from World War II, the evaluation of PSYOP effectiveness in Korea relied largely on anecdotal indicators such as prisoner interviews. Unfortunately, interviews were often ineffective due to the haphazard system used to conduct them in Korea. The report noted "the tendency of prisoners to try to please the captor by giving the desired answers." The PSYWAR branch within Eighth Army was also limited by its meager research capabilities. The branch evaluated the resources the enemy assigned to countermeasures such as jamming and punishment for listening to broadcasts as an indirect means to measure PSYOP radio effectiveness. The usefulness of such a process was limited by the intelligence gathered.[16]

Captain Herbert Avedon, who had served with the OSS and later as commander of the 1st L&L Company in Korea, conducted his own evaluation of PSYWAR. He compiled a comprehensive after-action review in August 1953 to address problems that he hoped would lead to changes in the training, doctrine, and organization of PSYOP forces based on lessons learned. According to the study, United Nations forces dropped 2.5 billion leaflets during the war. The total included 1,200 separate leaflet messages. Avedon found problems with PSYOP equipment, particularly airplane-mounted loudspeakers that could not be understood from safe flying altitudes and extremely heavy and immobile tactical loudspeakers. The problem of how to mount loudspeakers effectively on aircraft lasted into the Vietnam War. Avedon also reported a lack of coordination among PSYOP elements.[17]

Access to PSYOP-relevant intelligence had proved to be a major difficulty. Evaluating PSYOP effectiveness was the mission of the military intelligence staff, which usually had little interest in performing that duty. Most military intelligence organizations focus on order-of-battle data, such as topography and enemy force locations. Psychological operations require a much more complex set of data. Conventional forces frequently overlook such "human terrain" data. For instance, Avedon complained that "no one in PSYWAR radio operations seems able to come forth with reliable statements as to the size of the potential radio audience." As a result, it was difficult to gauge the effectiveness of the PSYWAR radio effort. Avedon also noted the lack of an overall master plan for conducting PSYWAR. His report itself was delayed because of renewed attempts to disband the army's PSYOPs when the war ended. By October, the military decided to retain a small force, and only then it disseminated the report.[18]

PSYWAR Bureaucratic Development

Upon the formation of the US Air Force in 1947, its leadership immediately sought to define the service's roles and missions. Air force leaders demanded dominance over other services in PSYOPs. In February 1948, the air force staff created the Psychological Warfare Division. This unit began developing plans for creating an air force PSYWAR capability. Initially, it designed a concept for a Special Operations Wing (SOW) to support theater-level PSYWAR staff and operations. Then, in January 1951, the Military Air Transport Service received the mission to organize, train, and equip SOWs. For operational security reasons the air force designated them as Air Resupply and Communications Wings within the Air Resupply and Communications Service (ARCS). The 581st Air Resupply and Communications Wing had responsibility for conducting leaflet drops as well as supporting unconventional warfare in Korea. The unit's 581st Operations Squadron operated a variety of platforms for delivering personnel and propaganda, including twelve B-29 bombers, four C-119 Flying Boxcars, four H-19 helicopters, and four SA-16 Albatross amphibious aircraft. The wing also contained the 581st Reproduction Squadron that worked along with the 1st L&L Company to print PSYOP products. From their base in the Philippines, the 581st ARCW rotated elements to Korea to support operations.[19]

One unusual ARCS element was the balloon program. Using meteorological research as a cover, ARCS conducted testing and evaluation to "clarify the capabilities that the proposed balloon flying squadrons needed" to drop leaflets in denied areas. In November 1952, the 1300th ARCW became the only active US Air Force PSYOP balloon squadron.[20]

The air force also developed an outstanding PSYWAR training program for ARCS officers. The program included an intensive, three-stage curriculum. Training began with a four-month course at Georgetown University's Institute of Languages and Linguistics, focused on PSYOP theory and area studies. The second phase "transitioned the students from theory to operations" and lasted as long as three months, depending on the officer's career track. A third phase included advanced language training, working with other agencies, and special operations, Ranger, and parachute training. By the time the air force terminated the program in May 1953, more than 500 officers had completed at least the first phase of the training.[21]

General Munro MacCloskey assumed command of the ARCS in late 1952, and the air force seemed poised to become the dominant service for PSYWAR. MacCloskey quickly initiated Operation Think, a program that challenged ARCS's 500-plus PSYWAR officers to develop programs and PSYOP campaigns. This proved to be a last flash of brilliance for air force psychological operations. By the fall of 1953, the air staff ordered ARCS units to "confine itself to projects requiring implementation only by the Air Force." The air force's brief primacy in the PSYOP field began to wane with the deactivation of the ARCS wings, effectively killing the air force's PSYOP program. By 1954, Air Force Reserve and National Guard units acquired most of the remaining capabilities.[22] ARCS had been a victim of interservice rivalry. General McClure had championed the primacy of the army in psychological operations and argued that the air force should be "essentially a supply agency for unconventional warfare activities."[23] He won that argument.

With the activation in May 1952 of the Psychological Warfare Center at Fort Bragg, McClure received the task of training and developing PSYWAR and Special Forces units and evaluating equipment.[24] Under McClure, American PSYOPs essentially received the organizational structure that would be used for the next three decades. The OSS naming system for teams used a digraph designation, in part to conceal their functions. For instance, *SI* was responsible for secret intelligence, the *SO* branch performed special operations, and the *MO* branch conducted covert PSYOPs under the cover of morale operations. This naming method carried through to the postwar conventions of the PSYOP and Special Forces units. It is not surprising that both PSYOPs and Special Forces carried over this method of designation from the OSS, given that many of the organizers had served in it.[25]

The contemporary naming rule for a twelve-man Special Forces A-team derives from the original digraph *FA*. By 1960, the *F* was removed.[26] However, PSYOPs units maintained a digraph designation into the 1970s. *AA*, *AB*, and *AC* designated command and control teams at the company, battalion, and group levels, respectively. In the 1955 version of US Army PSYOP doctrine, the command and control teams retained the staff sections generally found in a military unit, such as the *S1* for personnel, *S2* for intelligence, and *S3* for operations. However, all other elements were cellular in structure.[27]

Regardless of the various versions of TO&Es for PSYOPs, in general the first character designated the type of PSYOP mission and the second a subset of that mission. *F* referred to propaganda analysis and design, *G* to production and printing, *H* to mobile loudspeaker and propaganda teams, *I* to radio, *J* to heavy printing, and *K* to consolidation. The *HA*, *HB*, and *HE* teams were the primary operational units within a PSYOP company, providing mobile command, loudspeaker, and audio-visual teams, respectively. (See figure 2.1.)

The 1955 version of the army PSYOP doctrine assigned the Joint Chiefs of Staff with approval authority for psychological warfare plans, policies, and guidance. Any deviations required immediate notification of the JCS with the reasons for the change. Outside of a military theater of operations, the US National Security Council (NSC) remained the primary agency responsible for propaganda.[28] At the national level, the Psychological Strategy Board was created in 1951 to coordinate at the NSC level the activities of the State Department, CIA, and military services in countering the Soviet Union in the information war. Although it lasted only two years, the Psychological Strategy Board laid a foundation on which Eisenhower built his policy-making

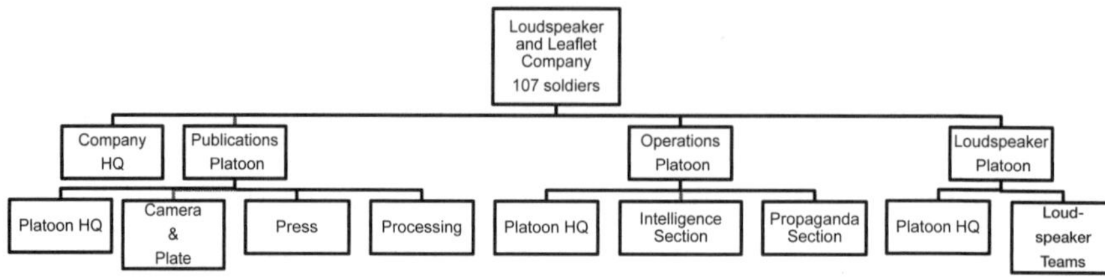

Figure 2.1. Loudspeaker and Leaflet Company

Operations Coordinating Board. He placed the board under the NSC "to coordinate implementation of cold war planning."[29]

Before his election as president, Eisenhower had "promised a coherent national security strategy that accorded paramount significance to psychological considerations."[30] After he won in 1952, the president established the Committee on International Information Activities. He also established two key agencies designed to streamline the control of PSYOPs "so a symphonic theme can be played which will be heard and enjoyed by the people of the world and our people." Eisenhower's New Look strategy assumed an indefinite Cold War, necessitating a shift to political and psychological means to advance American interests during the long struggle.[31]

The primary overt PSYOP agency responsible for fighting the Cold War was the US Information Agency, created on 3 August 1953. According to Eisenhower, its purpose was to "submit evidence to people of other nations by means of communication techniques that the objectives and policies of the United States are in harmony with and will advance the legitimate aspirations for freedom, progress, and peace." The director of the USIA reported to the president through the NSC.[32] The USIA operated Voice of America, as well as international cultural and educational exchanges. To maintain legitimacy, the organization could never admit to conducting PSYOPs. While the focus of USIA was overt, it was an often reluctant supporting element of the entire PSYOP effort. It also coordinated, through the NSC, messages and themes used by the military and CIA. The agency's overseas arm was the United States Information Service, which fell under the direction of the US ambassador in each country. In Vietnam, for example, the USIS "provided advisory and material support to the [Vietnamese government] propaganda effort and coordinated military training assistance."[33]

Often working in tandem with the USIA, USAID helped in building modern communications capacity to help disseminate American PSYOP themes in countries such as Vietnam. This included conducting technical studies to determine optimal broadcast facilities and locations, as well as providing aid in constructing stations. USAID also supported the pacification effort in Vietnam. By 1964, the agency had spent nearly $7 million to improve broadcast facilities there.[34]

The third leg of American PSYOPs was the CIA. Propaganda consumed 40–50 percent of the CIA budget in the 1950s, mostly in the form of gray propaganda. Throughout the decade, the CIA continued to develop its covert information capabilities. Eisenhower appointed Allen Dulles as director of Central Intelligence partly because he "shared the president's faith in the efficacy of covert action and psychological warfare," and Ike's administration used this capability frequently throughout the decade. As an example, the agency supported the publishing of books, such as a project code-named AEDINOSAUR for the publication of *Doctor Zhivago.* The CIA facilitated typesetting photographs of Boris Pasternak's original Russian manuscript and helped to publish and distribute the book. The agency then used various covert means to slip copies of the book to Russian tourists in an effort to spread the story of a failed revolution to the Eastern Bloc satellites. For instance, the CIA provided 800 copies of the Russian-language edition to contacts "for distribution to Soviet personnel and tourists in Western Europe" attending the 1958 Brussels World's Fair.[35]

Operation PBSUCCESS (1954) illustrated the mutually supporting interactions of covert, gray, and overt PSYOPs. President Eisenhower authorized PBSUCCESS in August 1953 to overthrow the government of President Jacobo Árbenz Guzmán of Guatemala. Árbenz became president in 1951 and instituted social changes that American policy makers viewed as supporting Communists. To carry out the mission, a CIA covert PSYOP team created news stories designed for the USIA overt operation to amplify. The program targeted the leadership of Guatemala while simultaneously reaching out to wider world opinion in order to justify the coup. The CIA targeted news media throughout the region, creating stories that could be used to support the overt program. At one point the chief of station for the operation requested all other stations "able to support efforts to discredit the Guatemalan government as communist-controlled" to report on their actions so that it could be "picked up for play in the Western Hemisphere." This chief suggested themes such as noting that the cost of Soviet-provided arms would be borne by the Guatemalan people and enhance Soviet control over Guatemala. He argued that "the Guatemalan case involves saving democracy from totalitarian aggression."[36]

The CIA also operated a covert radio station in Nicaragua and conducted leaflet drops in support of PBSUCCESS.[37] A declassified internal history published by the CIA described a plan to have a submarine land a cache of Soviet weapons. The CIA would then exploit the discovery of arms for propaganda purposes. Before implementation, however, "a ship carrying 2,000 tons of Czech weapons and ammunition arrived," and the subsequent international furor obviated the need for the CIA ploy.[38]

An after-action report of the operation, however, was less than laudatory. The initial leaflet drop "caused the opposition to spring into action right at the moment when the inner organization was necessarily the most active." This nearly caused the failure of the entire operation. The chief of project argued, however, that such criticisms may have been Monday-morning quarterbacking and that other aspects of the PSYOP program were useful.[39] In the end, the uproar over the coup brought the United States into disrepute. Another memo from the operational headquarters in Florida to the Project PBSUCCESS HQ argued for a full overt/covert program to salvage the US position in the face of worldwide condemnation for the operation. The author suggested that "in either case, the interrelationship between communist victory in Indochina and communist progress in Latin America should be kept well in mind."[40] The coup occurred during the concluding days of the Geneva negotiations to end French involvement in Vietnam. Viewed through the Cold War lens, all areas of the world were interrelated.

Vietnam Developments

Under President Franklin Roosevelt, American foreign policy had focused on decolonization. FDR did not favor a return to colonial rule in places such as Indochina after the war. However, as it became clear to President Truman that the postwar relationship with the Soviet Union required a new framework, the anticolonial policy withered. The Soviet refusal to leave northern Iran on schedule initiated this shift. The Allied forces had occupied Iran during World War II with the understanding that each would vacate the country within six months of the war's end. Rather than leave as the other Allies had, the Soviet Union remained and instituted a puppet government in northern Iran. Truman became convinced that the United States needed support around the world to contain the Soviets. Regardless of who bears the fault for the Cold War, such perceptions drove the change in American policy.[41]

As a result of this change, President Truman acquiesced when the French returned to regain control of Indochina shortly after Ho Chi Minh proclaimed the Democratic Republic of Vietnam (DRV). This set off the First Indochina War. As the war dragged on, the French people became war-weary. However, the Chinese Communist victory in October 1949 upped the ante by supplying arms, especially artillery, to the Vietminh in the North. In an effort to rally popular support against the Vietminh and to enhance the pacification struggle, France allowed the creation of the independent State of Vietnam as an associated state of the French Union. Initially, America refused to recognize the State of Vietnam or to assist the French in their struggle unless "France granted true independence to its new client state." The French formed the Vietnamese National Army to take over combat duties. By this point, the Vietminh had been driven out of most cities and were conducting a guerrilla war, and controlled much of the countryside. The United States increased aid to the French and eventually covered much of the expense for France's war in Indochina.[42]

The French included PSYOP units in the nascent Vietnamese army fighting the Vietminh. By the end of 1953, US support included the advice of America's premier advocate for psychological operations support for counterinsurgency war. Edward G. Lansdale, who worked in advertising, served in the OSS during World War II, conducting MOs. At the end of the war, he joined the air force and was assigned to the Office of Policy Coordination, where he first served in the Philippines coordinating operations to defeat the Hukbalahap Rebellion in central Luzon. (The Office of Policy Coordination was the predecessor to the CIA.) Lansdale helped to mold Ramon Magsaysay into a national hero and later president of the Philippines. For this, Lansdale became known as a man who could accomplish any mission. Although Lansdale was not a CIA employee, Allen Dulles, the Director of Central Intelligence, arranged for the air force to detach him to support CIA missions in Asia.[43]

Lansdale recalled that Dulles tasked him with visiting Vietnam in 1953 to advise and assess the French PSYWAR program. This was the beginning of a long and influential relationship between Lansdale and Vietnam. "General Cogny [Commander of French Forces in Vietnam] was strengthening his psychological warfare organization then and invited my comment" on PSYWAR teams, Lansdale later wrote.[44] The French formed the 1st Vietnamese PSYWAR Battalion from the 1st and 2nd Armed Propaganda Companies and authorized it to conduct mobile propaganda in the First and Second Corps Tactical Zones.[45] These companies functioned much as the Vietminh's Armed Propaganda Teams, operating in disputed areas to mobilize the populace to support, or at least to comply with, the government. The Americans and French, however, were not the sole source of training. Saigon Radio English Service reported that Le Van Hoach, the Vietnamese vice premier, accompanied a July 1953 group "to study psychological warfare used by the British forces in Malaya."[46]

To maintain control of its territory, the Vietminh and the Democratic Republic of Vietnam instituted an all-encompassing propaganda program. In addition to selective assassinations, they used more familiar methods to influence behaviors. The party urged writers to produce propaganda literature aimed at mobilizing the masses. The People's Army of Vietnam (PAVN) established the Army Office of Art and Literature in 1949 and created the journal *Van Nghe Quan Doi* (Army Art and Literature) for this purpose.[47] Military units used three-man cells to control the behavior of their soldiers. As one participant later described the process:

> Each evening after they had eaten, these cells had to examine together their actions and thoughts during the day. Every member of the cell had to reveal his fears of hunger, hardship and death as well as his thoughts of envy, lust or enjoyment. This system was said to be necessary to maintain discipline, but it weighed heavily on human dignity and the personality of the individual who was always forced into repentance.[48]

The Vietminh apprehended landlords and initially redistributed land. "Party leaders hoped that such measures would raise the enthusiasm of poor peasants and encourage their participation in the war effort." As historian William Duiker notes, however, collectivization soon followed redistribution. During this period, the Vietminh killed as many as 100,000 people in the North and imprisoned at least as many.[49]

For the Vietminh, the addition of Chinese aid was a double-edged sword. Based on captured documents, French intelligence "reported friction between Vietnamese Cadres and Chinese advisers." Additionally, Vietminh deserters often stated that their motivation to desert was "excessive Maoist influence." Still others reacted to Chinese control of army units and the implementation of "more stringent land reform regulations that focused on eliminating the economic and political influence of the landlord class at the village level." Ho attempted to use the growing Sino-Soviet split to the Vietminh's advantage. Chinese diplomatic recognition of the DRV was followed two weeks later by recognition from the Soviet Union. The quest for influence over the DRV—and more broadly the world Communist movement—characterized the relations of the two Communist giants throughout the Cold War. However, the seemingly cozy overt relationship between Moscow and Peking only reinforced Washington's fears.[50]

By 1952, French operations began to weaken the Vietminh's hold in the South, especially when the latter tried to stand in open battle. At one point, Le Duc Tho, a founder of the ICP and the senior Vietminh official in the South, argued for a shift away from emphasizing main-force units toward guerrilla operations. He warned that "the ideology of pure militarism and looking down on the guerrillas while divorcing the mission of the main force from the local area must be smashed." Armed Propaganda Teams operated in French-controlled villages to rebuild the political infrastructure. The Vietminh conducted proselytizing among the French, as well as among the Cao Dai and Hoa Hao sects. These last were religious groups with their own armies, which at times operated alongside the Vietminh against the French. After falling out with the Vietminh, these sects became an alternative nationalist movement, hostile to both the French and the Vietminh.[51]

After their 1954 defeat at Dien Bien Phu, the French rapidly sought to disengage from Vietnam. By that time, the Vietminh controlled over half the country. In part, due to pressure from both China and the Soviet Union—but also due to Ho's belief that final victory would come via referendum—the DRV also agreed to end the war. Signed just two weeks after Diem became prime minister of the State of Vietnam, the 1954 Geneva Accords temporarily divided Vietnam at the 17th Parallel and ordered a regroupment of Vietminh forces northward and French forces southward. The accords also provided for a period in which civilians could choose to move north or south of the line. The State of Vietnam did not sign the accords and thus did not consider itself bound by them. Contrary to subsequent North Vietnamese propaganda and popular understanding, the State of Vietnam had the right to refuse to adhere to a contract it did not sign, including the required referendum. The United States also refused to sign what was in reality a document to disengage France from an ongoing war. When it came time for nationwide elections, Diem refused to cooperate, saying: "We did not sign the Geneva accords; we are not bound in any way by these agreements, signed against the will of the Vietnamese people."[52]

Diem stepped into a maelstrom when he became president. Diem's residence at Catholic institutions in the United States and Belgium connected him "to international networks of priests and lay Catholics who could safely carry messages without fear of interception by the French police," and this helped him keep in touch with supporters in Vietnam.[53] However, he had been out of the public eye during a critical period. He had to assert his right to rule, having just returned from years living abroad. One of his first challenges was the Binh Xuyen, a criminal gang with an army that controlled parts of Saigon. His first task was to defeat the Binh Xuyen in order to gain legitimacy as a leader. Meanwhile, the regrouped French army remained in the South and worked against him. Diem subsequently defeated the private armies of the Hoa Hao and Cao Dai religious sects.[54]

After a brief return to the Philippines, Lansdale received orders to return to Vietnam as the CIA's Saigon Military Mission chief following the signing of the 1954 Geneva Accords. He reported directly to Allen Dulles, bypassing Emmitt McCarthy, the chief of station in Saigon. Lansdale met Diem and secretly furnished him with funds and assisted him in PSYWAR training for South Vietnamese units. The United States had to move quickly to put people in place before restrictions on manpower under the accords came into effect. Lansdale rapidly built a team and conducted psychological warfare for more than a year. Additionally, many credit Lansdale with encouraging as many as a million people to move south of the 17th Parallel, a sign to the international community of Ho's repression.[55] Lansdale hired "a Vietnamese counterfeiter to produce a bogus Viet Minh leaflet instructing citizens of the north on how to behave when the Communists took over; the leaflet told them to make a tally of their possessions so that the Viet Minh would know what to confiscate." This black PSYOP was so successful that it even fooled Vietminh officials. Operation Passage to Freedom, as the migration program was known, resulted in most of the Catholic population in the North fleeing to the South.[56]

Simultaneously, Diem also encouraged people to move south. At one point, he flew to Hanoi to talk with Catholic leaders about moving their parishioners. Despite his efforts, both American and Communist narratives focused on the US role, suggesting either that American propaganda was solely responsible for the mass migration, or that it was completely spontaneous. Communist propaganda portrayed Operation Passage to Freedom in sinister terms, claiming that the refugees had been "tricked by a U.S. psychological warfare campaign." Lansdale did effectively use disinformation and deception in this operation. However, many chose to leave either out of concern over Vietminh reprisals or for economic opportunities. Diem's own propaganda effort focused on these themes. Most of the estimated one million refugees traveled on French ships or aircraft. About one-third sailed on US Navy transports.[57]

In the confusing aftermath of the 1954 Geneva Accords, the French initially retained control of the South Vietnamese Army's PSYWAR branch. They used the branch's radio facilities to discredit Diem, whom they felt did not support French interests. To counteract French efforts, Lansdale requested the assignment of a US Navy officer who had been a classmate of the radio station commander during PSYWAR training at Fort Bragg. He hoped this navy officer, as the adviser, could stop the negative use of the radio station. However, Lansdale later admitted that he was able to adjust the editorial slant only slightly.[58] The main CIA station established its own transmitter at the same time, which the president's brother, Ngo Dinh Nhu, used to "proselytize army officers and to announce new adherents—some authentic, some notional—to Diem's cause."[59] *Proselytizing* was a term used by both sides during the Vietnam War to describe troop indoctrination, that is, attempting to make soldiers understand the purpose of the war and the nature of the enemy as the leaders wanted it known. It does not necessarily mean to lie, but it is clearly subjective in nature. For example, one VC letter stated:

> Dear Friends, Ngo Dinh Diem is turning the South into a furnace of war, carrying out a great mobilization plan under the name of "military obligation" in order to bring upon the Vietnamese people a war of mutual slaughter. Ngo Dinh Diem is the servant of the imperialists and the representative of the reactionary feudalists now plotting to sell Vietnam into slavery to foreigners. His government has no concern for you and thus cannot possibly represent your interests.[60]

Lansdale realized that the national army was the sole nationwide organization in the South Vietnamese government. Despite its shortcomings, the army contained the basis for a nationwide communication system and an "officer corps with some training and experience in leadership and administration," including a college education for nearly all field-grade officers. Thus, the South Vietnamese Army was the only force capable of extending the writ of the new government into areas from which the 1954 Geneva Accords required the Vietminh to vacate.[61]

Along with Lansdale, recently arrived CIA officer Rufus Phillips advised the PSYWAR companies beginning in 1954. The two companies had about a hundred soldiers each. Under leadership of the French and American PSYWAR-trained Captain Giai (G-5 of the army), the companies "produced and distributed written propaganda as well as educating and entertaining troops and villagers by putting on skits."[62] In preparation for reasserting government control over areas vacated by the Vietminh after Geneva, Phillips took the lead in developing an indoctrination program for the army. It focused on treating civilians respectfully and emphasizing civic action, providing medical and other services. The PSYWAR company, under Phillips's guidance, created a counterfeit Vietminh proclamation stating that two days prior to their departure all Vietminh *dong* (currency) could be exchanged for Vietnamese piasters "at the rate of one for one at Viet Minh headquarters in Qui Nhon." Widely believed, this proclamation disrupted the Vietminh departure as they tried to pacify angry crowds demanding money exchanges.[63]

Once Diem decided to hold a referendum, which he believed would give the nation legitimacy without Bao Dai, he conducted a propaganda program to belittle the emperor. The president attacked Bao Dai in print, on radio, and by loudspeaker truck. He invoked the image of three enemies "to depict Bao Dai as a craven puppet who had a long record of selling his loyalties to feudalists, colonialists, and communists."[64]

General J. Lawton Collins, the US special representative to Vietnam, kept pushing for Diem's removal over a disagreement with the president's handling of the sects. After the destruction of the Binh Xuyen group, President

Eisenhower was more willing to support Diem, and he told Collins to back down. The crushing of these alternate power centers strengthened Diem's position and, rather than alienating the people, rallied them to his side, giving him the momentum to take on the Vietminh's "stay-behind" organization. It was from this time that he began to popularize the term *Vietcong* in an attempt to undercut the Vietminh propaganda name with his own variation that more closely reflected the organization's true purpose.[65]

In 1955, the Diem government launched the To Cong anti-Communist campaign, arresting, imprisoning, or killing even marginal Vietminh supporters. After this bloody campaign, property owners could return to the villages due to increased government control. Under Lansdale's guidance, the Americans expanded psychological warfare training within the South Vietnamese Army. The hasty instruction program taught soldiers how to enter a village and greet civilians in a respectful manner. Additionally, Lansdale supported Diem's Armed Propaganda Teams. These were heavily armed, twenty-man squads trained in psychological warfare and equipped with US Navy loudspeakers, bullhorns, and some larger French voice amplifiers. Men, selected for their patriotic motivation, "carried leaflets, booklets and posters, and at times, phonographs, films, film projection equipment, and simple medicines." Higher-level training began with the establishment of a psychological warfare training center at Fort Cay Mai, Saigon, in 1956.[66]

The APTs operated on an area support basis rather than the direct support role that US PSYOP doctrine preferred. In a counterinsurgency, area support made more sense. It often takes time to develop the relationships needed for PSYOPs to change behaviors. Rather than shift areas every time a supported unit moved, an APT could focus its effort on one area. "The teams were successful in penetrating remote regions," according to Lansdale, "attracting crowds through the distribution of simple medicine (such as aspirin) or showing of movies—and then talking to the crowd to explain the peaceful mission of the Army, the aims of the Free Vietnamese government, and then distributing leaflets and booklets." One ploy APTs used was to offer to exchange a villager's old, faded photo of Ho Chi Minh for a fresh, new, color photo of President Diem. Lansdale stated, "The teams knew that if they entered the huts and pulled down the pictures of Ho Chi Minh, they would only anger the villagers."[67] Along with Lansdale, the USIA began to staff positions in South Vietnam. Among the first to arrive was Everett Bumgardner, another American with a long association with Vietnam. He ran the field operations office responsible for supporting the Vietnamese Information Service (VIS) in communicating to the nation.[68]

Throughout 1955, APTs supported South Vietnamese operations to secure former Vietminh areas vacated in accordance with the 1954 Geneva Accords.

Beginning on 8 February, they coordinated an airdrop of leaflets with radio speeches by Diem and loudspeaker broadcasts explaining the peaceful mission. The USIS assisted Vietnamese G-5 staff (civil-military) in developing tapes, leaflets, and posters for these operations. Teams later used bicycle messengers to deliver mimeographed newspapers to remote villages. Lansdale reported that a major effect of all this activity was an increase in raw operational intelligence. Maintaining a brotherly attitude toward civilians helped overcome Vietminh intimidation. However, the return of the original landowners was not an unalloyed success. Although it represented a change in control, it upset landownership patterns that had developed over years. The Vietminh had organized land reform, and Diem had now overturned it. The associated anger and instability opened the populace to antigovernment propaganda.[69]

Lansdale urged Diem to act more like a campaigning American politician, visiting the countryside, meeting the people, and providing greater visibility. However, Diem remained reluctant to become the glad-handing politico, perhaps for valid cultural reasons.[70] American officials viewed Diem's reticence as a mistake, and many advisers angrily denounced him. Diem reluctantly allowed himself to be filmed for weekly newsreels and recorded ceremonial radio broadcasts in his office. However, as one USIS officer stated, "where the Viet Cong cynically propagandized the peasants with promises of good things they would do for the people, Diem and [his brother] Nhu talked repeatedly about the people's duty to the government."[71]

Lansdale dominated the conduct of PSYOPs in Vietnam for more than a decade. His relationship with Diem was a key factor in maintaining US influence with the Republic of [South] Vietnam (RVN) and in affecting the views of American leaders. He had greater success because he was better at developing the relationship and at understanding the information and cultural climates in Vietnam. However, Diem was not Lansdale's puppet. Diem had his own ideas on how to influence the Vietnamese people and had wide experience at all levels doing so. He also understood Vietnamese culture in a way no American ever could.

Advising South Vietnamese PSYWAR

After Lansdale left Vietnam in 1957, the Military Assistance and Advisory Group, Vietnam (MAAGV) continued to assist South Vietnamese PSYOPs units. In August 1957, for instance, the RVN Armed Forces General Staff requested "that MAAGV reserve one place for Vietnam in the study tour program for 1958–1959 to send one field grade officer of the Office of Psychological Warfare to go to the U.S." to study methods of organization and operations of the Psychological Warfare Branch.[72] Additionally, the US training division scheduled Lieutenant Colonel Nguyen Van Chau, director of psychological warfare actions, to attend a PSYWAR seminar on Okinawa, Japan, starting on 25 May 1958.[73]

Meanwhile, the Army of the Republic of Vietnam (ARVN) Intelligence and Psychological Warfare School graduated 112 personnel that year. The following year, MAAGV programmed nine more slots for Republic of Vietnam Armed Forces (RVNAF) officers to attend the Psychological Warfare Officer Course at Fort Bragg. This six-week course began on 16 February 1959, leading into a nine-week Information Officers Course. Twenty-four Vietnamese officers and enlisted personnel were slotted to attend PSYWAR training in the United States the following year. On 1 January 1958, the ARVN combined the two original Vietnamese PSYWAR companies to form the Mobile Cultural Battalion; on 1 November 1959, it changed the name to the PSYWAR Battalion. Throughout this period, Everett Bumgardner and his USIS staff of local hires provided advice, support, and oversight to the civilian Vietnamese Information Service program.[74]

To understand better the psychological environment in which it might need to operate, the US Army contracted with the Special Operations Research Office (SORO) at American University to produce detailed PSYOP studies of various nations. Under Project PROSYMS, SORO compiled PSYOP-relevant data such as media availability, social and historical information, and demographics to aid in developing themes and target audiences for PSYOPs. Among the first published was a study of Vietnam in April 1959.[75]

The development theory guiding Lansdale and his associates derived from the Americans' New Deal experience. Some focused on big projects and a top-down approach to change the physical environment. In the context of the Cold War, this included worldwide New Deal–style development projects, such as the implausible TVA-like Helmand Valley Project in the parched deserts of Afghanistan that the United States was undertaking at the same time. Many experts of the day—Walt Rostow, for example—saw a linear process of development from primitive to modern. Rostow was an economist and a future national security advisor to President Johnson. He had recently published *The Stages of Economic Growth: A Non-Communist Manifesto.* Rostow saw economic growth as propelling societies through stages leading to the "age of high mass consumption." This was a process that could be aided by outside actors. In the case of Vietnam, he viewed American aid as critical to setting the preconditions for economic takeoff. These ideas infused American advisers with a can-do spirit that often overlooked the nuances of Vietnamese culture and misunderstood Diem's actions.[76] Many historians trace this American mission of "uplift" to the nation's founding and the idea of "American exceptionalism." They tie these traditions to Cold War efforts to "promote the transformation of Asian, African, and Latin American societies" by reinforcing the United States as a model for other nations to follow.[77]

To the Americans, Diem presented "himself as a progressive reformer who believed that U.S. aid and expertise would figure prominently in Vietnam's postcolonial future." However, Diem rigorously maintained his independence. The Republic of Vietnam was not merely a passive recipient of aid. The sometimes poor or confusing advice Diem got from a few American advisers during his consolidation phase "confirmed his suspicion that the Americans did not understand Vietnam's political realities."[78]

Diem worried that the South Vietnamese would develop a "colonial mentality" with an influx of Americans and that local officials would defer to the Americans on the scene. He warned that relying on US aid would be "'like sitting and waiting for fig to fall'—that is, it would foster passivity and discourage Vietnamese from taking up the hard work of development."[79] For all these reasons, Diem wanted to limit the role of the United States.

Based on his experience fighting Communists in the 1930s, Diem devised his own approach. He stressed the "importance of mass mobilization, indoctrination, and other activities designed to forge close ties between government and ordinary Vietnamese." However, he was willing to use harsh "repressive and authoritarian measures designed to root out and destroy clandestine enemy networks." In May 1959, Diem began to experiment with "Agrovilles" (the American name for *Khu Tru Mat*, or "populous zones") in an effort to deny the Vietcong's access to the population. Agrovilles were an attempt to spur development from above and to separate guerrillas from the population. Consolidating the population and denying the enemy a base were common tactics in counterinsurgency operations. Despite problems with implementation, the system helped Diem gain ground.[80]

By that year, the ARVN had destroyed much of the remaining insurgent infrastructure. In May, Diem issued Law 10/59 authorizing the death penalty for a host of crimes associated with supporting the Vietcong. While Communist propagandists had a field day attacking "the colonial-era practice of conducting executions by guillotine," the law further constricted cadre operations.[81] In conjunction with the law, according to Ngo Dinh Nhu, the decision to reinstate village elections would be "very advantageous in the psychological field."[82]

The VC strategy after 1957 focused on utilizing armed propaganda units, but normally they were no match for the well-armed ARVN. Organization in the South began to atrophy under the increased pressure. A document seized in July 1959 detailed the "Situation and Missions for 1959" in Nam Bo (the North Vietnamese label for the southern region of South Vietnam). It stated that "the enemy has created greater losses for the Party and for the popular movement than in previous years, and has been able to carry out a relatively greater number of plots. . . . Basically, the movement is in a defensive position in the face of the daily increasing strength of the enemy's attacks." Many on the DRV politburo grew restive as Diem's pacification program decimated the stay-behind forces in the South.[83]

Le Duc Tho and Le Duan, the Vietnamese Communist politicians, eclipsed Ho's dominant role within the party, in part due to the failure of the 1954 Geneva Accords to unify Vietnam. Le Duan returned to Hanoi from the Mekong River delta in 1958. By January 1959, he took a lead role in the party. He emphasized the dire situation, saying that the southern forces faced annihilation without party intervention. In all of Nam Bo, only an estimated five thousand party members remained. Dissatisfaction among southern cadres contributed to the 15th Party Congress resolution changing from a strategy of "political struggle" to "armed struggle." The resolution allowed an increase in assassinations and small-scale guerrilla attacks for propaganda affect in the South. Hanoi leadership adopted this strategy, understanding "that it could no longer continue to advocate restraint without losing the control and allegiance of the southern Communists as well as the reunification struggle to Diem."[84]

Le Van Chan, a former deputy secretary in western Nam Bo, claimed that Communist propaganda was very clever. He later said:

> They never propagandize Communism, which teaches that the land must be collectivized. If they did, how would the peasantry ever listen to them? Instead, they say: the peasants are the main force of the revolution; if they follow the Party, they will become masters of the countryside and owners of their land, and that scratches the peasant's [*sic*] right where they itch. . . . Say one word about collectivism, and he already is against you. This is a truth the Party has studied and learned to exploit.[85]

According to Le Van Chan, after the party meeting 1959 party leaders understood that peaceful struggle would not be successful. Contrary to predictions, with American assistance the southern regime became stronger and was instead destroying the party in the South.[86]

This change in policy also led to the creation of Group 559 to transport men and arms to the South and later to return trained regroupees to lead the Vietcong. Under the surface, VC propaganda teams began to develop support in some areas. Along with that came the capability of shifting "from propaganda, intimidation, and assassination to effective military action." By the end of the year, assassinations had doubled from the eleven-per-month average in 1958. Vietcong-initiated attacks averaged over one hundred per month, and the ARVN faced a series of setbacks as districts began to fall to enemy control. In the context of increased violence across South Vietnam, US officials began to look more critically at Diem, the man who until recently had been viewed as a miracle worker.[87]

The shift of resources to the South represented a transition to the total war advocated by Le Duan and Le Duc Tho. They led the "South first" faction, which was determined to unify the country regardless of the cost to reconstruction in the North. Le Duan rose from head of the Central Office for South Vietnam (COSVN) to general secretary of the Lao Dong Party by 1960. The Soviet-associated "North first" faction, which included Ho and Giap, became increasingly marginalized. Le Duan began to use the security services against that faction. Meanwhile, Ho's role increasingly became "that of senior diplomat and foreign policy adviser," relegated to playing the propaganda role as father of the revolution and the kindly Uncle Ho. Duan and Duc Tho, both monomaniacally focused on North-South unification, were crucial to the ruinous decisions to escalate the war in 1960, 1963, 1964, and 1968 and the invasion of 1972, as well as the successful 1975 campaign.[88]

To manage operations in the South, the party reestablished the COSVN. On 23 September 1960, the "Party Committee ordered all provinces to launch [a] general uprising." Early the next year, they formed the National Liberation Front for South Vietnam under the control of the People's Army of Vietnam.[89] This harkened back to Ho's preference for front groups, providing a veneer of broad-based support over a Communist structure. According to Le Duan, the front was to be a "'broad united front' to fight imperialists and their reactionary allies." Officially adopted in February 1961 under the Vietnam Workers Party Central Committee, they also subsequently created the People's Liberation Armed Forces (PLAF) to fight in the South under the PAVN and the command of party leaders. This indicated a return to pre-1954 military activity.[90]

The future justice minister of the NLF asserted that Diem's To Cong campaign, "jailing and executing thousands who had fought against the French," left no organized centers of power to confront him. Truong Nhu Tang described it as a negative power, not associated with a positive agenda. Rather than attempt to co-opt the former Vietminh, Diem crushed them as well. Outlining his own thoughts as those of the broader base, Truong said: "As it was, the South Vietnamese nationalists were driven to action by his [Diem's] contempt for the principles of independence and social progress in which they believed. In this sense, the Southern revolution was generated of itself, out of the emotions, conscience and aspirations of the Southern people."[91] Truong was one of those nationalists who might have been co-opted.

At the founding meeting of the NLF in December 1960, Huyen Tan Phat, a member of the Lao Dong Party and later president of the PRG, spoke after being selected to be chairman of the NLF; Truong was selected as justice minister. After the formalities, the evening's entertainment consisted of a performance by the Liberation Troupe. Truong reminisced, "The play turned out to be a broad satire on the ineptitude of the Saigon government troops . . . *Primitive Weapons Can Win against Modern Weapons.*"[92]

According to Le Van Chan, the NLF provided cover for northern military aggression behind a non-Communist façade. Furthermore, they attempted to deceive South Vietnamese peasants about the Marxist–Leninist ideology undergirding the National Liberation Front. Using "traditional means to achieve radical ends," the NLF conformed to Vietnamese expectations of their rulers as good mandarins, much the same as Diem had attempted with his officials.[93] However, according to Chan, peasants lived "in intimate contact with the Party and thus were aware that it was still the communists." The main target audiences for this deception were "city people, intellectuals, and foreigners."[94]

At the same time, the NLF sent "representatives abroad to show that the Front was not Communist, in order to isolate the [Diem] government internationally," according to Chan.[95] The image of a southern-based opposition helped Hanoi confuse foreign audiences about the nature of the struggle. While most Southerners were not fooled by the NLF, the propaganda line did provide cover in the West. Thus, they portrayed the external attempt at unification as an internal struggle. Truong Nhu Tang was one of the intellectuals willingly deceived by the NLF's fictions. He knew the Communists were involved, but he saw them simply as like-minded nationalists.[96]

In My Tho Province, south of Saigon, the insurgency had been nearly defeated when the party ordered the change. By 1960, the NLF began the "Destruction of the Oppression Campaign," raising violence levels significantly by targeting Diem's government officials. At the same time, they increased armed propaganda activities. One factor in this resurgence was a doubling of infiltration from the North by 1961 to five thousand per year, initially mostly southern regroupees.[97] This increased effort threw back the ARVN.

Meanwhile, problems with training, indoctrination, and manning plagued the expanding ARVN. After eventually giving in to Diem's demand to expand the army, General Samuel Williams, commander of MAAGV, worried that the American forces' total advisory strength in the country would soon rise above the allowable maximum. To avoid doing so, he brought soldiers in on temporary duty status. Under this program, several US Army Special Forces teams and three PSYWAR specialists from Okinawa arrived in Vietnam in May 1960. This rotational policy was meant to keep from obviously exceeding the force level authorized by the 1954 Geneva Accords.[98]

During this period, the Center for Vietnamese Studies in Saigon began conducting opinion surveys on behalf of the USIS to help evaluate PSYOP themes and products. These surveys were "designed to assess the impact and effectiveness of various information media on particular audience groups and to learn the relative importance of mass media and personal contact in international communications." These initial surveys must be viewed skeptically, because the systems to conduct them were still in their infancy. However, they do allow for some estimation of trends in public opinion among educated Vietnamese. Skill at conducting the surveys grew with time. Eventually, polling by a variety of agencies led to greater reliability and helped to provide a clear trend line of public opinion on a broader scale. One early lesson learned was the need to pretest products and surveys on Vietnamese people.[99]

The center conducted a survey of teachers in 1960 to assess their media preferences. This group was a critical intermediate target audience. By effectively targeting teachers, wide dissemination of credible messages could be accomplished through the teachers' perceived legitimacy in the eyes of their students. This survey was conducted while teachers were in Saigon to attend a workshop, so the sample was somewhat skewed. Still, from an analysis standpoint, knowing whether the 57.5 percent who reported listening to the radio every day was precisely accurate was less important than knowing that radio was generally a good method of reaching teachers. Providing teachers with radios could then help spread messages in remote villages. Ominously, the Vietcong also viewed them as a critical target audience, to the point of specifically targeting teachers for assassination.[100]

Conclusion

Political consolidation characterized Vietnam—North and South—in the period after 1954. The North imposed its collectivist vision on the society utilizing repression and propaganda. At the same time, Communist cadres in the South attempted to maintain an overt and covert presence in the face of efforts to root them out. Ngo Dinh Diem used skillful political maneuvering to dominate and control the army and government, while slowly expanding his authority outward using the much-criticized Agroville program. Violence and repression accompanied this expansion, though not to the extent found in the North. The United States increasingly provided military and PSYOP training and advice to the South Vietnamese government, but Diem remained an independent actor. His close relationship with Lansdale, however, helped ensure US influence. In South Vietnam, a nascent PSYOP structure already existed and had contributed to the formation of the nation. The expansion of training created an organization doctrinally connected to the United States. However, it was not a mere copy.

US military, USIA, and CIA PSYOP components each were well established and possessed well-developed doctrine by 1960. By that year, USIS personnel had over five years of experience conducting operations in Vietnam. Additionally, the personnel had wide experience in the art of PSYOPs. At the national level, structures were in place to implement and coordinate American presidential directives across a wide spectrum of psychological operations agencies.[101]

Perhaps the weakest leg was the military component. The dissolution of the US Air Force's PSYOP training program meant that the US Army was the sole proponent of PSYOP doctrine and training. Limited army resources and support left a gap. With the distribution of the 1955 field manual on psychological warfare operations, however, the army acquired a solid doctrine on which to build. Additionally, a training center now existed to educate soldiers on how to conduct such operations. Indeed, the US military was better prepared to conduct PSYOPs at the beginning of the Vietnam War than any previous war.

In the meantime, greater Vietcong activity put the South Vietnamese government under increasing pressure, and a major expansion of the war was on the horizon. The Americans' PSYOP effort confronted a strong, united, and singularly focused propaganda program emanating from the North. That organization had contacts with similar propaganda programs in other Communist countries and front groups throughout the world, giving it a ready echo chamber for dissemination. Well-meaning American amateurs now faced professional and experienced propagandists.

1. Benjamin Franklin's 1754 "JOIN, OR DIE" warning to the British colonies in America, exhorting them to unite against the French and the natives, shows a segmented snake. "S.C., N.C., V., M., R., N.J., N.Y., [and] N.E." Library of Congress, LC-USZC4-5315.

2. "4 Minute Men—A Message from Washington," town crier, with bell in hand, in front of Independence Hall. The 4 Minute Men program produced propaganda talking points during World War I. Library of Congress, LC-USZC4-2824.

3. Vietcong cadre addresses soldiers before battle. VAS001581, undated, The Vietnam Center and Archive, Texas Tech University.

4. South Vietnamese propaganda dissemination in early 1960s. VA066941, undated, Rufus Phillips Collection, The Vietnam Center and Archive, Texas Tech University.

5. Edward Lansdale meets with Ngo Dinh Diem. VAS002698, undated, Douglas Pike Photograph Collection, The Vietnam Center and Archive, Texas Tech University.

6. South Vietnamese Armed Propaganda Team. VA066416, undated, Rufus Phillips Collection, The Vietnam Center and Archive, Texas Tech University.

3 America's "Special War," 1960–1962

Beginning in 1960, the North Vietnamese intensified violence in the South. In response, US involvement in Vietnam escalated, especially after President John F. Kennedy's election that November. US policy makers focused on the defense of Vietnam, the next potential domino to fall after the neutralization of Laos in 1962. But it was not this domino that had them most concerned. Instead it was the last domino: the strategic Strait of Malacca, the linchpin to controlling all trade in East Asia. By the end of this period, the National Liberation Front offensive withered. At the same time, South Vietnamese PSYWAR capability grew considerably, advised and equipped by the United States. The US Army began a PSYOP advisory effort in Vietnam with the deployment of PSYWAR personnel on 27 April 1960. However, the main components initially remained the small US Information Service group (about six officers), the CIA station in Saigon, and a few military advisers. In the next three years, the system expanded considerably and helped to blunt the NLF attempt to undermine the government. Despite progress, 1963 opened with a decisive psychological defeat for the South. For many Americans, the Battle of Ap Bac in January 1963 raised questions about the regime, the ARVN, and US strategy in Vietnam.[1]

In the twilight of Dwight Eisenhower's administration, the Laotian crisis nearly brought about nuclear war between the superpowers. A struggle between Royalist and Communist factions for control of Laos resulted in the United States and the Soviet Union choosing sides within the Cold War framework. Kennedy adviser Clark Clifford recorded President Eisenhower telling incoming President Kennedy that "if Laos fell, then Thailand, the Philippines, and of course Chiang Kai Shek [*sic*] would go."[2] He "considered Laos of such importance that if it reached the stage where we could not persuade others to act with us, then he would be willing, 'as a last desperate hope, to intervene unilaterally,'" according to a 24 January 1961 conference memorandum. Kennedy almost immediately began to look for ways to defuse the Laotian crisis. The resulting Laotian neutralization pact made Vietnam the new fallback position for American security. Kennedy did not believe that the loss of Laos would make the defense of Vietnam more difficult.[3]

President Kennedy Shifts Strategy

Immediately after his inauguration, President Kennedy placed renewed emphasis on counterinsurgency. He ordered an evaluation of capabilities and needs, focusing on Vietnam. Prior to Kennedy's inauguration, General Edward Lansdale conducted a fact-finding trip to South Vietnam from 2 to 14 January 1961. Within his first week in office, Kennedy expressed his desire that the CIA get "guerrillas to operate in the North." This expanded program involved greater use of covert propaganda against North Vietnam. All this indicated a rapid shift in US policy toward Vietnam. Additionally, Kennedy began to transfer much of the worldwide responsibility for irregular warfare from the CIA to the Pentagon in reaction to the failed Bay of Pigs operation later that spring.[4] In another change, whereas Eisenhower sought to avoid a propagandistic tone in USIA programs, Kennedy brought in journalist Edward R. Murrow as USIA director and immediately changed to a more aggressive stance. His new mission statement included actively "influencing public attitudes in other nations."[5]

By 1961 there were thirty USIS personnel operating in Vietnam. Additionally, USIS had hired Vietnamese USIS representatives for each province to work alongside the Vietnamese Information Service. Funding rose dramatically, with the USIA budget for Vietnam growing to $750,000 by 1963 and to more than $2.7 million in 1968. Despite the eventual improvements, Lansdale noted that the North was "way out in front" in the propaganda battle at the time. He believed the North's messaging capabilities were stronger than those of the South. Comparing communications capabilities, North Vietnam had eleven radio stations, all based in Hanoi, whereas the South had twenty-two government-owned stations around the country. According to Lansdale, the Hanoi stations had stronger signals, however.[6]

Upon Lansdale's return from Vietnam, he produced a memorandum for the secretary of defense. He noted that without mobilizing their total resources, the South Vietnamese could do little more than postpone defeat in the face of increased VC activities. The nation required expanded psychological operations capability to assist in this mobilization against the NLF's information and military intensification. Despite Diem's imperfections, Lansdale argued, "we must support Ngo Dinh Diem until another strong executive can replace him legally."[7] He noted that President Diem felt isolated by American criticism and that he had begun to withdraw into a shell. In addition, Lansdale encouraged appointment of a new ambassador as well as other personnel changes in Vietnam because of a perceived inability to work with and influence Diem. In Lansdale's mind, Diem was the indispensable man. As he wrote, "The next time we have become 'holier than thou,' we might find it sobering to reflect on the DRV [North Vietnam]. Do the Soviets and the Chinese Communists give Ho Chi Minh a similar hard time, or do they aid and abet him?"[8]

Lansdale noted that although the NLF had made gains in 1960, it had "neglected doing sound political work at the grass roots level and broke one of Mao Tse Tung's cardinal rules. Many people in the South now under their thumb are unhappy about it, but too terrified to act against these new rulers."[9] Treating the people harshly had created psychological opportunities for exploitation, which he discussed further in his recently drafted "Basic Counterinsurgency Plan" for South Vietnam. Lansdale wrote of the need to "foster a spirit of national unity and purpose among all elements of the Vietnamese society [and] strengthen the people's confidence in and respect for the RVNAF as a security force vis-à-vis the VC."[10] He also urged the United States to assist in raising South Vietnam's international stature.

The "Basic Counterinsurgency Plan" envisioned intensifying psychological operations to keep the populace informed of what the RVN was doing on their behalf in order to "strengthen their feeling of participation in government and thus their loyalty to it." Among the goals Lansdale set were expanding communications facilities and an improved public relations and strategic communications policy for the RVN. He also argued for a more robust counterpropaganda program to expose the fallacies of the DRV program.[11] In many ways, the plan reflected Kennedy's vision and the liberal consensus that focused on managerial and technical solutions to development problems.[12]

On 3 February, in the wake of Lansdale's memo, President Kennedy signed National Security Action Memo (NSAM) No. 2, directing the development of counterguerrilla forces. "In consultation with other interested agencies," the president ordered the examination of the "means for placing more emphasis on" that mission.[13] Furthermore, the president wrote National Security Advisor McGeorge Bundy three days later regarding a strategic PSYOP message in this connection. He asked whether a recent counterguerrilla case study by Lansdale could be used as part of a strategic communications program. In his letter to Bundy, Kennedy noted that he thought that Lansdale's study "would be an excellent magazine article for magazines like the *Saturday Evening Post.* Obviously, it could not go under Lansdale's signature. . . . It seems to me they would find it interesting and it might serve as an example of what can be done."[14] If Bundy felt that publication was not a good idea, Kennedy urged that US ambassadors in Asia, and the CIA, still provide the story to reporters for use on background. He ordered him to "make sure that this type of material has good distribution." It was published anonymously in May, discussing the creation of village security in the Mekong Delta.[15]

Lansdale and Diem

Lansdale also had advice for Diem. He wanted the South Vietnamese president to become more of an American-style populist leader. "Perhaps the wisest move would be to call in the younger people among the opposition. It would be best if you talked to them personally," he wrote to Diem that January.[16] Lansdale's advice harkened back to his experience in American journalism and advertising, as well as his efforts in the Philippines on behalf of Magsaysay. However, he did not seem to grasp traditional Vietnamese conceptions of what a leader should do. Despite providing typically sage advice, Lansdale embodied the problem of cultural misunderstanding that encumbered the American PSYOP effort throughout the war. Although Diem listened to people like Lansdale, his Civic Action Program reflected his own ideas more than an American formula. Built around "communitarianism, self-sufficiency, and community development, Diem stressed that village residents should participate in and contribute to all Civic Action projects," in line with his ideology of "Personalism."[17]

In contrast to most historians' characterization of Personalism as a discredited cult created by Diem, it was a deep and broad movement in French-Vietnamese intellectual circles. Personalism arose out of the work of the 1930s French Catholic philosopher Emmanuel Mounier's critique of liberal capitalism and of Marxism. Diem thought this philosophy best adapted traditional Vietnamese beliefs to a changing world and could provide an ideological counterbalance to the lure of Marxism. Diem's brother Ngo Dinh Nhu was the chief proponent of the ideology in Vietnam. The movement sought to balance "human material needs" with what Mounier called "spiritual considerations." Nhu and Diem fused this with Confucian ideals stressing personal duty, rather than rights, in order to create a third force that could mobilize the people. The historian Edward Miller argues that Diem was not the hidebound traditional mandarin he is often portrayed as. According to Miller, Diem "was much more interested in how he might use particular Catholic and Confucian principles to craft what he believed was a distinctively Vietnamese vision of development for Vietnam."[18]

John Richardson, who replaced William Colby as the CIA chief in Saigon in June 1962, recognized the need for active peasant loyalty, saying that he was "receptive to Nhu's exposition of an esoteric doctrine called Personalism." Colby, the CIA's Far East division chief, suggested that Nhu write an article on the Strategic Hamlet Program and on the brothers' "social and politico-economic program" for the journal *Foreign Affairs*. Colby and Richardson hoped it would counter Diem's image as ruling a "right wing, authoritarian, bigoted administration."[19]

Despite later taking part in the coup that overthrew Diem, General Tran Van Don, the future minister of national defense, stated: "Diem's early actions were tremendously popular. In the army, all ranks were enthusiastic at our unified spirit after the suppression of the sects and the rallying of their members to our side." Even in the countryside, Diem was popular initially. Don related how farmers told him that "in ten years of leadership by Ho Chi Minh, he never once took the trouble to visit us, but after only ten days, Ngo Dinh Diem has already come to see us and listen to our problems."[20]

The new US administration quickly moved forward to implement its vision for low-intensity capabilities, in contrast to Eisenhower's reliance on nuclear deterrence. In search of results, Kennedy's men pushed back against bureaucratic inertia. On 9 March, Bundy sent NSAM No. 29 to the secretary of defense and the director of Central Intelligence. It called for increased guerrilla operations against Hanoi. The president ordered "that we make every possible effort to launch guerrilla operations in Viet-Minh territory at the earliest possible time" and ordered a prompt reply on what steps were feasible.[21]

In April 1961, Lansdale continued to urge that the United States emphasize psychological and political support to Vietnam. He pressed for Vice President Lyndon Johnson to "visit Saigon and announce U.S. determination to support Vietnam's desire to remain free." Indeed, Johnson visited Vietnam in May 1961 and did just that. Lansdale also urged the creation of a national reconciliation program such as the one he had organized in the Philippines. It proved to be the precursor to the successful Chieu Hoi (Open Arms) program that Diem announced in 1963. Chieu Hoi was a program designed to entice VC supporters to lay down their arms and support the RVN.[22]

Throughout this period, PSYOPs were seen as an integral part of fighting what Kennedy called "brushfire wars" as well as the larger ideological struggle between the United States and Soviet Union. In a series of NSAMs in June 1961, Kennedy continued to expand his emphasis on counterinsurgency. He ordered the US Department of Defense (DOD) to determine precisely its counterinsurgent force needs.[23]

That spring, Kennedy also ordered an increase in military forces in Vietnam. Reinforcements included transferring sixteen H-34 helicopters and four C-130 cargo airplanes from the DOD to the CIA for use by Civil Air Transport, the forerunner of Air America. Most of the unconventional capabilities in the active US Air Force had been downsized and transferred to the Air National Guard in 1954. However, the air force retained a limited capability composed of two troop carrier squadrons, one of them based on Okinawa. Working in conjunction with the CIA, this unit had supported operations aiding Tibetan rebels against the Chinese. In line with President Kennedy's emphasis on counterinsurgency, the air force expanded and reorganized these squadrons into Air Commando Squadrons (ACSs). Elements of these squadrons arrived in Vietnam with C-47 aircraft in November 1961 equipped with belly-mounted loudspeaker systems for PSYOP use, part of Operation Farm Gate. Farm Gate included aircraft to support ground-attack and commando operations, as well as other PSYOP aircraft. Kennedy authorized this move in October, along with an expansion of the US military effort in Vietnam.[24] Defoliation aircraft, part of Operation Ranch Hand, arrived at nearly the same time. The defoliant operation became a potent theme in NLF and DRV propaganda against the Americans' actions in Vietnam. The primary goal of the spraying was to clear the jungle along roads and around bases, although later missions included crop eradication. The first aerial spraying under Operation Ranch Hand occurred on 10 January 1962 using the herbicide Agent Purple.[25]

As noted above, after-action reviews from the Korean War described the poor performance of aircraft-mounted loudspeaker systems. However, the lesson had apparently gone unresolved. Tests of the C-47-mounted system in Vietnam found that due to the Doppler effect the belly-mounted speakers "kept changing pitch as the aircraft approached and departed, leaving no more than two or three intelligible words out of a complete sentence." As a consequence, these aircraft were relegated to leaflet drops while a technological fix was devised. The fix took two years and eventually required side-mounted loudspeakers.[26]

As his first year in office ended, President Kennedy became concerned that the military was not doing all it could to utilize Civic Action projects in support of psychological operations. Within the limits of military necessity, he ordered the DOD to "encourage local forces to undertake civic action projects as an indispensable means of strengthening their society's economic base and establishing a link between army and populace." In the view of the administration and military doctrine, Civic Action projects were perceived as seamlessly supporting the PSYOP effort to influence behavior at the tactical level and to help maintain support both domestically and internationally for the mission. Good deeds, while good in themselves, were done for the psychological benefit to be accrued.[27]

Revised American PSYOP Doctrine

The 1954 Geneva Accords limited the United States to about 700 military advisers in Vietnam. Although not strictly bound by this number, the United States tried to at least appear to comply. Seeking to halt the perceived military decline in South Vietnam, Kennedy approved the special presidential military adviser General Maxwell Taylor's November 1961 recommendations for increased US assistance and the quadrupling of advisers to 3,200 men. This increased commitment included "two helicopter companies, a squadron of fighter aircraft, communications, intelligence, and other U.S. elements to the war effort." Kennedy also liberalized the rules of engagement and combat roles, allowing advisers to accompany ARVN units in a wider array of operations. To oversee this rapid growth, Kennedy established the US Military Assistance Command, Vietnam (MACV) under General Paul D. Harkins in February 1962. The first PSYOP Mobile Training Team (MTT) arrived that same month.[28]

The US Army issued an update to its PSYOP doctrine and force structure in *Field Manual 33-5, Psychological Operations* (January 1962). Pursuant to FM 33-5, the definition of *psychological warfare* was now a subset of psychological operations and referred to the use of propaganda "in time of war or declared emergency." The revised 1962 doctrine expanded and clarified the 1955 manual. Although there was little change in the TO&E for PSYOP units, the new doctrine placed much greater emphasis on the role of intelligence in PSYOPs. This change was clearly a response to problems observed in the Korean War. In line with Kennedy's preference, the US Army highlighted PSYOP support to counterinsurgency operations. However, this emphasis was largely theoretical due to limited recent experience with such operations.[29]

Under the revised doctrine, the renamed Broadcasting and Leaflet Battalion as well as the Loudspeaker and Leaflet Company remained the basic US military PSYOP units. The doctrine built the battalion staff around the standard "S" staff sections, but the Radio Broadcast Company, Reproduction Detachment, and Consolidation Company remained cellular in structure. The separate L&L Company structure remained unchanged.[30]

FM 33-5 explained the importance of measuring PSYOP effectiveness but was realistic in noting inherent difficulties. It suggested having prisoner panels pretest products to ensure they were appropriate for the mission prior to dissemination. It called for analyzing captured documents and enemy reactions such as "tightening discipline against troops who pick up propaganda leaflets." Interrogating prisoners also provided indicators of effectiveness. Without such metrics, measuring progress was nearly impossible. Unstated was the need for a clear statement of psychological objectives and measurable supporting objectives. Accordingly, the doctrine stressed the criticality of intelligence in developing the accurate psychological profile required to help the PSYOPs officer "identify his targets, reveal their vulnerabilities, and indicate the effectiveness of his effort." However, without a change in the TO&E, PSYOP units lacked sufficient intelligence personnel to effectively do this.[31]

PSYOP units continued to be dependent on a military intelligence system preoccupied with studying the enemy's order of battle rather than the human terrain—that is, the complex array of people active within the operational area. This focus made sense in a conventional war because the enemy force was typically the "center of gravity." However, the center of gravity during an insurgency may include such factors as enemy propaganda and tax-collecting capabilities. If the intelligence structure does not ask the right questions to gather information on what may be important, then leaders may miss the center of gravity and waste their efforts. Although the enlisted Military Occupational Specialty that comprised most PSYOP positions was "96B, Military Intelligence," personnel were functioning in an operational, rather than an analytic, role. Additionally, most officers viewed PSYOP assignments as a career detour. Prior to assignment, many had no prior experience in intelligence or psychological operations generally. These factors limited the internal analytic capabilities of PSYOP units.[32]

In line with Kennedy's emphasis on counterinsurgency, the US Army Special Warfare School at Fort Bragg began offering a course for counterinsurgency special warfare staff officers in March 1962. The course's goal was to "provide commissioned officers and civilian personnel with a general knowledge of the latest doctrine and techniques of unconventional warfare, psychological operations and counterinsurgency operations." Additionally, advisers were scheduled for a twelve-week basic course in Vietnamese. All these moves aimed at overcoming the advisory communications problems noted in after-action reviews of operations up to that point. Unfortunately, the expansion of the advisory effort continued to outpace the ability to overcome the obstacles.[33]

As the US Army struggled to generate a force sufficient for the challenge, the CIA stuck to its foundational doctrine for its North Vietnam operations. Like many organizations, the CIA was a prisoner of its institutional memory. Based on OSS operations and the early Cold War experience, the CIA initially focused on agent operations in North Vietnam. Despite poor results infiltrating teams into Communist-controlled areas, this technique became the standard CIA response to any demand for action as "the way we do things."[34] This aspect of CIA operations in Vietnam would have tragic consequences; the covert PSYOP program it engendered was to have mixed results. Beyond infiltration, the agency established covert radio stations, conducted leaflet drops in the North, and supported the advisory effort in the South.

The Special War

With this increase in US interest, America's "Special War" had begun. Radio Hanoi and associated platforms used this term to describe the influx of American forces, specifically Special Forces, using counterinsurgency tactics.

Much of this activity took place in the Central Highlands, training various Montagnard (highland) tribes. The North relentlessly attacked the program using divisive themes, concerned that it threatened to cut off access to the populated coastal regions.

Quoting the Liberation Press Agency (LPA—the VC press arm), Radio Hanoi accused Diem of indiscriminate attacks on Highlands villages, asking ARVN soldiers from that region: "What would you think if the village where you were born was strafed and bombed by Diem's planes?"[35] Later, in a broadcast in the Rhade language, Radio Hanoi accused the Americans of destroying the village of An Lao. The announcer then plaintively asked: "Have you ever realized that you have been fooled? You must realize that you have chosen a very dangerous path." These divisive themes played on the honor of the Highlanders and aimed at building a common interest with the Vietcong. They warned "those who have been fooled by the U.S.-Diem clique and have joined the enemy that unless they return to their families and their villages, they are digging their own graves." The broadcasts attacked the United States and specifically US Special Forces for constructing roads and airfields and establishing camps and intelligence networks in the Highlands. *My-Diem* (US–Diem) was a term used repeatedly to portray Diem as a lackey of the United States who was dependent on its support. Picking up on atrocity themes reminiscent of World War I, Radio Hanoi in mid-August claimed that the My-Diem cut off "the heads and limbs and opened up the abdomens of the people." Furthermore, it claimed that "the U.S.-Diemists have caused much sadness and suffering to the compatriots, although they shamelessly pretend to be generous, in distributing goods to the compatriots."[36]

A few days later, Radio Hanoi accused Diem of using "lying propaganda and demagogic tricks" to force "the compatriots to join the concentration camps" and build strategic hamlets.[37] The Agrovilles became a failure, in part due to the resurgence of VC violence after 1960. Diem then implemented the Strategic Hamlet Program. This program was similar to the Agroville project, but it combined village improvements, self-defense, and Personalism's concept of self-development in order to expand the government's control of remote villages. The program reflected South Vietnamese goals, not American military objectives, and generated critical commentary from the American advisers. Diem saw Strategic Hamlets as a solution to the three enemies: "communism, underdevelopment, and disunity." The goal was to cut the ties between guerrillas and villagers, forcing the former to resort to coercion to gain local support. This would, he hoped, poison Front relations with the villagers.[38]

Albert Pham Ngoc Thao, a close friend of Truong Nhu Tang, impressed Diem with his ideas on counterinsurgency warfare. As a result, Diem sent him to "Malaysia to study the counterinsurgency techniques that had been used so successfully against the Communist guerrillas there." According to Truong, it was Thao—in actuality an underground Communist agent—who convinced Diem to expand the Strategic Hamlet Program rapidly in hopes of ensuring failure, and "under his supervision the strategic hamlets created even more hostility among the peasants than had the Agrovilles before them." At the time, Thao's connection with the Communists was not known to the administration in the South. He served Hanoi as a source for news on how he was subverting the Strategic Hamlet Program.[39]

Despite Thao's access, the NLF suffered military setbacks due to the program. Inability to compete on a material basis led the Communists to attack the results of that progress. In the propaganda war, the VC information service disparaged the Strategic Hamlet Program, which it considered integral to America's "Special War." Responding to Diem's information services call for Highlanders to move to Strategic Hamlets, Radio Hanoi accused him of wanting to exploit the wealth of the region. NLF propaganda made much of the hardships incurred by peasants during the rapid expansion of the plan in an effort to mobilize them against it.[40]

Among a litany of attacks, Radio Hanoi said that Diem had "sown dissension among highlanders and seduced them into joining their ranks and struggling against their compatriots" and had "forbidden the Vietnamese in the delta to sell rice, salt, dried fish, and so forth, to highlanders." In a strange twist that makes one wonder if it could have been a covert broadcast meant to discredit Hanoi's propaganda line, it even attacked Diem for allowing "some highlanders to be elected National Assembly deputies to show that highlanders and Vietnamese are equal."[41]

In yet another attack by Radio Hanoi on Diem's Strategic Hamlets, the broadcaster claimed that "our compatriots in the South have already destroyed some hundreds of these strategic hamlets." Hanoi understood the serious threat to the revolution that the program represented. Indeed, Ho Chi Minh personally remarked during a November 1962 politburo meeting that "we must figure out a way to destroy them." Diem came to believe that he had to expand the program rapidly to outpace the NLF's ability to destroy them. That August, Radio Hanoi had already claimed the destruction of 256 hamlets. Although that may well have been true, the expansion of Diem's program far exceeded the actual numbers destroyed.[42]

In VC-controlled areas, the party organized its own version of strategic hamlets, which were labeled "combat villages." Within this concept, land reform became one of the lures to engender popular support. The party viewed land reform as an "integral part of people's war." Party leaders considered it to be a "strategic task which must be carried out regardless of cost, in order to produce an impact on the peasantry and in order to set the peasantry in opposition to the government and the landlords." Initially this counterprogramming helped maintain support in rural areas. Later, as taxes rose and became more arbitrary, support waned.[43]

To counteract growing VC influence in the villages, the CIA initiated the Census-Grievance program. Starting in Kien Hoa Province in 1961, Census-Grievance teams entered villages ostensibly to conduct a census while eliciting grievances from the people. Rufus Phillips worked with Lieutenant Colonel Tran Ngoc Chau in developing this program. The kinds of questions asked, according to Phillips, were: "Have there been recent problems in your hamlet?" "Who caused the problems?" and "What would you like the government to do to help you?" This helped in determining the level of VC activity as well as the quality of government administration. In addition, the teams did not punish VC members; instead the census teams attempted to convert them.[44] Even though an actual census was conducted, it was merely a pretext to conduct interviews of villagers. Also, Census-Grievance teams gave the province chief a way to monitor popular sentiment and to obtain a reliable head count, gauge what it was the villagers wanted, and determine the level of corruption present. Because everyone was being asked questions, the system provided anonymity in hopes of eliciting honest answers and useful intelligence despite asking delicate questions.[45]

In contested areas that the Census-Grievance teams could not infiltrate, the South Vietnam Ministry of Information deployed Armed Propaganda Teams. Major Nguyen Be formed fifty-member APTs to move "into a village and reactivate its local government and services [before] moving onto another after several-weeks stay."[46] Major Be's efforts were part of the initial CIA attempts to thwart the Vietcong's infrastructure such as it was. APTs were the forerunners of Revolutionary Development Teams that were deployed later in the war.

Meanwhile, President Kennedy had settled the Laotian crisis that had bedeviled his first days in office. To defuse tensions with the Soviet Union and remove Laos as a flashpoint, Kennedy sanctioned an agreement that neutralized Laos. All external parties were required to refrain from interfering in the country. Signing the Geneva Agreement on Laos dated 23 July 1962, the communist Lao Dong Party General Secretary Le Duan saw it as a model to settle the Vietnam issue short of war. The agreement also created a coalition government—something that President Eisenhower had cautioned against. However, in the view of the Kennedy administration, the agreement indeed avoided a broader regional war (and a troublesome US–Soviet flashpoint), even though it opened South Vietnam to infiltration by Communists. To the US administration, it was an acceptable risk, as the insurgency in the South "was not yet seen as an imminent threat."[47]

The Laos agreement on neutrality had been signed just two weeks prior when Radio Hanoi used the legitimacy conveyed by the State of Vietnam's former prime minister, Tran Van Huu, in a broadcast to the South. Radio Hanoi quoted from a National Liberation Front story claiming that Huu "declared his approval of South Vietnam pursuing the path of neutrality." Huu reportedly "voiced hope" that reunification could be achieved. Earlier, during the negotiations over Laos, Radio Hanoi quoted an Associated Press (AP) story stating that NLF leaders proposed settling the war by "the establishment of a national coalition government, general elections, neutrality, nonparticipation in military alliances, withdrawal of U.S. aid, and the holding of an international conference like in Geneva."[48]

Radio Hanoi noted the tendency on the part of intellectuals in the South toward neutralism in the wake of the 1962 multilateral agreement on Laos. In addition, using neutrality as a theme was a "way to buy time for revolutionary forces and for the party in the South." This softer propaganda line reflected a rhetorical shift in the face of renewed strength among the South Vietnamese administration. This tactic resurfaced during the war when the NLF came under greater military pressure.[49]

South Vietnam also began expanding its PSYWAR forces in light of the combat situation on the ground. On 1 May 1962, the ARVN's Joint General Staff activated two more PSYWAR battalions. This move provided one battalion each for the I, II, and III Corps Tactical Zones (CTZs), which geographically divided South Vietnam for military operations. (A fourth battalion was added later.) Each battalion consisted of three PSYWAR companies.[50] Radio Hanoi, in its broadcasts to the South, continuously attacked the expanded PSYWAR actions, which they characterized as "uttering contemptible slanders in the hope of lessening the prestige of the [NLF]," especially the South's campaign criticizing "North Vietnam's aggressive and subversive activities."[51]

Despite increased infiltration of soldiers and matériel through Laos, the burgeoning civil war within South Vietnam turned a corner in favor of the Diem administration. By 1962, the South Vietnamese military and administrative leadership was improving. After seven years in office, Diem's cohort of young nationalists had replaced many of the holdovers from the French colonial occupation. Along with increasing US military aid, this new generation of leaders brought about civic progress. These factors helped decrease the violence associated with increased Communist infiltration, thwarting any hope of decisive victory by the North through military action. The Vietcong took a major beating during this period, although it was not clear from the reporting at the time. The official PAVN history of the period states: "Just as the political struggle combined with armed struggle began to develop in late 1960–early 1961, the enemy launched a vicious counter attack." It continued:

"Using large numbers of troops, superior mobility and heavy fire power, the United States and its puppets constantly attacked our bases in the mountains, mounted sweeps and blockades of the contested areas, and seized and occupied portions of our liberated areas in the lowlands." By the end of 1962, the South had defended itself against Hanoi's initial attempts at subversion and conquest. Instead, the North "turned to a strategy of eroding America's will to support Diem by means of protracted low-intensity warfare."[52] (In his 2014 book reflecting on this history, the USIS field operator Frank Scotton described the activities of the ARVN 4th PSYWAR Company, operating near Qui Nhon City during this period. Teams conducted nighttime loudspeaker missions in conjunction with ambushes. Mortar-launched flares lit the way home while coordinated loudspeaker messages urged the Vietcong to surrender.[53])

On 12 May 1962, the CIA's Covert Action Station in Vietnam reported that a covert radio station run by the ARVN's Psychological Warfare Directorate had begun broadcasting from the Quang Tri area. This radio station claimed to be an NLF-affiliated station based in North Vietnam. The target audiences for this covert station were North Vietnamese military officers and middle-class urban residents.[54] This was possibly the spurious NLF Liberation Radio that the Saigon station chief William Colby had started. It broadcast on an adjacent frequency to the real Liberation Radio and generally sounded like the original "except for certain false segments that, it was hoped, cast aspersions on the VC in the minds of the listeners."[55]

In a blow to the CIA commando program, Radio Hanoi reported in July that "twelve U.S.-Diem spy-commandos, including two frogmen, were brought before the military court." They had been captured nearly a month prior to the announcement. In a letter of protest read on air, Radio Hanoi accused the United States of crying "stop thief" for complaining about infiltration from the North while conducting similar activities against the North. It alleged that the American-controlled 1st Observation Group (presumably ARVN, it was actually a CIA-sponsored organization that evolved into Strategic Technical Directorate) conducted this operation after the commandos attended a fourteen-week course in Taiwan. When captured, the observation group's equipment bore US markings. Furthermore, Radio Hanoi connected this operation to a C-47 cargo plane shot down the prior year bearing Civil Air Transport markings. According to this broadcast, "the sending of frogmen to North Vietnam for sabotage is no novelty for the U.S. imperialists and Ngo Dinh Diem." It referred to a 24 May 1961 *Wall Street Journal* report on American plans to conduct "sabotage and subversion in . . . North Vietnam."[56]

However, Radio Hanoi's internal broadcasts were not so cavalier regarding such incursions. A *Nhan Dan* editorial titled "Let Us Continually Raise Our

Vigilance and Smash the U.S.-Diemist Sabotaging Schemes" told domestic listeners: "In the past few years the U.S.-Diemists sent, on many occasions, spies and rangers into North Vietnam." The editorial continued by alleging that the commandos "joined with agents, who had been left behind, and other counterrevolutionaries in spying on our military strength, on national defense plans, and our political and economic situation and in carrying out activities of sabotage."[57]

However, at this time the North faced internal rebellion over the consequences of its collectivization program. The arrival of RVN commandos exploited popular disenchantment, "particularly in minority-dominated areas" in the North. In some areas, these ethnic minorities may have joined forces with "disillusioned southern regroupees to stir up trouble." Fears that the commandos would fan these flames forced Hanoi to intensify operations in order to suppress dissent. The regime increased repression and arbitrary arrests to deal with this, which "involved forced relocation of rebellious tribes."[58]

To counter US support for Diem, the North Vietnamese and VC propaganda agencies began to cultivate a symbiotic relationship with American elements that might be opposed to Kennedy's actions. In July 1962, Radio Hanoi read a letter from the Vietnam Mothers' Association to women in the United States calling on them to join in "urging the U.S. Government [to] halt at once the armed aggression and terrorist raids in South Vietnam." The letter listed cases of American bombings and mentioned the use of "noxious chemicals to destroy crops and vegetation." It closed by expressing thanks to the sixteen "American intellectuals and Mrs. Eden Mirke in New York and other peace and justice loving people in the United States for their support to the South Vietnamese people's struggle." Radio Hanoi also wished success to the American World for Peace Movement.[59]

Despite the ARVN's struggle throughout 1962 to undo the VC advances during the previous year, many villages remained under Communist control. In contested villages, the daily pressure of Communist Party organs contrasted with the "sporadic appearances of government officials or soldiers." Regardless of the villagers' preferences, this left them with the choice between accepting VC rule or facing reprisals.[60]

However, determining who controlled a village was not as simple as counting weapons or miles of fortifications. Such numbers, though useful for some purposes, provide little information on the real center of gravity in an insurgency, "the balance of political power in the countryside[,] or the political loyalties and views of the peasantry."[61] Problems measuring progress not only affected Americans and the government of South Vietnam. In an interview conducted with a VC village cadre who later rallied to the Government of Vietnam (GVN), a man said, "Generally speaking, I didn't dare report truth for fear of being criticized. . . . I reported that the villagers were determined to support the revolution when in fact they weren't." In the end, he "almost had to arrest" people to force them to attend indoctrination sessions and doubled the attendance figures in reports.[62] Until the advent of the Hamlet Evaluation System (HES) by MACV in 1967, both sides used subjective reports to estimate loyalty and village control. Controversial and open to criticism on several fronts, the HES at least provided a variety of useful data for determining who controlled a village, which helped to focus the information program.

The Battle of Ap Bac

Despite an internal understanding of the problems faced, the North Vietnamese propaganda line remained constant, buoyed by American press reports. Radio Hanoi quoted one story by United Press International (UPI) reporter Neil Sheehan, who accompanied an ARVN operation in the Ca Mau Peninsula. The primary source for his story was a "high-ranking military adviser with considerable experience." According to this American officer, the troops "showed no enthusiasm in seeking out the Viet Cong [and] angered the already alienated peasants by swooping up most of the chickens and ducks which ran among the grass huts."[63] This mirrors ARVN 7th Division adviser Lieutenant Colonel John Paul Vann's accusations later made after the Battle of Ap Bac, forming the basis of the myth of South Vietnamese and ARVN failures preceding the overthrow of Diem.

Radio Hanoi's domestic service prepared listeners for the coming year in a 27 December 1962 broadcast. Despitetheir attempts to put on a good face, 1962 had not gone well. The conflict "more clearly had the character of an extremely desperate, protracted and arduous struggle against foreign aggression," according to the report. Amid clearly exaggerated descriptions of victories, the broadcast indicated flagging interest, saying that the struggle movement was "not as intense and large as in the previous year." It claimed credit for the destruction of 1,436 strategic hamlets in 1962, although almost all of these "were repaired at once by the U.S.-Diemists" and therefore they could only claim 300 completely destroyed—less than 10 percent of the total. The attempt to achieve rapid victory had failed, and the propaganda line reverted to emphasis on the protracted war. Even Wilfred Burchett, the well-known leftist Australian journalist, acknowledged that the government held the "strategic and tactical initiative" at this point.[64]

Days after that disappointing broadcast, the North's luck took a dramatic turn. Shortly before noon on 3 January, Radio Hanoi broadcast a report of a large-scale battle the previous day near the hamlet of Ap Bac. At this point, a crossroads in the propaganda battle occurred. The report claimed that NLF guerrillas had killed three Americans and shot down five helicopters. Describing the fight, the announcer said that "some 1,200 Ngo Dinh Diem troops supported by aircraft and amphibious armored cars began the raid. More than 500 paratroopers were later thrown into the battle as reinforcements."[65]

The target of the attack was a VC radio transmitter for controlling operations in the Plain of Reeds area southwest of Saigon. The importance of this target meant that defending the village was not a matter of choice for the Vietcong. Intelligence indicated about 150 guerrillas on the objective, to be attacked by 300 airlifted soldiers. They were supported by blocking positions held by a mechanized company and several less-qualified Civil Guard battalions. In reality the village was held by a company each from the 514th Provincial and 261st Regional Battalions, or about 350 men total.[66]

The same day that the ARVN 7th Division launched its attack on Ap Bac, the ARVN launched Operation Burning Arrow in Tay Ninh Province. This operation targeted COSVN headquarters west of Saigon, a higher-priority target. However, no US reporters accompanied the Tay Ninh mission, and because no Americans were killed, Burning Arrow was barely mentioned in the American press at the time. That operation expended most of the helicopter support then available in South Vietnam, leaving just ten H-21, twenty-passenger "Flying Bananas" available. As an economy-of-force mission, Ap Bac did not have priority for air strikes, airlifts, or reserves.[67]

The first lift landed about 0700. Fog delayed the second and third lifts, meaning that the outgunned first company confronted the friction of battle alone. By 0935 the second and third lifts had landed, but the force now confronted a tougher situation associated with five downed choppers. The flight leader of the third lift, a US pilot, ignored the recommended landing zone and set down just 200 yards west of the VC-occupied treeline, well within machine-gun range. Four craft were lost within five minutes, another shortly after. Several canals barred the arrival of the mechanized unit of M-113 armored personnel carriers to support the downed helicopters because the banks were too steep to safely enter the water. Crossing these canals took hours, and as a result the armored personnel carriers did not arrive on the battlefield until about 1330.[68]

As the fight at Ap Bac evolved, Colonel Huyen Van Cao, the division commander, selected a drop zone to insert the reserve force. This unit constituted the only reserve force available due to the operation in Tay Ninh Province. It is likely that the Vietcong intercepted radio communications and knew the final airdrop plan. The drop took place at twilight. Unfortunately, due to an error in calculating the release point for the jumpers, many landed well within

VC machine-gun range. The battle ended by the evening, and by morning the ARVN was in control of Ap Bac. The VC force had departed during the darkness.[69]

On 4 January, Peking's New China News Agency reinforced Hanoi's messaging about the battle, reporting that US-Diem forces had suffered a "costly defeat." It quoted from a UPI report, probably by Sheehan, and asserted that "the Ngo Dinh Diem army and its so-called U.S. military advisers suffered 'one of the most costly and humiliating defeats'" of the war. It claimed that attackers outnumbered defenders ten to one and suffered sixty-five killed and five helicopters lost.[70] The number and type of Vietcong involved were downplayed for propaganda purposes, hyping the David-versus-Goliath narrative. Quoting from the UPI story, New China stated that "American advisers were disappointed and angered that the South Vietnamese troops should fail in one of their biggest tests after more than a year of training."[71] Recounting the battle, Radio Hanoi quoted an American pilot as saying, "It has been a bad day right from the start." Radio Hanoi broadcast the same story the following day in English, claiming 150 total ARVN casualties. These casualty numbers closely match Lieutenant Colonel Vann's with good reason: the original UPI story contains his estimates. One study put the numbers at 3 Americans killed with 6 wounded and 25 ARVN killed with 100 wounded.[72]

It is from here that the battle grew to mythic proportions. American reporters, such as Sheehan, arrived shortly after the fighting to report on the battle, largely basing the narrative on Vann's telling. That story may have been colored by Vann's attempts to deflect the results of his poor advice and American mistakes onto the Vietnamese soldiers. Communist propaganda feasted on the distorted US press reports and constructed a narrative used to divide South Vietnamese audiences, recruit new members, and dishearten American supporters of the Diem government. They were successful in each area. In retrospect, a minor, botched tactical engagement became a metaphor for ARVN failure. The misrepresentation of Ap Bac, in turn, gave the NLF the psychological initiative going into 1963.[73]

Many reporters had preconceived notions of the war at this early phase and wrote stories to support that view. Selection bias clearly played a part in the reporting. John Paul Vann, adviser to the ARVN 7th Division based south of Saigon, fed into this bias. Despite a strong dedication to the war that led him to volunteer to serve in Vietnam for nearly a decade, Vann was a self-absorbed man with serious character flaws who pulled the wool over the eyes of reporters. He deflected blame for Ap Bac by discrediting the ARVN to reporters. Contrary to Vann's insistence that Diem had ordered his officers to cease attacks and avoid casualties, Ngo Dinh Nhu, in the wake of Ap Bac, criticized what he described as his brother's insistence on "frontal attack and constant maintenance of the initiative in aggressive troop actions."[74] However, Vann's narrative influenced the flawed retelling of the Battle of Ap Bac and dealt a serious blow to the Diem government at a critical time. As Rufus Phillips later noted: "The main problem with Ap Bac was its characterization as evidence that the entire Vietnamese army was incompetent and lacked the will to fight and that everything was Diem's fault."[75]

Diem, however, saw the battle as merely a speed bump on the way to victory. He deemed the Central Highlands to be the strategic center and allocated resources accordingly. Ap Bac did not rate that high. Colonel Ha Mai Viet, ARVN armor specialist, concurred, writing that the problems in a small, regimental operation were not representative of larger ARVN capabilities. The ARVN was already conducting large, complex operations with some success. Certainly, Ap Bac was a tactical defeat, but terrain issues, Clausewitzian friction, and the difficulty of coordinating widely dispersed combined-arms operations using three languages accounted for most of the problems—not haplessness or cowardice.[76] Nevertheless, Diem's opponents used Ap Bac to attack him for "what they called the 'dictatorial' and 'erroneous' policies," and they cultivated negative propaganda prior to the coup that would oust him. The battle became symbolic of the ARVN and GVN's failures, supporting preconceived narratives. Throughout the war, Ap Bac remained a propaganda theme for North Vietnam and the budding American antiwar movement.[77]

Conclusion

Shortly after the battle, Radio Hanoi noted that the "raiding operation" on Ap Bac cost three American lives. It began a campaign to target American soldiers in the wake of this failure. The station remarked that Hanoi's Voice of Vietnam English service had recently asked American GIs in South Vietnam: "For whom and for what purposes have these men sacrificed their lives?" Playing on themes of loneliness, it also claimed that the men died for the "arms dealers and warlike rulers of the United States" who dragged GIs from their families.[78] Hanoi also instituted an "Ap Bac Emulation Campaign," which urged VC fighters throughout South Vietnam to struggle to emulate the PLAF victory in that battle. Emulation campaigns were a standard strategy in Soviet propaganda and indicate the continuity brought by Ho Chi Minh. A detailed indoctrination program from the People's Revolutionary Party (South Vietnamese Communist Party) listed weekly lectures and award ceremonies.[79] The People's Revolutionary Party (PRP) was the communist core of the National Liberation Front. That fall, the Liberation Press Agency announced the results of that campaign, claiming that between April and August

Front forces “killed, wounded, and captured 50,000 U.S.-Diem troops, including 372 Americans.” The total number of US and ARVN casualties in Vietnam was actually less than a third of that, including 78 Americans for the entire year. The LPA also reported destroying 3,300 strategic hamlets.[80] These exaggerated results led Le Duan to assert that the Americans “could not defeat us.”[81]

The Battle of Ap Bac was a psychological defeat for the South. It colored the perceptions of the ARVN and President Diem for many Americans, despite the positive trend lines in pacification. The coming spring witnessed a series of protests and Buddhist unrest that further shook the nation and ultimately led to the coup to overthrow Diem. In spite of these troubles, the PSYOP structure and adviser support had improved greatly during the previous year.

4 Buddhist Crisis and Coup, Spring–Fall 1963

Despite the negative perceptions of the government that festered among some American advisers after the Battle of Ap Bac, President Diem continued to fight the counterinsurgency. In February, he ordered implementation of the Chieu Hoi (Open Arms) program to encourage desertion from the NLF. Initially, the program was quite successful. However, as unrest spread that summer, the number of so-called ralliers dropped off. The collapse of support for Diem among key officials in the Kennedy administration encouraged supporters of a coup. Any revolution on the scale Diem envisioned was bound to create anger by those who stood to lose power. The overthrow of Diem in the fall of 1963 led to a near collapse of governance in rural areas and marginalized the South Vietnamese PSYWAR program. It also sapped the wellspring of government legitimacy in the years to follow.

In line with Lansdale's encouragement to create a VC desertion campaign, Diem announced the Chieu Hoi program during Lunar New Year in 1963, just weeks after the Battle of Ap Bac.[1] This program encouraged Vietcong to rally to the government side. That April, the ARVN's PSYWAR directorate began publishing a monthly magazine aimed at indoctrinating its soldiers about the program. That publication, titled *The Republic of Vietnam Armed Forces*, focused each issue on a specific theme, such as "Communism in South Vietnam" or "The People of Vietnam." The supporting psychological objective of the magazine was to educate soldiers about the Chieu Hoi program and to build support for it. Thus, each issue tabulated the numbers of returnees and weapons since the inception of the program. Issues also contained Diem's 17 April 1963 proclamation for all Vietnamese "to return and uphold the just cause of the fatherland and to contribute their efforts along with those of all our people in order to build, in a militant spirit, the new society and civilization where every citizen will be able to develop totally and in full freedom." Though aimed at the Vietnamese, the magazine was also published in English- and French-language editions.[2]

Supporting this campaign, the South mocked the Vietcong and North Vietnamese as Russian and Chinese clients. This utilized nationalist, anticolonial rhetoric in hopes of undermining a central attraction of the NLF. On 23

May 1963, Saigon's *Vietnam Press* (in English) noted large-scale VC attacks using Russian and Chinese weapons. The story reported recent propaganda from the North stating that Hanoi would increase support for the Vietcong. The paper charged that Communist authorities sought "by every means to sabotage these efforts toward peace, whose success would indeed hinder the expansionist designs of international Communism." The report claimed that the targets of this effort were the strategic hamlets, "which are designed to give the population a better life in peace and security and allow each to reach the full development of his faculties in dignity and justice."[3]

During the summer and fall of 1963, US PSYOP advisers and air commando aircraft increasingly supported ARVN operations. The Chieu Hoi program and appeals to surrender were the major themes used. Through testing, advisers determined that aerial loudspeakers were most effective at night, which also reduced the risk factor for aircrews. They also found that angling the speakers off the side of the aircraft mitigated the Doppler effect, which distorted and garbled the broadcasts. The first operational use of the reconfigured aerial loudspeakers took place in June 1963. In an effort to remove noncombatants from the battlefield, aircraft informed Montagnard tribesmen in contested areas surrounding the Kon Brai outpost in Kontum Province "that after a certain date anyone found in the area would be killed." Tribe members made taped messages that the aircraft repeatedly broadcast over the area. Within five days, 2,400 Montagnards had come to the outpost.[4]

Chemical Warfare Propaganda and Commando Operations

The US Air Force also continued its defoliation operations, with an emphasis on clearing roadways to prevent ambushes. Although such operations were possibly tactically sound, they presented constant fodder for propaganda. In May, Radio Hanoi broadcast a Soviet response to the use of defoliants. The Soviets reported sending a message to the British regarding "U.S.-Diemist use of chemicals as a means of war in South Vietnam."[5] In a June broadcast to Europe, Radio Hanoi again protested the use of toxic sprays. They underscored the fact that US-Diemists "use chemical substances as a war means in South Vietnam to ravage the crops and vegetation, decimate the cattle, impair the people's health, and force the people into concentration camps described as strategic hamlets." To support this narrative, Radio Hanoi quoted Rachel Carson, the American biologist and author of *Silent Spring*, as saying that "indiscriminate use of such chemical substances can have harmful effects on birds, animals, and even human beings." According to Carson, the danger came from Americans' use of chemical substances "as a war means, especially when they are being sprayed repeatedly and indiscriminately upon densely populated and agriculturally rich areas." Radio Hanoi compared such use to Adolf Hitler's tactics. Use of toxic substances was among the North's longest-lasting and internationally most effective themes.[6]

By 1963, many in the CIA began to question its sponsorship of the South Vietnamese agent and commando insertion program in the North. Political limitations placed on the CIA meant it could never do more than agitate people in the North. Although the original CIA interest was to gather intelligence in North Vietnam (and, as President Kennedy commented, give the North "a taste of their own medicine"), the CIA's interest in using propaganda to incite resistance forces gathered steam. This was a delicate matter, however, due to the Eisenhower administration's failure to support the Hungarian resistance during the uprisings in 1956, when Western powers stood by in the face of a rebellion that some charged CIA propaganda had incited. One idea was to create notional teams to harass the North. In this sense, "notional" meant fictional teams intended to deceive Hanoi.[7]

Herbert Weisshart, a covert political action specialist, arrived in Saigon in March 1963 to organize a notional resistance movement for the CIA. Weisshart had experience providing PSYOP support to imaginary teams in China and later helped to draft Operational Plan (OP) 34A, which transferred covert operations in Vietnam to the military. Appealing to Vietnamese mythology, Weisshart created the Sacred Sword of the Patriots League (SSPL). In 1428, the mythological Sacred Sword had been delivered to the tale's hero by a turtle swimming in Hoan Kiem Lake in Hanoi. Using the power of this sword, the Vietnamese defeated the invading Chinese. Weisshart hoped to utilize this myth—filled with nationalist appeal—to fuel dissension between Vietnamese people and Chinese Communists. He recalled that "it would provide an ostensible sponsor for real teams on the ground and, if all went well, would provoke paranoia in the DRV hierarchy."[8] To support their propaganda efforts, both sides used appeals based on Vietnam's militant history of resistance. For example, at various times both sides used the term *Lam Son*—the fifteenth-century guerrilla Le Loi's anti-Chinese base—to identify their military operations. To support the SSPL deception, Weisshart began programs like Operation Loki.[9]

Under Loki, kidnapped North Vietnamese fishermen were brought to Cu Lao Cham Island, otherwise known as Paradise Island, ten miles off Da Nang. They were indoctrinated at an SSPL resistance camp notionally located in North Vietnam. The men were held for several days and then sent home with PSYOP-themed gift baskets that included information products, soap, and other useful goods. Reportedly, some fishermen actually sought to be kidnapped to receive medical treatment and, according to the Studies and Observations Group commander Colonel John K. Singlaub, the chance to gain "an average of over 20 pounds" during their ordeal.[10] Also, some captives knew exactly where they were, having visited the island during the course of their lives. Eventually, SOG—the successor to the CIA in these operations—had to limit the number of times a man could be kidnapped. Additionally, Radio Hanoi very quickly betrayed the details of the program, accusing Diem's navy of illegal acts for capturing the fishermen. As with the commando insertion operations, Hanoi was quick to exploit the propaganda value of such incursions in the North. Prior to the start of Loki, in August 1962, Radio Hanoi broadcast a report about navy ships that fired shots at some fishing boats and "compelled one of these boats to sail south" with five fishermen aboard. In light of this pushback, it is unlikely that the official purpose of the campaign was achieved. Nevertheless, Operation Loki may have provided other, intangible benefits later in the war when the program was used to disseminate disinformation.[11]

In the midst of this uproar, on 10 June 1963, Radio Hanoi announced the detention and trial of additional South Vietnamese commandos captured six months earlier. According to the report, a "U.S.-Diemist ranger spy gang" had landed in Ha Tinh Province, about 130 miles north of the demilitarized zone established per Geneva in 1954. It listed names, equipment, and training, noting that the captives "proved to be efficient agents . . . in carrying out the task of acting as ranger spies" but that they were hunted down and arrested. The men received sentences ranging from ten years to death. Hanoi sentenced another team on 12 July 1963. This may have been Team Tarzan, which parachuted north of Dong Hoi in January 1963. The DRV forces caught that team shortly after insertion. In both cases, it seems clear that the lag between insertion and the announced capture had more to do with spreading disinformation and using the team as double agents than with implying the effectiveness of commando operations. Team Tarzan had remained in radio contact for about three months, despite communications anomalies indicating that they had been co-opted immediately.[12]

As part of the increased emphasis on the CIA's SSPL program, infiltration teams in the North began to assume a secondary mission: psychological warfare. The first dual-mission team, code-named Team Easy, parachuted into the North on 11 August 1963. While collecting intelligence, the team distributed SSPL leaflets. Other teams carried special-purpose, single-use mortars to disseminate leaflets after the sabotage mission was complete.[13]

Buddhist Crisis

While these events continued out of view, reports about South Vietnam flashed across Teletypes in newsrooms worldwide. Throughout the late spring and summer of 1963, Diem had been dealing with a growing Buddhist crisis at home. The crisis began that May, when Buddhists began flying religious flags in Hue. This dispute followed a visit to the city by a papal representative for the twenty-fifth anniversary of the installation of Diem's brother, Ngo Dinh Thuc, as bishop. For that ceremony the papal flag was flown, and afterward local officials had forbidden the flying of any flag but the national flag. This angered Buddhist leaders. At the Buddhist celebration on 5 May, many openly carried the Buddhist banner, leading to a police response. Protests erupted throughout the South.[14]

The international press seized on this turmoil to weaken support for the Diem government. Unwittingly, this supported the North's propaganda efforts. Subsequent North Vietnamese histories make clear the extent to which the Buddhist movement received aid and advice from agents of the North who operated within the Buddhist movement. Although the North had no direct leadership role, they attempted to agitate and manipulate the movement to suit their own purposes. In the end, the crisis had the desired effect and may have helped to lead Ambassador Henry Cabot Lodge to begin working with coup-plotters as the crisis festered that August.[15]

To the American press corps in Saigon, the Buddhist crisis was further evidence that Diem lacked public support, was unreasonably repressive, and deserved to be overthrown. However, the unrest must be seen more as a political struggle than a religious one. The movement's leaders hoped to reignite Buddhist nationalism as the basis for Vietnam's postcolonial identity. They sought to institute a devout Buddhist rule, which had certainly not been the norm throughout Vietnamese history. Spurring them on were DRV agents seeking to diminish Diem. A later history admitted that Communist agents had been active in the protests and within "mass organizations and in the Buddhist Church."[16] The 1963 Buddhist crisis, as Diem biographer Edward Miller wrote, "was sparked not merely by fears of religious persecution but also by growing South Vietnamese anxieties about Diem's nation-building agenda." An example of this was resistance to Madam Nhu, Diem's sister-in-law, and her attempts to achieve equal rights for women. The social anxiety created exploitable opportunities, which the North Vietnamese agit-prop program seized.[17]

The Diem regime sought to drive a wedge between the Buddhist leaders and the people. On 12 June, Diem broadcast an appeal to any Buddhists influenced by "plots and tricks designed to postpone a solution and to repress Buddhism," saying that he and the constitution stood in support of national Buddhism.[18] Regardless, the protest events increased psychological pressure across the nation. By 10 May 1963, CIA officers had met with the dissident Buddhist leader Tri Quang. They described him as "self-confident, dominating, committed, and slippery, but able to make a joke and take one at his own expense. He would neither admit nor deny that his goal was to destroy Diem."[19] Repeated self-immolations, whereby Buddhist extremists literally set themselves on fire, and violent actions associated with unrest that summer, divided President Kennedy's cabinet.

Compounding the narrative of the GVN's decline, the foreign press continued to attack Diem, stoking rumors of a coup. Critics of Diem wrongly believed that all the anger at Diem stemmed from his repressive measures. On the contrary, many of the disgruntled believed that Diem "was being too soft, not too tough." Seeing American press reports indicating that Diem was in disfavor with US government officials led some in the South Vietnamese military to side with coup plotters as a way to ensure that the war against the Vietcong maintained American support. Many South Vietnamese officials viewed negative stories in papers such as the *New York Times* as evidence of official US government policy.[20]

Among the sources the US media relied upon for this narrative were Albert Pham Ngoc Thao and Pham Xuan An, both of whom were covert VC agents. As a source and translator for foreign reporters, Pham Xuan An in particular was in a position to feed the desired line to people with a bias against Diem. He subsequently claimed never to have done any such thing. However, the notion that he acted as a highly awarded agent of the North but never used his access to spread propaganda to support a key psychological objective is implausible, to say the least. Additionally, the leader of the Buddhist protest movement, Tri Quang, was possibly a Communist agent as well. (His brother certainly was.) Regardless, he seemed to be on a quest for political power and a larger role for Buddhism in Vietnam. Despite repeated attempts by Diem to placate the Buddhist agitators within the movement, Tri Quang continued to attack. At a minimum, the North agitated the movement and attempted to steer it in a way beneficial to the revolution.[21]

Supporting its effort to degrade American support for Diem, Radio Hanoi excerpted an open letter to President Kennedy from 650 American professors published in the *New York Times Magazine* in July. The letter condemned the US government for violating the 1954 Geneva Accords on Indochina by continuing to pursue a policy keeping Vietnam divided so that "at least one segment of it remains under western control." Despite US aid to the ARVN, South Vietnam "was no nearer victory," according to the letter. The professors claimed that guerrillas had captured and increasingly used "American-made rifles, machine guns, and mortars to shoot down American helicopters operated by U.S. fliers." Three days later, the Liberation Broadcast Service (LBS) carried the same story in Vietnamese to South Vietnam.[22]

The following month, Hanoi remarked on a letter from Reverend Donald

Harrington, secretary of the Ministers' Vietnam Committee, that was signed by "some 15,000 American clergymen of various faiths." Northern propaganda often used large numbers of ostensible supporters to lend legitimacy to messages. Sent to President Kennedy on 14 August, the letter urged "an end to American financial and armed support for the Ngo Dinh Diem government in South Vietnam," according to foreign reports. The letter especially protested against

> 1) U.S. military aid to those who deny religious freedom to South Vietnamese Buddhists; 2) the immoral spraying of parts of South Vietnam with crop destroying chemicals and the herding of many of the South Vietnamese into concentration camps called "strategic hamlets"; 3) the loss of American lives and millions of dollars to bolster a regime universally regarded as unjust, undemocratic, and unstable; 4) the fiction that this is fighting for freedom.[23]

A story in the *New York Times Magazine* likewise covered Harrington's letter. This letter parroted, amplified, and gave legitimacy to all the North Vietnamese propaganda themes then in use, further isolating Diem from his American supporters.[24]

November Coup

August arrived with rumors swirling through Saigon of a coup against the Diem regime. Some accused the president's brother, Nhu, of stoking rumors of a clampdown and running inflammatory stories in the *Times of Viet Nam* to denounce the Buddhist movement. He used deception and disinformation to get potential plotters to reveal their loyalties and to create fear in order to control them. The unrest over the summer diverted resources from the war and to a degree set back the military.

President Kennedy had earlier decided to replace Ambassador Frederick Nolting with the former Republican vice presidential candidate Henry Cabot Lodge. Commenting on the personnel change, Radio Hanoi blamed the "Ghost of Ap Bac," saying that in the face of recent defeats "the bellicose Americans have been forced to admit their inability to implement their plan of rapid aggression in South Vietnam and to prolong this aggressive war."[25] When Ambassador Lodge arrived in Vietnam to replace Nolting, he made a point of seeking out the press and met with the influential journalists David Halberstam, Neil Sheehan, and Malcolm Browne for private dinners. While Lodge used the press in a mutually supporting attack on Diem, President Kennedy assailed David Halberstam, of the *New York Times*, accusing him of "running a political campaign" and being "wholly unobjective."[26] Despite

Lodge's prejudice, Ngo Dinh Nhu attempted to manipulate Lodge after his arrival in Saigon. He seems to have been behind positive stories in the *Times of Viet Nam* and smiling posters of Lodge almost every day.[27]

On 21 August, the Saigon domestic service carried a statement from Diem regarding his imposition of martial law to "vanquish Communism." While this was his stated reason, Diem was also trying to preempt a possible coup against his government and to crush the Buddhist uprising. The domestic station then changed its name to Voice of the Armed Forces of the Republic of Vietnam, lending support to the idea that this was a military operation in support of Diem.[28] At midnight, ARVN Special Forces under Nhu's direction raided Buddhist pagodas involved in the unrest. The RVNAF leadership supported the martial law, though they later denied support for the raids. An RVNAF statement on the declaration of martial law accused those sowing division in the past few months of cooperating "with the feudalists, the colonialists, and the communists to rush our country toward danger."[29] Meanwhile, the Liberation Broadcast Service blamed the violence associated with martial law on the United States. According to the service, in the days following the declaration "thousands of Buddhist monks, nuns, believers, students, and professors, and other people from all walks of life were killed, tortured, and imprisoned by the devils."[30]

After the ARVN seized Tu Dam Pagoda in Hue, the Saigon domestic service stated that "everyone appears satisfied with the energetic act of the army aimed at protecting security, maintaining order, and serving the people." This support proved, according to the broadcast, that ARVN actions were directed against a "group of people who take advantage of the faith of the compatriots to work for the benefit of the Communists." Reportedly, forces seized explosives and a gun from one pagoda. This and other reports made clear "that most of the pagodas have been used as shelter for reactionaries and not for truly religious persons," according to Voice of the Armed Forces of the Republic of Vietnam.[31] General Tran Van Don, the ARVN chief of staff, acknowledged to CIA contacts "that he and several other generals . . . had planned and advocated martial law," but he denied any prior knowledge or sponsorship of the pagoda raids.[32] The anger among ARVN leaders arose from the claim that the ARVN had conducted the pagoda raid rather than from the implementation of martial law. ARVN leaders ultimately used that raid to justify the coup against President Diem.

When Diem cracked down in August, many American reporters thought it would be perceived in South Vietnam the same way as an antireligious crackdown in America would have been. However, the situation was complicated. Many Vietnamese wanted order and viewed the Buddhist venture into politics as sowing disorder—a good example of how viewing events through the lens of "hearts and minds" can lead to an incorrect analysis of events. Saigon claimed that despite "deadly blows to the Communist rebels," which placed the nation on the verge of victory that year, a new group of traitors sought to "undermine the gallant fighting spirit of the republican army."[33] They accused Communist rebels, camouflaged as Buddhist bonzes, of propagandizing against the government.

Following the pagoda raids, *Vietnam Press* issued an appeal from the Buddhist Sangka Association leader, Thich Thien Hoa, to lead the religion back to its normal place in Vietnamese society. According to him, "cruel provocateurs" sowed "confusion, hatred, division, and trouble between the government and the Buddhists." These provocateurs sought to tarnish the image of the government abroad by manipulating the "reputation of virtue and by shedding the blood of true Buddhists"[34]

Details on those whom Thich Tien Hoa believed to be behind these provocations came to light on 27 August. The South Vietnamese domestic service broadcast the names and VC connections of several leaders in the movement. These included North-born Thich Quang Chau, head of the Buddhist Follower Struggle Movement. His brother was a security officer in the Vietminh. Another was Thich Cu Quang, who had worked with the VC-sponsored Buddhist Association and indoctrinated students against the government. These Buddhist front groups acted much like the united fronts that Ho had championed, mobilizing the masses to support movements orchestrated for other ends.[35] Despite these details, the crackdown sparked resistance to Diem. Frank Scotton found that religious moderates on all sides in Qui Nhon City reacted with dismay to the pagoda attacks. Meanwhile, the characterization of the crisis as reflected in American press narratives spotlighted Vietnamese shortcomings and "treaded on nationalistic sensitivities" while directly attacking "their cultural code of politeness, indirectness, and public restraint." This only further inflamed the situation.[36]

Continuing the theme that the Communists had hijacked Buddhism, the Saigon station broadcast a Vietnam Press Agency report that the people disliked these "maneuvers carried out by communists, feudalists, and colonialist henchmen who hide themselves behind the religious screen in an attempt to overthrow the legal government and sabotage the freedom of faith."[37] Later, the *Times of Vietnam* claimed to possess "captured Viet Cong documents" and have information from "at least one Viet Cong officer" proving that they planned to take control if the CIA-sponsored coup took place.[38] This echoed a rumor then current in Saigon that elements in the CIA were supporting coup plotters. Events later proved this rumor to be true. Meanwhile, Diem's support among key military leaders waned.

According to the former CIA station chief William Colby and his contacts,

Diem "might actually have achieved what they sought in the raids on the Pagodas: the suppression of the Buddhist challenge to the authority of the Government—in short, a repetition of their success against the politicized sects in 1955."[39] A memo from the CIA acting deputy director of intelligence admitted that, along with the military progress of the previous year and the increasingly successful Strategic Hamlet Program, the author said he did "not believe one can rule out all possibility of winning the war under a Ngo administration." It was not a ringing endorsement, but an endorsement nonetheless. The analyst based the idea on the positive trends over the previous year and presumed short-lived effect of the backlash resulting from the Buddhist crackdown. However, political support in Washington, especially for Ngo Dinh Nhu, collapsed as a result of the pagoda raids.[40]

As the negative image presented by press reports on the Buddhist unrest and martial law took hold in America, a Saigon radio station reacted. Despite the generous aid provided by America, the broadcasts claimed, public opinion there had been "poisoned by the ever-increasing propaganda and slanderous arguments of the Communists." According to the station, the "Communist line of propaganda and slander on the Buddhist issue" had echoed throughout the free world.[41] After declaring martial law, Diem sat for an interview with *Vietnam Press*. He laid bare the divisive propaganda plan that he saw the Vietcong implementing. According to Diem, the main goals were "discouraging people of good will, cutting off the people of Vietnam from their allies, alienating the people from the government, sowing doubt and suspicion everywhere," adding to the burdens of the hot war that the nation was waging.[42]

As Diem reasserted civilian control in September, he answered announcers' questions regarding the speed of the expansion of the Strategic Hamlet Program. He acknowledged that a dilemma existed between stressing strength versus speed of implementation. He said that one theory argued that "the strategic hamlets should be built slowly, one after another," stressing solidity. A second theory argued that "an extensive network of hamlets should be built quickly and simultaneously with the aim of forestalling the foreseeable reactions of the enemy." That tactic would prevent the Vietcong from using neighboring villages as a base to attack strategic hamlets, including using innocent villagers in the first wave, "with women, children, and old people on whom you would hesitate to fire." To counter such a deadly stratagem, Diem approved the second option, taking the risk of spreading the hamlets rapidly in order to provide mutually supporting security. He acknowledged that it was an imperfect solution. However, it had helped spread government control during the previous year. Nevertheless, Diem was unaware that the head of the program, Albert Pham, was actively undermining it.[43]

The NLF increased actions against the Strategic Hamlet Program at the same time. On 11 September 1963, the NLF's Liberation Press Agency announced an intensification of the "Ap Bac Emulation Drive," urging all units to exceed their tasks. The agency stated that the NLF "appealed to all officers and men of the regular force and the militia [Vietcong] to vie with each other in reducing and exterminating the enemy force while improving and developing our own force, destroying strategic hamlets, and stepping up further the people's guerrilla warfare."[44]

1963 PSYWAR Advances

Despite Communist rhetoric, the US general Victor Krulak had a very optimistic image of the war effort after a September visit to South Vietnam at the behest of President Kennedy. Based on wide-ranging discussions with American advisers, he concluded that the trend was positive despite the press reports. During this period, the ARVN claimed 1,220 Vietcong killed in April 1963 and 1,242 in June.[45] A contrary view of this September trip by fellow visitor Joseph Mendenhall of the US State Department prompted the president to ask, "You two did visit the same country, didn't you?" Later, an October report on Secretary of Defense Robert McNamara's trip to Vietnam again contended that the war was going well and that the United States might be able to "withdraw one thousand military advisers by the end of the year." However, McNamara's team also heard contradictory testimony from US military advisers that the Buddhist crisis and martial law had caused the mission to stumble.[46]

Following the Buddhist crisis, the I Corps Tactical Zone advisory effort focused on fielding recently trained PSYOP teams of the 2nd PSYWAR Battalion.[47] This battalion itself was occupied with "preparations for the national day and the II CTZ anniversary celebrations on 25 and 26 October 1963."[48] That fall, ARVN PSYWAR teams around the country began to experiment with nighttime psychological operations. This entailed sending out teams to harass VC Armed Propaganda Teams, in coordination with airborne loudspeaker and leaflet operations over villages in which Vietcong operated. One unit operating over Long An Province found this tactic particularly effective and reported that "future night PSYOPS missions are being planned."[49]

Meanwhile, in II Corps Tactical Zone, a village chief used an aerial loudspeaker to urge Montagnard tribesmen to return to their hamlets. Families had fled to the forest due to VC threats, and the broadcasts called on them to return home. During the first week 253 Montagnards returned, but it was several more weeks before all of them could be accounted for. The use of the aerial loudspeaker in conjunction with leaflet drops became increasingly common during this period. Additionally, the ARVN tested a novel capability in Binh Duong Province on 15 October. A newly rallied cadre member directed a loudspeaker aircraft to his unit's position, whereupon he made a live broadcast, calling on his comrades below by name to rally to the government. The rallying of several of his associates proved the value of rapid response loudspeaker broadcasts.[50]

Not all loudspeaker missions were as successful, however. Frank Scotton reminisced on a rally conducted in Quang Ngai Province around the same time. The province chief was scheduled to speak to a crowd at the village of Vinh Tuy where an American/Vietnamese medical team was providing aid. Arriving late, he rose to speak to the now restless throng on the theme of government aid as a benefit of rallying to the government. The plan called for his speech to culminate with a loudspeaker broadcast from an overhead C-47, followed by a leaflet drop of safe-conduct passes. Unfortunately, communications with the aircraft failed. During the middle of the speech, the crew broadcast "Mr. Province Chief, finished yet?" followed by a leaflet drop. Some of the bundles did not open properly, crashing like bombs among the crowd during the speech, Scotton later wrote.[51]

Diem had made progress in unifying the nation and fighting an increasingly Northern-manned insurgency. In the first half of 1963, American advisers observed dramatic improvements in governance and the skill of the Strategic Hamlet Program militia in repelling the Vietcong. Despite continuing problems in the Mekong Delta, Rufus Phillips, US adviser to the Strategic Hamlet Program, wrote that "the strategic hamlet program has so well proven itself in those areas where it has been well executed that there is every reason for optimism and confidence." It was an admittedly imperfect and controversial solution, but on balance it was working better than its detractors admitted. With better internal administration and the arrival of increased US military aid, the South had thrown the NLF onto the defensive.[52] However, events during the summer and fall had caused Diem to lose the support of key advisers to President Kennedy.

North Vietnamese documents support the view that Diem was winning. An increase in ARVN operations per month led one report to conclude that "protracted and large-scale operations launched unremittingly against any given region were more numerous and fiercer than in the previous year."[53] Prior to the November coup, the North admitted that the South had gained control over more than two-thirds of the rural population while establishing more than 3,500 strategic hamlets. The increased strength of the South Vietnamese government had made it increasingly difficult for the Vietcong to extract taxes, supplies, and soldiers. The North further claimed that more than 40,000 cadres and soldiers had entered the South by the end of 1963. In contrast to the popular image of a homegrown insurgency, it is clear that the North was taking over the war effort as a result of the gutting of the Vietcong and the decrease in Southern volunteers. According to the North Vietnamese official history of the war, troops from the North "represented 50 percent of the full-time armed forces in the South and 80 percent of the cadre and technical personnel assigned to the command and staff organization in South Vietnam in 1963."[54]

Postcoup Collapse

Despite these improvements in South Vietnam's strategic situation, reporters such as David Halberstam and Neil Sheehan claimed that RVN efforts had in fact dissolved after Ap Bac. What began as a myth soon bred a narrative peddled in the American press, one that was retailed in Communist propaganda and fed a revival of the insurgency after Diem's death.[55] His death unleashed a period of instability at a critical moment, which allowed the insurgency to grow into a structural threat to the nation. Despite his manifest shortcomings, Diem was the indispensable man. Even Northerners were incredulous at the US government's support for the coup. Colonel Bui Tin, a propagandist and the PAVN officer who helped accept the Southern surrender of Saigon in 1975, later wrote that "although we criticized Ngo Dinh Diem publicly as an American puppet, Ho Chi Minh adopted a soberer appraisal. He realized Diem was a patriot like himself but in a different way." Diem was a nationalist leader like Ho, "who lived an honest and clean life and . . . was unmarried," indicating a life dedicated to the nation.[56] After the coup, some Northern leaders saw an opportunity to exploit disintegration in the South, viewing Diem as "one of the strongest individuals resisting the people and Communism."[57] As the French journalist Pierre Darcourt wrote, "Under Diem there was an idea, a social and economic cadre, a politique and a faith. There was a man who knew how to make himself obeyed and in front of whom, people tremble. The directives he gave were executed." Not so with his replacements.[58]

The official radio station in Saigon broadcast a herald of change on the afternoon of 1 November 1963. The first reports of a coup crackled across South Vietnam's domestic service at 3:41 PM. General Duong Van "Big" Minh ordered loyalist "troops to lay down arms," and the station changed its name to Voice of the Armed Forces. General Minh continued: "Dear compatriots, as of this hour, the army has resolutely risen up to liberate you from the dictatorial yoke." Minutes later, the station deceptively announced that Diem had agreed to resign, and three hours later it announced the declaration of martial law.[59] On 2 November 1963, President Diem surrendered and was later murdered by coup plotters. The coup, along with the assassination of President Kennedy later that month, proved to be critical events in the brief history of South Vietnam.

The arrests of Diem regime loyalists that followed the coup also disrupted military operations as commanders at all levels were replaced. The army halted its aggressive patrols and left a vacuum for the Vietcong to fill. Post-coup justifications by Americans and Vietnamese asserted that the country was falling apart and that Diem had to go. Conversely, a later CIA analysis pinpointed the coup as the foundation of the rapid security declines.[60] A later North Vietnam assessment concurred: "The balance of forces between the South Vietnamese revolution and the enemy had changed very rapidly in our favor. . . . The bulk of the enemy's armed forces and paramilitary forces at the village and hamlet level have disintegrated, and what is left continues to disintegrate."[61]

North Vietnam faced hard choices as well. By 1963, the Le Duan faction had displaced Ho Chi Minh for control of North Vietnam. This faction was contemptuous of Ho and his focus on building socialism in the North first. With the ascendance of Le Duan, a shift to the "general offensive and uprising" strategy began. This meant expending all resources necessary and releasing operational constraints in order to unleash revolution in the South. Resources that could otherwise develop the Northern economy were sent instead to an open-ended war of choice in the South.[62] Le Duan shifted assets southward, hoping to rapidly collapse the regime there and then build a coalition government containing members of the NLF. This government would unify with the North, it was anticipated. The new strategy also meant the imposition of an increasingly totalitarian police state in the North to maintain control in the face of self-imposed hardships. Despite such internal North Vietnamese rivalries, in its English-language broadcasts Radio Hanoi continued to tout the role of President Ho Chi Minh in leading the nation from one success to another.[63]

Meanwhile, North Vietnam and the NLF rapidly exploited the psychological opportunities presented by the coup. Radio Hanoi laid the blame for Diem's ouster squarely on the United States for seeking to avoid defeat, by "ousting Ngo Dinh Diem, tightening their control over the South Vietnam military machine, and stepping up their aggressive war."[64] Shortly after the coup, Liberation Radio (i.e., the LBS) likewise blamed the removal of Diem on the "U.S. imperialists . . . the sworn enemy of the Vietnamese people." US use of chemical poisons against the Vietnamese people proved this point, according to the station.[65] To support the claim of US government complicity in the coup, Radio Hanoi International quoted a 2 November editorial in the *New York Herald Tribune*, asserting that "this revolt is our [America's] revolt."[66] But in broadcasts to the South Vietnamese, Radio Hanoi claimed that the reason for the coup was the "heroic fighting spirit" that forced the imperialists to "suffer bitter defeats," checking the United States's Special War.[67]

Simultaneously, Northern propaganda shifted from supporting Buddhists against the Catholic Diem to using his death to show Vietnamese Catholics the folly of supporting the new, post-Diem government. The divisive objective remained the same, but the exploitable audience changed. Using the theme "Lesson for Vietnam Catholics," Radio Hanoi quoted Reverend Ho Thanh Bien, vice chairman of a North Vietnamese front group, the National Liaison Committee of Vietnamese Patriotic and Peace-Loving Catholics: "During the past protracted war of resistance the Vietnamese Catholics learned a bloody lesson: Some priests, brought up by the imperialists, followed them in 'opposing communism purportedly to defend religion.'"[68] Saigon's *Vietnam Press* remarked on the new theme: "The Viet Cong have not missed the chance to make use of this situation in order to sow division between religions and create disturbances all over the country so as to facilitate their scheme of invading the free Southern area." *Vietnam Press* disparaged this attempt as blatant VC propaganda.[69]

In the days after the coup, the NLF went on the military offensive as well. It claimed to have "fought hundreds of battles in many parts of South Vietnam and destroyed over 100 military posts and many 'Strategic Hamlets.'"[70] Additionally, the Liberation Broadcast Service declared that forty-two military posts, eight villages, and two towns were liberated. It reported that many Southern soldiers had defected to the revolution in the days after the coup.[71] The LBS also stated that, in Tan An Province on the night of the coup, people "destroyed many 'strategic hamlets,' and encircled military posts"; in another case, they shot the chief of one post, seizing arms.[72] Information from ARVN PSYWAR teams concurred with the nature of the postcoup uprising. For example, an NLF group had "removed iron spikes and barbed wire" from a hamlet, and the Vietcong called on dwellers to abandon another strategic hamlet using "megaphones and leaflets," per PSYWAR Team One.[73]

Using an inevitability-of-defeat theme, Radio Hanoi further claimed more than 1,000 attacks in the period after the coup, saying that for the Americans it was "the bloodiest seven days in the history of its war" in Vietnam. To substantiate this, *Nhan Dan* quoted Bernard Fall, the Howard University professor and noted author. According to the UPI, Fall had declared that "the collapse of the Diem regime will not prevent the war against communist guerrillas from bogging down in stalemate. . . . Nothing has really changed. The Communists are no weaker than they were yesterday."[74] In addition to inevitability themes, the Communists used divisive themes meant to demoralize the Southerners. The LPA asserted that the United States initiated a series of coups against "puppet leaders . . . in Latin America, Thailand, Korea, Turkey, and many other places." These LPA claims hyped the supposed duplicity of the United States and may have been aimed at the coup plotters themselves, showing them how tenuous America's support could be.[75]

In the immediate aftermath of the coup, South Vietnamese PSYWAR units focused on explaining what had happened. The principal themes consisted of reasons for the coup d'état and the new military government's policies. For example, the ARVN 23rd and 25th Divisions dropped a total of 640,000 leaflets using these themes during the first week alone. According to one MACV report, the new government also attached ten-man teams to each district in Darlac Province to counter VC propaganda regarding the coup.[76] After the coup, American advisers reported a renewed Vietnamese acceptance of PSYOP advice, in contrast to Diem's perceived desire for complete control of the program. However, rather than displaying a newfound enthusiasm for PSYOPs, the new leaders were more dependent on the United States and thus more pliant to its suggestions, regardless of the appropriateness of messaging within Vietnamese culture.[77]

Indeed, an analysis of reports covering the five months between July and November 1963 paints a picture of the intense PSYOP activities being conducted by the end of Diem's rule. In fact, Diem had backed the program to the hilt—as long as it pursued objectives he deemed to be appropriate. The vast majority of nearly 1.2 million leaflets airdropped during this period were in support of the Chieu Hoi program. In measurement terms, this resulted in an average of 1,400 ralliers per month prior to the coup. Ralliers dropped to 650 per month afterward. The 88 reported precoup loudspeaker missions (many sorties went unreported) appear to have been effective as well. One key task was to encourage refugee movement, which successfully returned thousands of Vietnamese to government control prior to the coup.[78]

On the ground, ARVN PSYWAR and Vietnamese Information Service teams operated across the country, conducting face-to-face operations, showing films to thousands of viewers, and fielding and testing novel equipment usages. The ARVN now included three PSYWAR battalions with US adviser support integrated at all levels. The VIS functioned nationally, supported by USIS representatives. However, this structure was now essentially leaderless with Diem gone.[79]

Even at this early stage of the war, two clearly different approaches to propaganda operations in Vietnam appeared. Both the North and the South utilized international broadcasts to influence world opinion. However, the North had a ready echo chamber in the Warsaw Pact nations and China to spread its narratives. South Vietnam lacked this advantage, largely facing a hostile press. Circumstances also constrained each side. The North and the Vietcong were forced to rely heavily on face-to-face communications in the South. Though potentially highly effective, this technique had limited range and coverage. The LBS was a mobile radio station with intermittent broadcasts, and the NLF had no ability to conduct leaflet drops. Radio Hanoi provided better population coverage.

The North and its organs, in line with the agit-prop model and united front tactics, displayed an opportunistic ability to seize on events, build effective themes, and agitate their audiences. From neutrality to religion, toxins, bombings, and ethnicity, they exploited hot-button issues as they arose. Their long-term goals were to divide the RVN from the people, the RVN from world opinion, and the American people from support for the war. Although the North had some success in the short term, jumping from theme to theme led to haphazard and often contradictory messaging. In the long run, it may have been counterproductive.

Indeed, as both nations faced a widening war, the South had some clear advantages, especially in logistical backing for widespread use of loudspeakers, radio, leaflets, and magazines. However, its face-to-face operation was not as consistently effective, nor was it supported by strong ideological underpinnings, as in the North. The RVN had to create a system almost from scratch in the midst of combat operations. Less reactive to events, it tended to act as an information conduit, issuing surrender appeals and orders for population movements. However, it did include nationalist themes and the Chieu Hoi program, which had a degree of success. US aid, training, and equipment rapidly expanded South Vietnam's radio broadcasting capabilities. In many ways, the South adopted the mass-market system of the Americans. Despite positive trends, the effect of governance disintegration, along with the growth of the Vietcong, proved too much in the upcoming phase.

The US Army continued to develop its counterinsurgency doctrine, techniques, and technologies despite the coup. Psychological operations were seen as a key component of that doctrine. General Edward Rowny received the task to form the Army Concept Team in Vietnam (ACTIV) on 6 November 1962. The ACTIV organization evaluated "new methods of countering insurgency in actual combat."[80] This program, which operated through April 1972, was the largest US Army research and evaluation project of the war. Though initially focused on helicopters, the program eventually expanded to cover a wide array of topics, including PSYOPs.[81]

At a May 1963 conference in Hawaii, Secretary of Defense McNamara ordered ACTIV to study how to increase tactical PSYOP responsiveness and overcome the high illiteracy rate in Vietnam. The first of the studies pertaining to PSYOPs involved a 1963 evaluation of heliborne loudspeakers "for use in the psywar and civic action aspects of counterinsurgency warfare." This development was critical when considering the failure of airplane-mounted loudspeakers, which had yet to be resolved at the time of the helicopter tests. Initial trials proved the system to be intelligible from 3,000 feet even when traveling at 45 knots—an essential capability to increase aircrew survivability.[82] In addition to the aerial loudspeaker test, ACTIV tested audio-visual Tri-Lambrettas. Ninety of these three-wheel scooters equipped with projectors and loudspeakers were ordered for South Vietnam's PSYWAR/Civic Action teams at the sector level. These scooters arrived in December 1964 for distribution in early 1965. Finally, ACTIV requisitioned seventeen 1,000-watt loudspeakers and thirteen 250-watt ones for helicopter use.[83]

US intelligence reports by 1964 indicated that the insurgency was increasingly being taken over by the North. One CIA study determined that 5,800 Northern fighters had infiltrated in 1962 and a further 4,000 in 1963. More important, these newcomers provided key skills and filled leadership positions in the Vietcong. The numbers increased in 1964 and included more Northern-born personnel. However, NLF propaganda regarding the nature of the war may have impacted the senior policy makers that President Lyndon Johnson had inherited following JFK's assassination in November 1963. Despite access to these CIA reports, many advisers continued to believe that the Southern insurgency was self-sustaining, and "hence actions in Laos or North Vietnam would not have dramatic effects on the war in South Vietnam."[84] President Johnson based his policy on a flawed understanding of the war. Furthermore, ideas from the academic community, such as the Harvard professor Thomas Schelling's expositions on game theory, led Johnson's advisers to believe North Vietnam "could be restrained through limited actions that made heightened conflict seem contrary to their interests." Schelling's 1960 book *Strategy of Conflict* posited a world of rational actors susceptible to the right pressures in international conflict.[85]

Game theory assumed that all decision makers acted rationally. However, viewing North Vietnamese leaders as rational was problematic. By this point, Le Duan may have been acting rationally, but he was playing an entirely different game than Johnson. The North Vietnamese ignored messages Johnson meant to convey through actions. Johnson viewed Vietnam within the Cold War context and had to pull American public opinion along while balancing his domestic priorities and keeping the "dominoes from tumbling." Thus, Johnson struggled to find the precise propaganda policy pressures to influence the North, maintain American popular support, and keep the war at a low simmer at home so he could focus on his domestic agenda.[86]

SOG Takes the Reins: Covert PSYWAR

By 1963, the psychological effort began to take precedence in the covert program against the North as the CIA slowly realized that its agent operations had been a failure. At the Honolulu Conference in November, William Colby, the CIA's Saigon station chief, expressed his doubts about the entire agent operation and urged phasing it out. Colby argued that most agent teams had been captured or killed and wanted to end the program. He preferred "infiltrating ideas, rather than agents and explosives." Reporting from Radio Hanoi supported Colby's assessment of the current program. The CIA hoped to discontinue infiltration completely by 1965 and thereafter focus on covert radio and leaflet operations to influence the North. It was in the context of these CIA failures that pressure grew to transfer operations to the Pentagon. Secretary of Defense McNamara reportedly took the view that if CIA agents and PSYWAR operations were a failure, then the fault was with the CIA. Colby bluntly stated, "It isn't working, and it won't work any better with the military in charge." MACV expressed a willingness to phase in responsibility for the operations with the "CIA footing the bill until 1 July 1964." However, authorities envisaged keeping CIA specialists inside the new organization.[87]

During the late fall of 1963, the Kennedy administration decided to transfer responsibility for the PSYWAR program against the North from the CIA Covert Action Station to the MACV Special Operations Group. The new organization became the Studies and Observations Group, a generic name designed to provide cover. In a 14 November cable, MACV, the Covert Action Station, and Ambassador Lodge all concurred in this change. The MACV had already prepared, along with the CIA station, what became Operational Plan 34-63 to run these operations. The plan echoed nearly all the current CIA operations. The cable noted that expanding the operations, as envisioned, required accepting "a reduction of plausible denial to a level of discrete overt operations."[88] The transfer placed CIA officers in the new organization but reduced the agency's support for SOG's PSYOP effort over time. SOG's Psychological Studies Branch functioned under the designation "OPS33." Among its programs was a counterfeit Radio Hanoi that broadcast on adjacent radio frequencies, in hopes people would "surf" in. SOG produced radios that were incapable of receiving the correct Radio Hanoi frequency but were able to receive the spurious one. SOG dropped the radios, or agents left them after raids in the North, expecting people to find and use them.[89]

By late November, the Liberation Broadcast Service reported about the spraying of poison chemicals in My Tho, asserting that "fruit trees were damaged and a number of domestic animals were killed" (as well as fish).[90] On 10 December, Radio Hanoi's English service read a letter purportedly from a

Southerner to a Northern friend regarding the use of chemical weapons. The author noted that the "U.S.-Diemists are more barbarous than anyone could imagine." He said that "American aircraft were sent to spray toxic chemicals over our province during the Tet holiday. The poison affected more than 20 villages particularly those around the chief town of Ben Tre." Interestingly, this was broadcast prior to the Tet holiday; more than a month after the coup, Radio Hanoi had not yet devised a new term to replace "U.S.-Diemists." Despite these propaganda flaws, the chemical weapon theme supported all the major psychological objectives of North Vietnam. It aimed to sap American domestic support, feed international anger, and divide the South Vietnamese from the GVN and the United States. Considering the limited need to embellish the story, and popular revulsion with chemical weapons and confusion over their terminology, it was a potent theme.[91]

Several weeks later, the paper *Tien Phong* (Vanguard), an organ of the Vietnam Labor Youth Union, attacked "the bellicose Johnson group and its new henchmen in South Vietnam" for continuing to spray toxic chemicals "as a means to intensify their aggressive war there." It also denounced a counterpropaganda campaign in the South Vietnamese press, claiming that "the spraying of the chemical poison is part of their 'land reclamation program' and that 'the poison is not harmful to crops and animals.'" Despite this use of chemicals, Radio Hanoi continued to profess that victory was certain: "No reactionary forces, no violence, or up to date weapons and techniques can quell the will to fight and to win" on the part of the NLF.[92] It based this theme of inevitability on "famous victories such as Ap Bac," among others, and the effective "political and military struggle and agitation among the enemy troops."[93] The "inevitability of victory" became a dominant theme as actual victories grew in the weeks after the coup.

In early January, North Vietnam conducted a meeting to condemn "the United States and its hirelings" for continuing to spray noxious chemicals in South Vietnam. Professor Nguyen Yuan Nguyen, vice president of the Vietnam General Medical Workers Association, denounced the United States and its "henchmen for using chemical poison as a war means to suppress the patriotic movement in South Vietnam."[94] Accusations included dropping gas bombs and firing gas rockets. The Liberation Broadcast Service claimed that the United States resorted to poison gas because "it is clear that, even though they have modern weapons and better equipment, the enemy troops continue to suffer defeats. Thus, they must use poison gas against our troops."[95]

In order to better understand the human terrain in the South, USIS and its Vietnamese counterparts began an in-depth psychological study of Long An Province, near Saigon. This was a contested province with a large VC presence. Ambassador Lodge had tasked USIS with this project. The USIS conceived of the study as a "hamlet by hamlet survey in the area destined for a clear and hold operation." Long An had both the benefit and drawback of being near Saigon, providing easy access but limiting its usefulness for understanding the situation elsewhere in the nation. Everett Bumgardner, chief of USIS field operations in Vietnam, conducted interviews with students at a few Long An schools in December 1963 in preparation for the full study. Although not a random sample, it provided some baseline data for future surveys. According to Bumgardner, the "students considered themselves (and their families) to be merely bystanders in the Viet Cong–Government war" and focused on "safety and self-preservation." The students respected the behavior and discipline of ARVN units and soldiers and stated that the Vietcong were "afraid of ARVN." One student asked an important question, though: "We see and talk to the Viet Cong all the time. Why doesn't the government ever come to our hamlet and speak to us?"[96] This feeling of abandonment left villagers susceptible to VC influence. The report mentioned that some villagers were reluctant to talk to the interviewer because security forces, following a previous visit, had informed the district chief of critical remarks and punished the informers. This raised a problem that all sides in the war had to contend with. Villagers were reluctant to express their true feelings to outsiders due to fear of reprisals.[97]

Amid a developing PSYOP effort to influence behavior, work continued on assessing enemy propaganda to help gauge the effectiveness of that program. USIS also conducted one such analysis of VC propaganda activities in Long An Province. It found that the VC propaganda teams typically consisted of four to five armed men who arrived in a hamlet between seven and nine o'clock in the evening. These were usually area residents who stayed in the village until dawn conducting propaganda sessions, of either a "house-to-house type of contact or a group meeting and lecture involving all inhabitants."[98] They distributed handbills and leaflets at this time, criticizing the Strategic Hamlet Program and the Southern government, using threats as well as persuasion. The Vietcong also attempted to jam loudspeaker broadcasts by requiring villagers to beat on pots and pans to make the broadcast unintelligible. They frequently issued death threats to motivate villagers, but normally they carried them out only when a specific psychological goal could be achieved. This cruelty often backfired and ran counter to the normal Maoist concept of building support among rural populations. The report noted that in one village people were initially sympathetic to the Vietcong, "but when one of the village representatives was murdered by the VC, the inhabitants reportedly turned against the VC. This resentment is not unmixed with considerable fear, however."[99]

The NLF's agit-prop teams spotlighted any grievance to exploit. For the masses, agitation fomented anger and frustration, relying on simple slogans to direct that anger at a target. The goal was to divide groups and create psychological vulnerabilities that could be exploited by VC cadres. "Land to the tiller" was a good example of an agitation slogan. Consistency of effort never constrained them. As noted in the discussion about religious agitation, one moment the Communists attacked the Catholics as despots, while in the next they warned Catholics not to trust the Buddhist government. Cadres identified a dissatisfaction and agitated people to increase their resentment of the "other." Then, cadres provided slogans and emotional appeals meant to rally the people to action. They used both printed product and radio broadcasts as supplements to the direct approach in South Vietnam. The Long An study concluded that similar person-to-person propaganda, focusing on positive deeds by the GVN's PSYWAR/Civic Action teams, was necessary.[100]

The South Vietnamese government attempted to seize the initiative in direct communications. By sending VIS teams in villages to conduct face-to-face meetings, the government sought to gain support and better understand the needs and desires of villagers. This program was in addition to the CIA Census-Grievance teams and PSYWAR teams. Normally the VIS teams, along with a Vietnamese representative of the USIS, produced a report to help illustrate what was happening in rural areas. One study concluded that only people in remote areas were influenced by VC propaganda, whereas "the people living in the district, near the market always side with the Government." For instance, in one hamlet about 80 percent of the people supported the government. But in four VC-controlled hamlets farther away, the ratio was reversed.[101]

Visiting PSYWAR teams also conducted assessments of village needs. For example, per PSYWAR Team One, Ap Xuan Hua village requested a classroom, a maternity hospital, and "medicine for first aid kit they already have." According to the team leader, after the 3 November visit the villagers enjoyed the distributed publications and "were impressed that the American with the team spoke some Vietnamese," asking the cadres "how long he had studied the language." This was likely Frank Scotton, who organized and oversaw the interviews for the initial Long An survey. This face-to-face effort grew in the coming year, but government instability limited its effectiveness. The skill and ideological commitment of the teams was mixed as well.[102]

North Vietnam and VC propaganda tended to downplay their matériel strength, in one instance saying that "our guerrillas and people using rudimentary weapons counterattacked the enemy troops, killing three and forcing the enemy to conduct their operations in the open since their troops dared not approach the hamlets."[103] They connected this victory to the "Ap Bac historic battle" and the subsequent Ap Bac "emulation" movement.[104] The

David-versus-Goliath meme supported both the international and domestic propaganda programs.

One of the most important victories of the propaganda war took place twenty-five miles west of Saigon in late November 1963. A PLAF force overran the US Special Forces camp at Hiep Hoa, capturing several Americans, including Special Forces Sergeant First Class Kenneth M. Roraback. According to Radio Hanoi, "75 weapons were captured, including some 50 individual rifles and carbines, 8 automatic rifles, 7 machine guns, a 'sizeable' quantity of ammunition, 8 shotguns, and 2 mortars. The booty was reported by AP to be 'enough to equip a battalion.'"[105] The Liberation Broadcast Service later mocked attempts to recover the missing men in a broadcast to the South. According to the story, "the U.S. imperialists and their lackeys foolishly sent thousands of their troops" to areas throughout Tay Ninh Province to search for Roraback. To support this search, "the enemy sent planes equipped with loudspeakers to fly over" the areas.[106] Roraback was executed in 1965 by the NLF.

Both sides utilized counterpropaganda and inoculation to prevent enemy messages from being accepted by the populace. The GVN's Ministry of Security cautioned citizens against "deceitful" VC propaganda, saying that NLF agents attempted to "launch a propaganda campaign for the purpose of inciting the people to fill up the ditches and destroy the defense works in the strategic hamlets and quarters."[107] It was likewise with the defoliation program. Originally the NLF feared this program greatly. However, as historian Eric Bergerud found in Hau Nghia Province, once "it became obvious that most areas of the province would not be sprayed, Front cadres used the issue vigorously for propaganda, portraying it as an 'inhuman, cruel and barbarous act.'"[108]

At the same time, instability in the government as a result of the November coup caused the PSYOP effort to lose focus. The need for consistent messaging could not be met in such a situation. The rapid turnover in South Vietnam's PSYWAR leadership exacerbated this. A last major factor was the Johnson administration's internal struggle to keep the situation under control while it sought a way forward. These concerns dominated the PSYOP effort for most of 1964.

In a memorandum for the president, Assistant Secretary of State Robert Manning presented the 1964 information program for Southeast Asia. He designated three audiences: Southeast Asia, the international arena, and "the case to be made to our own people and Congress." As Manning saw it, the Johnson administration faced a shift in public opinion as the nation focused more attention on the war. Until that point, it had remained an insignificant issue for most Americans. However, he asserted that Americans' memory of the Korean War, France's defeat in Indochina, and the brutal nature of the current war, "mixed with the odor of confusion and frustration that seeps out of Saigon, are poor material on which to build understanding and confidence." Like the North Vietnamese emulation campaigns, Manning argued for giving "publicity to Americans who have performed feats of gallantry and heroism in Viet-Nam." He suggested working with Milton Caniff, the artist behind the popular *Steve Canyon* and *Terry and the Pirates* comics, to have the characters "operate" in Vietnam. He felt that the comic strips were "ideally suited to dramatize the conspiratorial and savage nature of the Communist assault on Viet-Nam and why and how the U.S. is trying to combat it."[109]

Conclusion

In the three years since President Kennedy had come to office, American involvement in Vietnam had grown considerably. The NLF and the North's attempt to seize control of the South had been blunted, and both the South Vietnamese psychological warfare capability and the American advisory effort greatly expanded. Despite these factors, the Battle of Ap Bac had given the NLF a propaganda boon at a critical time. The propaganda benefited from widespread repetition of the theme of ARVN collapse in the American press. Followed closely by the negative press emanating from the Buddhist crisis, Diem's support among political elites in the United States and in South Vietnam collapsed. The resulting coup unleashed two years of instability, more coups, and military defeats. The rotating series of leaders, each valuing loyalty over competence, further delegitimized the government. Despite improvements in PSYWAR capabilities, there was no way to spin defeat into victory. The burgeoning PSYOP program was left to tread water in the face of this tide while the North's program surged. By the end of 1963, the United States was increasingly confronted with hard choices.

Map 1. South Vietnam, late 1963 (chapter 4 locations)

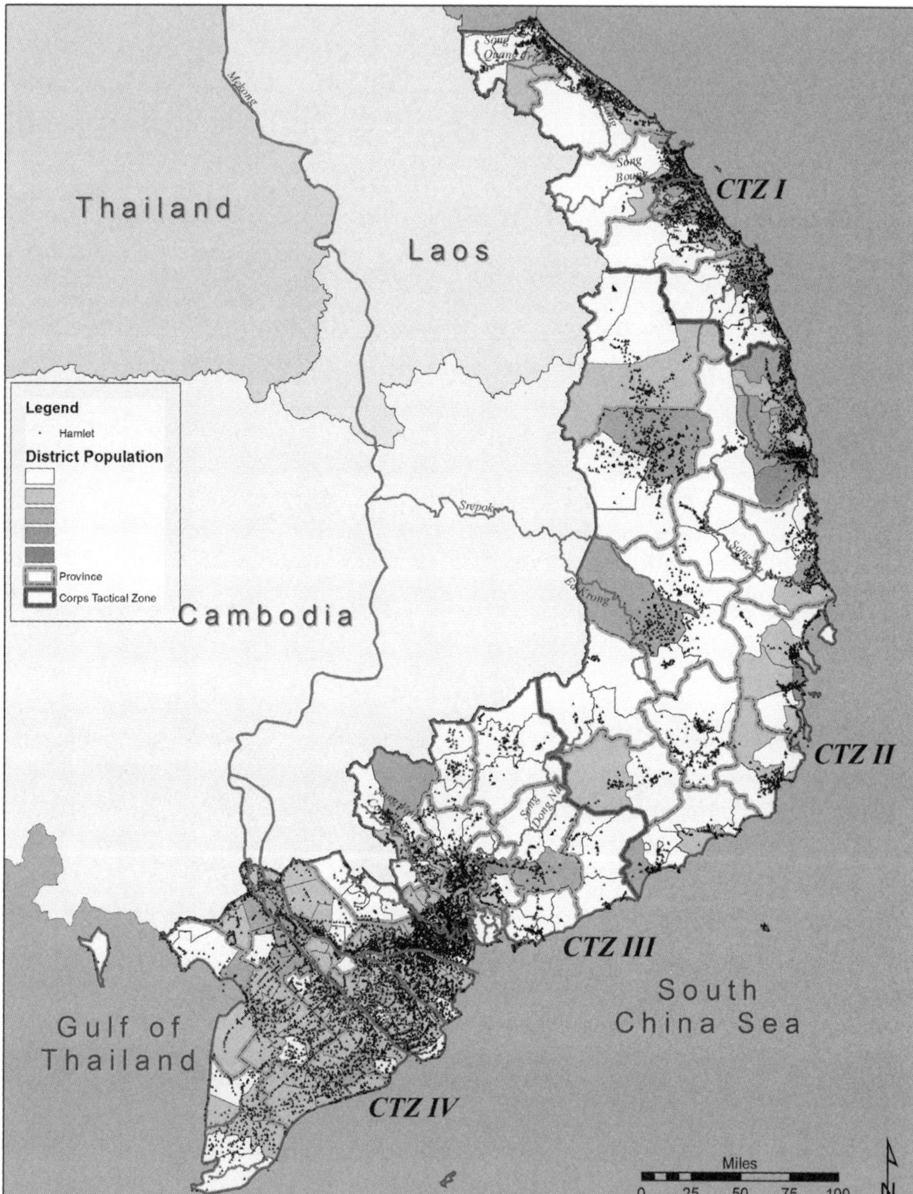

Map 2. South Vietnam hamlet density overlaid on district population

5 "Only Hard Ones Lie Ahead," 1964

As the postcoup disarray took hold in 1964, the South Vietnamese government's power waned. The United States was left with hard choices to make. Even as the PSYWAR structure continued to grow and improve, government instability hampered messaging. The covert program targeting the North also faced a succession of agent operation failures that called into question the utility of this program. The transfer of the program from the CIA to the US Department of Defense also led some to recommend a radical change. At the same time, as US troop strength in Vietnam grew, the North began targeting troop morale and sought to divide US and ARVN troops. The initial attempts were poorly executed, but they improved over time.

Expanded Advisory Effort

As 1964 opened, there were indications that the burgeoning joint US/ARVN PSYOP program was reaching its physical limitations. On 7 January, MACV notified the Joint Chiefs of Staff that leaflet printing alone was utilizing all available printing capabilities in Vietnam.[1] An investigation determined that, due to a bureaucratic oversight, no one had requisitioned the printing plants scheduled for delivery to Republic of Vietnam Armed Forces PSYWAR units. As a result, ARVN and USIS presses in Vietnam, as well as locally contracted presses, were completely overwhelmed. Subsequently the US Army Broadcast and Visual Activity Pacific (USAB&VAPAC) unit received authorization to ship a heavy mobile press from Okinawa to fill this gap. Additionally, the USAB&VAPAC presses on Okinawa provided backup printing capacity.[2]

ARVN's three PSYWAR battalions had responsibility for enemy and civilian information programs and troop morale. ARVN planned to add a fourth battalion in 1965, providing one for each Corps Tactical Zone. Until then, however, the 1st PSYWAR Battalion had to cover both III CTZ and IV CTZ. Each battalion had an authorized strength of 367 soldiers and consisted of three 60-soldier tactical companies. The PSYWAR companies were composed of six mobile loudspeaker/leaflet teams, essentially acting as APTs. Each PSYWAR battalion had eighteen such teams, as well as a cultural platoon that

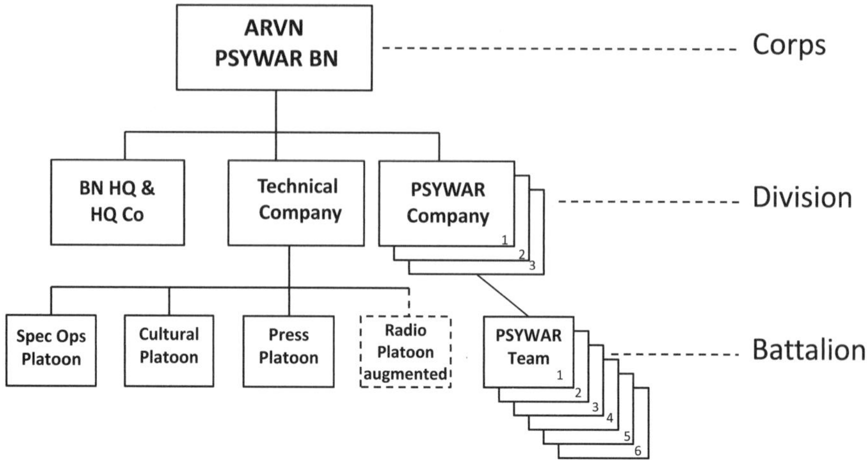

Figure 5.1. ARVN PSYWAR Organization

used entertainment to spread propaganda.[3] Each technical company provided the battalion with a cultural platoon, a print platoon, and a special operations platoon conducting covert propaganda activities. According to a capability briefing, South Vietnamese Civil Affairs units acted as a support tool for the overall PSYOP effort. Each PSYWAR team—the basic tactical element of the battalion—provided support to an infantry battalion, utilizing airdropped leaflets, loudspeaker support, and face-to-face leaflet/product distribution.[4] (See figure 5.1.)

Despite issues at the tactical level, the ARVN PSYWAR force and the US advisers and support units continued to expand and test the limits of PSYOP capabilities in support of counterinsurgency operations. On 31 December 1963, the South Vietnamese prime minister had announced the transfer of responsibility for the Chieu Hoi program from "the former Civic Action Department . . . to the Ministry of Defense."[5] Meanwhile, the MACV J3 (Operations) office split into two sections, Operations and Special Warfare. The Special Warfare Branch controlled PSYWAR operations. On 1 January 1964 it became the PSYWAR/CA branch under J3. According to the MACV's *Command History*, this change "reflected recognition of the inter-relationship of these functions" and mirrored existing ARVN staff levels with the MACV advisory channels.[6]

In early 1964, 150 ARVN PSYWAR/CA personnel and US advisers, along with the British MACV advisory mission, met in Saigon. This meeting indicated the importance of PSYWAR and the ARVN's Psychological Warfare Directorate to the new government. At the conference, Defense Minister Tran

Van Don chastised officers who argued that a lack of equipment hindered the PSYWAR mission. He thought their excuse was a poor one, covering for their "lack of initiative." "Leaflets, loudspeakers, moving pictures, and so on are very fine and effective," according to Don, "but head to head talks are still the best and most effective means of propaganda, which the Viet Cong never fail to employ."[7] Despite this praise for the enemy, the ARVN and the GVN had teams operating in this "head to head" war throughout the country. However, the tactical teams needed to do more.

A study by US Assistant Secretary of State for Public Affairs Robert Manning had reiterated that the VC propaganda program was capable of exploiting developments quickly. For example, he discussed the rapid shift after the November coup. Manning wrote that the Vietcong used a "strategy of news" in their propaganda, rather than depending simply on bombastic phrases. This news was presented as straight facts and often given credibility by echoing Western news reports. However, depending on the target audience, hyperbole was present, along with clearly embellished, subjective news to support a given theme. And rapid thematic shifts and fraudulent numbers hurt the program. Over the long haul, this lack of candor undercut the credibility of an otherwise sound program.

A consistent theme, exemplified on a leaflet presenting the reasons for VC success in Ap Bac, was the dominance of willpower over material. Noting that VC strength was "10 times or 20 times less than that of the enemy" and its equipment "100 times inferior to that of our enemy," the leaflet explained the reasons for victory. "Even with superior number and weapons, the enemy soldiers lack the will to fight," the leaflet claimed, meaning "that in battle they do not measure up to us in strength and weapon."[8] Another frequent theme was desertion of enemy troops. On 2 January, Radio Hanoi claimed that 40,000 ARVN personnel had deserted during the previous year. The most recent defector mentioned was a PSYWAR officer, Sub-Lieutenant Thai Tran Trong Nghia, who defected on 22 December to "join the struggle against the U.S. aggressors and their henchmen in South Vietnam."[9]

North Vietnamese Propaganda Reactions

The North Vietnamese propaganda campaign grew after the coup in South Vietnam deposing Diem, and its effectiveness seemed to increase as well. It is difficult to precisely correlate the two, but clearly the postcoup disorder aided the North's program. Hanoi also issued orders to intensify the ground war to take advantage of the new situation. The 1964 North Vietnamese Party plenum culminated with the announcement of Resolution 9. This directive shifted the nation to expanded war in January 1964, nine months prior to the Gulf of Tonkin incident that brought the United States actively into the war.[10]

To control internal debate about the change, party leader Le Duan created a police state that answered to him. Domestically, Le Duan and the politburo member Le Duc Tho were "determined not to tolerate the practice of expressing views in an unorganized manner on the party's line, policies, resolution, and instruction," according to Tho. Le Duan promoted the Cong An (Security Police) and Bao Ve (Military Security) to "surveil not only Communist cadres, intellectuals, and high-level military officers but also the general population" in the struggle against "counterrevolutionaries." This resulted in the detention of rightists, especially those linked to the Soviet Union.[11] Despite this, Le Duan retained Ho as a front man after 1964, while Duan committed the nation to violently liberate South Vietnam in the face of American opposition. Internal distrust provided a psychological opening, which SOG successfully exploited in the coming years.[12]

The North-first faction that Ho represented generally sided with the Soviet Union, whereas Le Duan favored China. However, North Vietnamese leaders used tensions between the two Communist powers against each other in the quest for aid. This shift gave North Vietnam leverage in the Sino-Soviet split. But each nation placed expectations on North Vietnam as well. Beijing pressured North Vietnam "to avoid peace talks at all costs and to reject Soviet aid and advice," and Moscow pushed it to keep the war from upsetting Soviet foreign policy elsewhere.[13]

Until this time, the Soviet Union had remained only indirectly involved in Vietnam, in pursuit of Nikita Khrushchev's policy of "peaceful coexistence." This required limiting the war in Vietnam to an internal struggle. The previous Soviet promise of support was simply a propaganda ploy designed to enhance the Soviet position relative to China.[14] But Hanoi required support from both powers in order to unify Vietnam. By 1964, the Sino-Soviet split became an open rivalry and powerfully impacted Moscow's debates about Southeast Asia. This year was the low ebb for Soviet influence over the DRV while pro-Chinese sentiment overwhelmed North Vietnamese officials. Chinese propaganda hailing the North Vietnam–China friendship swept the country. However, the divisions were not readily visible to the casual observer, as both China and the Soviet Union sought to appear to be the strongest supporter of North Vietnam and/or the NLF.[15]

Unlike the empty claims and downhearted tone of the previous year-end message, the LPA hailed the successes of 1963 and looked forward to the coming year. On the first anniversary of the Ap Bac battle, the agency emphasized that "this was one of the greatest victories of the South Vietnam patriotic forces which smashed a major enemy raid in Ap Bac . . . killing or wounding 450 enemy troops, including 12 American officers, shooting down 6 and hitting 10 aircraft, setting afire 3 armored amphibious cars, and damaging 1 gunboat," proclaiming it to be a "historic battle" in which the United States was "once again defeated by the people's patriotic forces." All these numbers had increased considerably in the year since the battle. Furthermore, the LPA now claimed that more than half of the Southern population was under its control.[16]

To support these claims, Radio Hanoi employed *Newsweek* reports on the failure of the war effort. The broadcast made the suspect assertion that 75 percent of the strategic hamlets "were overrun in November alone" and that in the crucial Mekong River Delta "the guerrillas are now stronger than they were when Vietnam was French." According to *Newsweek*, the US military officials thought the South would win if the ARVN maintained a kill ratio of "three Viet Cong to one Vietnamese soldier." However, the story quoted Bernard Fall as saying, "The South could not win with less than a kill ratio of 50 to 1." Fall says flatly, "The war has never gone so badly," per the broadcast.[17] Despite the obvious exaggerations, the year had ended well for the revolution.

To divide ARVN soldiers from American advisers, in January Radio Hanoi read an open letter to "U.S. puppet soldiers." It accused the Americans of bringing "debauchery" to Vietnam and said that "the gangsters and murderers work hand-in-glove with the monopoly capitalists, Congressmen, and law officers" who were responsible for "bringing these problems with them." Among the accusations: Americans were responsible for disemboweling nearly 3,000 people, "their livers plucked out, cut, and eaten," as well as the rape of more than 4,000 women. The story quoted several international left-wing newspapers reporting on the war. It appealed to ARVN troops who were "deceived and forced by the U.S. imperialists and their henchmen" to fire on their compatriots. It also pleaded for soldiers to "return to the people's side" to be "worthy of the fatherland." This was the real target audience; even though the letter was addressed to US troops, it was read in Vietnamese. More than 87,000 South Vietnamese soldiers had deserted, the broadcast claimed. Just days earlier the number had been 40,000 in broadcasts to the same audience. The statements, if taken in isolation, might be effective. However, any regular listener would catch this misinformation, destroying the broadcaster's legitimacy on other issues. Again, inconsistency regarding figures plagued the North's propaganda.[18]

Much of the Communist propaganda effort was channeled through "front groups and organizations, making use of the social movement propaganda approach" rather than depending on mass media. The Vietcong saw mass media playing a supporting role. Leaflet quality was poor, but an American study once again noted that it reacted quickly to exploit events. Although the North expended "considerable effort" to make its messages appear credible, the speed of the program exposed lapses in truth.[19] The Front also used the threat of death as a way to prove it controlled the area, undermining inhabitants' "faith in the GVN and its ability to protect them." NLF propaganda urged youths to join the Front or at least to avoid serving in the ARVN. At this time, service in the Vietcong offered the opportunity to operate locally, a benefit in an agricultural society. Young men could avoid the ARVN draft, the Front informed them, by "cutting off their trigger fingers." ARVN PSYWAR teams "reported several cases of young men in one hamlet with missing trigger fingers, another example of the effectiveness of VC person-to-person propaganda." Further indicating the effectiveness of the North's propaganda, residents of another village were conditioned to turn down PSYOP/CA gifts.[20]

In contrast, PSYWAR Team Three reported that a VC Armed Propaganda Team arrived in strongly progovernment Binh Tuy Hamlet to distribute antigovernment propaganda using a megaphone. The APT "scattered leaflets and hoisted VC flags in the hamlet," exhorting the "people to destroy strategic hamlets and not to cooperate with the government." However, the villagers there refused to accept the leaflets calling for the withdrawal of Americans, or to abandon their strategic hamlet.[21] The PSYWAR team leader Nguyen Ba Trung reported that his team visited each hamlet family, presenting gifts and pursuing grievances. They visited 125 families, distributing copies of *Huong Que* (Rural Spirit), *Free World,* copybooks, and *Progress of Viet Nam* to each, as well as various leaflets. *Free World* was a digest of Western news reports. Trung wrote that "no government teams had visited the hamlet since the coup," so the presence of members of the PSYWAR team gave residents "great satisfaction," and it received a "cordial welcome." Even here, though, the team identified a few VC supporters by noting their reactions to the PSYOP product, such as tearing up leaflets.[22] Other teams reported similar results at this time. In such cases, the PSYOP products had been useful in opening a dialog and in determining loyalties.

One peasant described the propaganda activities in VC-controlled areas to a visiting PSYWAR team. The Vietcong organized theatrical groups and used dancing and singing performances to "win the children over to their side by taking advantage of these performances to distort the [South Vietnam] Government's actions and news." The peasant said that local students had identified with the groups because their peers were involved.[23] The following day, the PSYWAR team found homes in the hamlet empty, except for "children, women with babies, or aged people." The adults they did encounter kept strictly to the Vietcong's "four no" slogan: "no knowing, no listening to, no seeing, no having." This slogan indoctrinated inhabitants to profess ignorance in the face of government representatives' questions and to refuse help regarding the actual conditions in the surrounding area. The people in this contested village refused to use the word *Vietcong*, correcting the PSYWAR team with the "terms 'Liberation' or 'Resistance' government." Inhabitants expressed no interest in magazines that the team dispensed. When the cadres later returned, the magazines were left untouched.[24]

Pham Van Tung, leader of PSYWAR Team Two, wrote that the Binh Cu strategic hamlet south of Saigon contained "many underground VC cadres who regularly organized study and propaganda among the people; meanwhile, the government cadres rarely come to this hamlet, [which] makes the people more confused." The lack of government presence ceded control of the village, according to Tung.[25] Another team leader believed that a "lack of nurture from our side and regular contact with the enemies" led hamlets to shift loyalty away from the GVN. Given the haphazard nature of operations during the period, many exploitable opportunities opened for the Front.[26] (See map 3.)

An interview conducted by the team with an old woman from the Binh Tay Hamlet pointed to the shaky ground on which the relationship stood for all sides. She told of a recent visit by a medical team from the ARVN 3rd Paratrooper Battalion. She noted that the men prepared their own meals and "didn't ask the hamlet inhabitants to furnish them anything, even fish sauce or pepper." Their righteous behavior won the sympathy of the residents. However, a later team, including five nurses and some soldiers, was not as well respected. After giving aid during the day, the nurses "organized music entertainment and trifled with the soldiers" at night, disturbing the peace of the hamlet. Moreover, the woman reported that they used residents' cooking utensils as drumsticks, breaking one. The next day, the nurses left quietly without cleaning or replacing the kitchen tools. Their flirtatious behavior led the old woman to cry that she "detested and didn't have any sympathy for these girls."[27] In essence, their careless acts undercut the positive psychological value of the medical aid.

Southern Instability

On 30 January 1964, three months after Diem's overthrow, another coup erupted in Saigon. This time, General Duong Van Minh was overthrown, continuing a revolving door of regimes in South Vietnam. General Nguyen Khanh assumed power. As before, political upheaval meant confusion in the PSYOP program. Saigon Radio began to attack foreigners' and internal schemes to neutralize the nation, especially those put forward by the French. However, Rufus Phillips referred to the station's charges as "preposterous." Khanh used the charges to keep his political enemies off balance. France

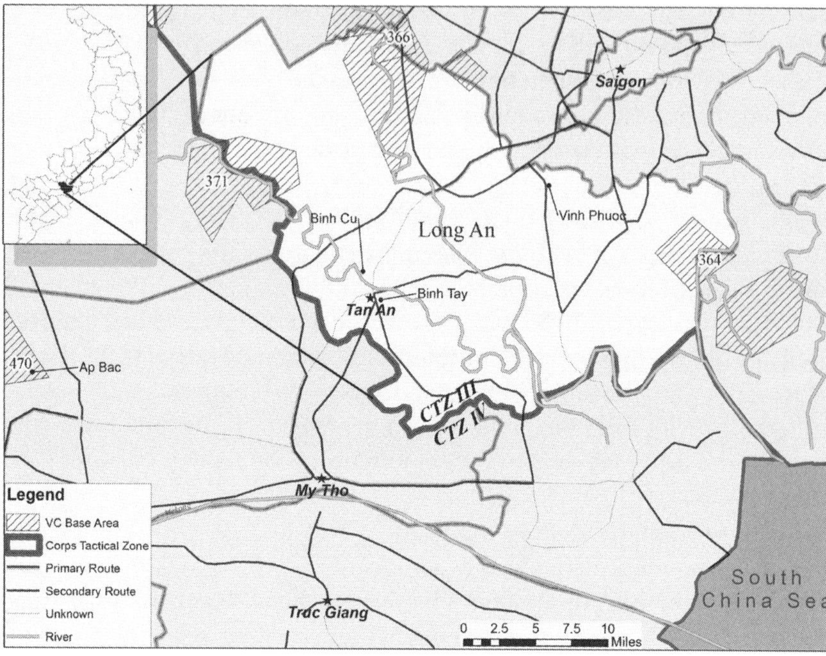

Map 3. Long An Province, South Vietnam, 1964

had offered plans, but the specific charges of South Vietnamese complicity were wrong. The station denounced the "colonialists' scheme to cooperate with the communists to subvert the political situation in the Republic of Vietnam." According to the report, the Vietcong sought neutralization because it could not win on the battlefield.[28] In referring to previous calls to neutralize South Vietnam, as Laos had been, the government declared: "The front's objective . . . was not the liberation of [the] South Vietnamese population but rather the extension of communist domination over all Vietnamese territory, which is contrary to the 1954 Geneva cease-fire agreement."[29]

Liberation Radio used the occasion of the latest coup to drive a wedge between Americans and Vietnamese. Hyping terrorism as a tool to divide supporters of the GVN, Hanoi broadcast a warning to stay away from jobs supporting the Americans. Citing a 26 January bombing at Tan Son Nhut Air Base and a later one at the Kinh Do Theater that killed fifty-six, the station quoted a former officer in the French Union as saying, "I have the impression that the God of death is waiting for the Americans wherever they set their feet." The officer urged his compatriots to avoid contact with Americans in order to stay safe.[30]

Meanwhile, President Lyndon Johnson announced his own plan to neutralize both North and South Vietnam. "I think that the only thing we need to do to have complete peace in that area of the world now is to stop the invasion of South Viet-Nam by some of its neighbors and supporters," stated the president at a press conference in response to a question regarding the French general Charles de Gaulle's neutralization proposal. Johnson declared that he would consider a plan that neutralized both the North and the South.[31] Despite its previous support for neutralization in the South, Hanoi ridiculed Johnson's call for neutralization of both states. The people of the North would never accept such a plan that was simply meant to thwart "the tendency for peace and neutrality which is spreading in South Vietnam and the ever-increasing desire of the people in Southeast Asia for an end to the dangerous U.S. war in South Vietnam."[32]

General William Westmoreland arrived in Vietnam in early 1964 as MACV deputy commander and became commander in June. He had been interested in PSYOPs since at least the Korean War. In a "Combat Information Bulletin" issued in 1952, he described a desertion program he had initiated in the 187th Airborne Regimental Combat Team's area, in which he employed a combination of loudspeakers and artillery to encourage desertions. "In my opinion," he wrote, "this type of operation certainly is deserving of further study and should this [Regimental Combat Team] encounter a similar situation in the future I would certainly like to try a similar scheme again." This interest in PSYOPs carried over when he became commander of MACV.[33]

SOG Psychological Studies Branch

The management of covert operations against the North transferred from the CIA to the Pentagon on 1 February 1964. SOG's Psychological Studies Branch consisted of four sections: research and analysis, print media and mail, radio, and exploitation. It expanded on William Colby's operational concept, emphasizing deception operations. The limited objectives of SOG included diverting North Vietnamese military resources to internal security, adversely affecting the Northern economy, impeding infiltration to the South, and creating the "impression that an active, unified, internal opposition" existed in North Vietnam."[34]

Herb Weisshart became the chief of the Psychological Studies Branch, the only CIA officer with managerial status in the new organization. Besides Weisshart, the CIA was reluctant to send its best people to MACV-SOG.[35] The CIA had wanted to suspend these sorts of operations and did the least it could in helping the new organization get off the ground. The first US Army PSYOP personnel assigned to MACV-SOG included twenty-two soldiers sent on temporary duty status from Okinawa in January. They were graduates of the PSYOP course, but none had experience conducting covert PSYOPs in denied areas. Additionally, Weisshart said, "they lacked area knowledge" of Vietnam.[36] Captain Fred Stables arrived from Okinawa in February with the second US Army PSYOPs element and received assignment to SOG's print media operation. Stables's team was officially assigned to III Corps to provide cover for entering Vietnam "because the national policymakers did not want a PSYWAR unit in country or the Vietnamese government at that time did not want us there." He recalled questioning at the time how "a democratic government [could] properly implement covert PSYOP operations because we're not a long-range planning kind of a country or government. We always were operating with our hands tied behind our backs simply because of our democratic process."[37] Although he referred to "democratic government," he likely referred specifically to the problem of maintaining secrets in the United States.

Fomenting regime paranoia provided a potential rationale for agent operations. If tied to a larger strategy, the loss of men could be justified as a cost of war. However, the operations themselves were futile. Despite attempts by some in SOG to create such a strategic goal, limitations placed by the Johnson administration stood in the way. Putting an exclamation point on the futility, Radio Hanoi quoted a UPI story reporting on additional commando trials in March. UPI "announced that these plots have been thwarted by the effective security and police system in North Vietnam," while Radio Hanoi reported that twelve trials of seventy commandos had been held. "The Americans said that if captured we would be tortured," said one prisoner, "but the truth is entirely different. The objective of the North Vietnamese Government is to change bad people into good people, not to use torture to kill."[38] Reporting on the eighteenth group of commandos put on trial, Radio Hanoi announced seven death sentences. Local militias captured the teams shortly after they parachuted into Quang Binh in North Vietnam. Among the group's tasks were "to collect information and carry out espionage activities, establish spy bases, kidnap or assassinate cadres and army men, and to carry out psychological warfare among the people by scattering leaflets and false news against the North."[39]

In addition to the radio and leaflet operations previously mentioned, SOG sent poison-pen letters, which were intended to cause harm to either the recipient or the purported sender or to disseminate disinformation. An example would be a letter from a low-ranking official complaining about actions of higher-level officials or making accusations designed to spur intergroup dissension or even to cause the arrest of individuals. These are particularly useful in manipulating paranoid state security services. In one of the more standard PSYOP actions, SOG dropped "gift kits" in the North that contained thread and cloth with the South Vietnamese flag's colors. This was meant as divisive propaganda to force a reaction that would separate the people from the government. Poor people with no access to cloth would likely use the gift. Government officials were likely to get upset at the display of the South Vietnamese colors and ask people to discard the cloth. Mothers might be upset at relinquishing clothes they had made for their children and take their ire out on the government, for example.[40]

The Saigon newspaper *Dan Quyen* reported in April on bombings apparently conducted by commandos in Hanoi. On 7 April 1964, a "special station" broadcast news of bombings at the Hanoi Great Theater and the Metropole Hotel using plastic explosives. This special station was likely referring to one of SOG's covert efforts. Another device "seriously damaged the Khai Tri House, which has been used by the Viet Cong as a site for press conferences in the North," reported the paper. *Dan Quyen* stated that these were conducted in retaliation for bombings in Saigon and claimed that "the sabotage taking place in Hanoi proves that anticommunist activities in the North are strong and noteworthy." It is unclear whether the events actually happened or were instead fictionalized responses to the psychological effect of the commando arrest broadcasts.[41]

A further report on a "Ranger Spy trial" in Quang Binh Province identified one culprit as Nguyen Dong, a PSYWAR officer. He confessed that three Americans had taught him the necessary skills and tasked his group with PSYWAR in the North. The broadcast attempted to portray this as a unilateral American mission. In fact, it was a South Vietnamese operation conducted by the Strategic Technical Directorate, advised and aided by Americans.[42] Dong testified that they were humanely treated upon capture. As propaganda, these messages targeted future teams, portraying them as dupes of Americans and urging them to surrender rather than die for a lost cause. Punishments of the team members seemed calibrated to induce surrender.[43]

Radio Hanoi defensively replied to President Johnson's charge that the North was "directing and supplying the patriotic guerrillas in South Vietnam." Living far across the Pacific Ocean, "the Vietnamese people have never been and never will constitute a threat to the U.S. people," implored Radio Hanoi. Meanwhile, it accused the United States of establishing a colonial base to enslave the Vietnamese people.[44] The station also mocked Robert McNamara as "absent-minded" for claiming that North Vietnam had supplied Chinese arms to the Vietcong. Instead, it posited that a UPI story, asserting that "guerrillas seized enough weapons from the government troops in February 1964 . . . to equip two companies," reflected the true source of the weapons.[45]

McNamara's March visit to Vietnam fed the propaganda war. President

Johnson had instructed him to enthusiastically demonstrate American support for Vietnam by "seeing Khanh in the newspapers with McNamara and Taylor holding up his arms." The result was unfortunate. Photos of the taller McNamara and Taylor flanking the diminutive Khanh were a propaganda disaster that the NLF was able to exploit, representing "visible proof that he was an American puppet."[46]

Radio Hanoi also attacked Southern efforts to eradicate malaria as being a cover for espionage activities. According to the North, "under the pretext of spraying mosquito-killing chemicals to help the people eradicate malaria," the teams gathered intelligence on the NLF and "tried to win the sympathy of and buy the foolish people." Furthermore, Hanoi claimed, the antimalaria teams "carry linen, clothes, salt, and other things for distribution in areas they passed through." The station called on listeners to prevent the US and South Vietnamese governments from achieving the antimalaria program's "dark and dirty objective."[47] Spinning US aid efforts as negative, the LBS also attacked the Americans' rice aid to the country. A common theme of Northern propaganda was to attack any positive actions by the GVN and the United States. The North could not compete in the battle for material well-being in its districts and so sought to derail the debate.[48]

As an indication of the leftward movement of the Communist Party and isolation of "rightist" Communists like Ho Chi Minh, Radio Hanoi proclaimed in March that "our party is a true Marxist-Leninist party," reading from a *Hoc Tap* (Studies) editorial to domestic listeners. Meanwhile, the broadcaster attacked those rightists who failed to "realize that strengthening dictatorship over the enemy and developing democracy for the people are two inseparable aspects of our dictatorship," again per *Hoc Tap*. The rightist attitude manifested partly in "overbearing Mandarin attitudes" and a "tendency to give the improvement of the people's living conditions priority over capital accumulation." Radio Hanoi and the LBS functioned as the propaganda distribution means of the agit-prop system in the South. Reading articles from *Hoc Tap*, the party's ideological organ, helped ensure unity of effort across a widely dispersed and functionally independent propaganda organization.[49]

PSYOP Adaptations

With the effective transfer of covert propaganda to the US military, work continued on developing a functional structure to coordinate PSYOPs in Vietnam. On 16 March, Johnson indicated his "own interest in the strongest possible information and psychological warfare program" and gave the USIA director Carl Rowan authority to expand his agency's control. Johnson assured Rowan "that no worthwhile undertaking shall be inhibited or delayed in any way by financial restrictions." The only stipulation dealt with VC defection programs, on which Johnson deferred a decision.[50]

Saigon Radio began a series of government leaders' speeches to the people in an effort to rally support for the Khanh regime. Deputy Prime Minister Nguyen Ton Hoan spoke on "safeguarding freedoms" in the second of these broadcasts. Speaking as a "revolutionary cadre who is continuing the struggle for the nationalist ideal," Hoan argued that the role of government was a means to the end of protecting freedom.[51] Prime Minister Nguyen Khanh initiated his weekly radio address by calling on his compatriots to "carry out a complete and total revolution and to achieve national liberation."[52] As in the North, the Southern government was attempting to cover itself with the nationalist mantle.

Among the lessons learned by that point was the importance of coordination to produce effective PSYOPs. As noted, radio served this function for the North, coupled with an integrated party political structure extending from the politburo to the hamlet. The South had to develop its structure on the fly. Combined PSYWAR/CA Operations Centers established in each CTZ by 1964 linked the US and ARVN systems. Eventually this program was extended down to the division and sector levels. Significantly, this combined PSYOP structure was never implemented at the national level due to South Vietnamese concerns. This failure hampered coordination as the war progressed; however, at least at the lower levels, a connection developed. These centers generated PSYOP plans and coordinated military and civilian PSYWAR/CA activities and messaging. They also allocated scarce resources, such as aerial loudspeaker sorties, and conducted propaganda analysis.[53]

Corps Operations Centers acted as the message coordinator for a dispersed and diverse effort. The overall structure was one of centralized guidance, with decentralized implementation. The program's ultimate success depended "upon the operators in the field who establish contact and relations with the people." In a sense, it was similar to the North Vietnamese system, the key difference being the North's insistence on integration at all levels to ensure that, even with decentralized action, the politburo's vision would rule.[54] Such control never existed in the South.

Always willing to use Western news stories to support its propaganda, Radio Hanoi had enlisted foreign reporters directly into its program by the spring of 1964. The Australian leftist journalist Wilfred Burchett had a long history of battlefield reporting. He had written dispatches from North Korea between 1951 and 1953 in order to highlight "the Korean people's struggle against the United States." The following year he reported from behind Vietminh lines at the Battle of Dien Bien Phu, followed by commentary on the anticolonial struggles in Algeria and Africa. In 1964, he gave a lecture in Hanoi and showed a Liberation Film Studio documentary on his recent trip to the "liberated areas in South Vietnam." Burchett said that "the U.S. and the puppet troops" exhibited low morale, and he expressed "confidence in the final victory of the patriotic war of resistance being waged by the South Vietnamese people against the U.S. imperialists and their stooges."[55]

North Vietnamese International Propaganda

But it was not only friendly reporters who provided material for propaganda support. *Life* magazine unwittingly gave powerful testimony that Hanoi exploited. Captain Jerry Shank, the USAF pilot of a T-28 ground-attack aircraft, was killed on 24 March 1964. His wife subsequently released letters critical of the war, letters he had written in the months prior to his crash. *Life* magazine published these, and *Nhan Dan* echoed them for their propaganda value. According to Radio Hanoi, Shank wrote, "Morale is at a big low over here, especially among the combat crews. I don't know what the United States is doing here." He was ashamed of America, writing that he believed "we can no longer save face over here, for we have no face to save."[56]

On 25 May, Radio Hanoi read a letter that Ho Chi Minh had sent to the *Minority of One* magazine in America. The publisher had asked Ho for his views on the war. Ho thanked the magazine "for affording me an opportunity to talk with the U.S. people on the present situation in South Vietnam." Reflecting on his time spent in the United States, and furthering the psychological objective of dividing the American people from government policy, Ho said that the "Vietnamese people never confuse the justice loving U.S. people and the U.S. government, which has committed numerous crimes against them in the past 10 years." Ho claimed that America's "Special War" was actually a "war of aggression waged by the U.S." and a betrayal of the Declaration of Independence. These themes all supported the North's developing psychological objective: decrease American support for the war.[57]

In the spring of 1964, North Vietnam began publishing the *Vietnam Courier* as an English-language propaganda organ for distribution primarily in the United States. This paper closely mirrored Radio Hanoi broadcasts in content. The target audience included Americans who were interested in the war but suspicious of US policy. Eventually the paper became available in many college libraries, providing a North Vietnamese spin on events. Like much Northern propaganda, it relied on a subjective truth to present an image of a deceitful American government to the reader. The *Vietnam Courier* reacted rapidly to events, producing detailed stories within days of a particular incident. For instance, the 5 November 1964 issue contained details and photos of the 1 November attack on Bien Hoa. However, the issues sent to Cornell University did not arrive at the university for approximately eight weeks, meaning that Bien Hoa was old news by that point. However, when viewed from a propaganda perspective, the late arrival might have been a net positive. The target audience would be unlikely to double-check the facts and exaggerations in the stories. Thus, in presenting only one view of a far-off event, Hanoi's spin would be the one with the most lasting and poignant impact on the target audience. Typically, the *Vietnam Courier* carried a mix of news from the battlefront and stories of emotional impact regarding life in North Vietnam. On 10 June, for example, the lead story covered Ho Chi Minh's message to the American people originally printed in *Minority of One*.

Printed on thin paper, the *Vietnam Courier* typically used British, rather than American, spellings for words and contained numerous misspellings. A typical blurb contained an advertisement for Wilfred Burchett's book *My Visit to the Liberated Zones of South Vietnam*, echoing his messages on Radio Hanoi. In September, perhaps in an effort to increase readership, the *Vietnam Courier* reported on the 5 August downing of US Navy pilot Everett Alvarez during the initial response to the Gulf of Tonkin incident. By carrying news of prisoners, the *Vietnam Courier* hoped to lure readers. The paper retailed for sixty cents, and, interestingly, most issues contained advertisements for machinery exports. It is hard to imagine American readers placing orders with Hanoi for machinery.[58]

Over the course of the war, at least a hundred college libraries in the United States subscribed to the paper, from the University of California at Berkeley to North Texas State University in Denton, Texas. An analysis of current library holdings shows that 65 percent of the extant copies are in American libraries, indicating that this was the paper's main target audience. Such distribution provided ready access to an important target audience to support the key psychological objective of dividing the American populace from government policy and forcing a change in that policy. The *Vietnam Courier* was straight propaganda, rather than agit-prop. Its aim was to inform people who might be in a position to agitate others.[59]

This approach contrasted with another magazine that became increasingly available in the United States at this time, especially through college libraries. *Vietnam* was geared toward making an emotional connection with a broader target audience. Similar to *Soviet Life* and *Look* magazines, the glossy color spreads showed life in the North and among the NLF, presenting both a heroic and a tragic image.[60]

China and Russia frequently echoed North Vietnamese themes in their own international broadcasts, lending credibility to them and providing wider dissemination. In one instance, the New China News Agency reported on a message received from the "student committee of Haverford College," near Philadelphia. Printed in the March issue of *Vietnam Bulletin*, which was published by the British Vietnam Committee, the statement condemned "U.S. Government intervention in the affairs of Vietnam" and the use of South Vietnam as a "testing ground for U.S. weapons and tactics." The students echoed North Vietnamese propaganda denying a role in the insurgency, describing it instead in Marxist terms as a "struggle between oppressors and oppressed." The committee attacked US government actions in South Vietnam, writing that Vietnamese "crops have been poisoned, their freedoms abolished, their men tortured, their women raped, and their children maimed." They asked fellow Americans to "condemn the United States for its guilt in the unjust treatment of the Vietnamese people" and to urge the US government to withdraw its troops. This incident shows the reinforcement that Western front groups provided to the international communist press's support of North Vietnamese psychological objectives. The North would issue a propaganda theme, which would be parroted in Western protests. Reports of the protest, harping on the original theme, would then play in Soviet, Chinese, and other friendly media. The story would recirculate through various North Vietnamese organs as fresh news, reinforcing and giving further credibility to the original story.[61]

South Vietnamese POLWAR Structure

In the battle for control of the population, refugees were a prime target audience. While the GVN sought to encourage these people to return to government-controlled areas and accept aid via refugee camps, the VC propaganda endeavored to impede this move. Refugees in Quang Duc Province reported being told by the NLF that "if they return to GVN control, they will be covered with gasoline and burned alive." MACV took this as an indicator of the "seriousness with which the Viet Cong view the refugee program."[62] Meanwhile, in Hau Nghia Province near Saigon, security had reached a level where Southern officials "visited insecure areas during hours of darkness to speak to the people, show movies, and present music."[63] Teams in Phuoc Long Province reported positive results at the tactical level after they had provided immediate assistance following a VC attack. The effort also successfully exploited the negative reaction to a VC attack on Bun Ho village.[64]

In April, PSYWAR teams reportedly began distributing colorful envelopes as pill containers during Medical Civic Action Programs (MEDCAPs). They portrayed Republic of Vietnam and United States flags along with an inscription reading: The Government of the Republic of Vietnam and the people of the United States of America wish you a healthy and prosperous life. Teams had also begun conducting PSYWAR training for Civilian Irregular Defense Group (CIDG) Strike Force soldiers and reported that the ARVN and Montagnards were working well together in this task. The CIDG troops were mostly Special Forces–trained Montagnard militia acting as a blocking force in the Central Highlands. The GVN established a PSYOP committee, reporting to the National Pacification Council, in mid-April to supervise field activities. By this time every ARVN division had a PSYWAR company attached to it.[65]

The trend looked positive, but the postcoup instability still severely hamstrung the PSYOP program and the pacification effort in general. Rather than the decrease in VC-controlled hamlets seen in the final months of Diem's rule, there was a rapid loss of hamlet control to the Front after the coup. Despite increased PSYOP sorties and tactical improvements and innovations, as well as reportedly spotty but sometimes "outstanding Psychological Warfare support of provincial pacification activities" by the Vietnamese Information Service, the negative trend continued for most of 1964. Capabilities improved, but tactical progress was no substitute for a stable government and a clear ideology. To make matters worse, some provinces such as Binh Duong had no PSYWAR program at all.[66]

Echoing Manning's earlier statement of priorities, sixteen officers representing State, Defense, the CIA, and the USIA met in Honolulu in May 1964 to refine the goals of the information program. The key campaigns were improving South Vietnamese morale in order to diminish VC influence and expanding the program in North Vietnam to warn citizens "of the risks imposed upon them by the Hanoi regime's continued aggressions." But the last goal, to "better inform the American people and our allies in Europe and elsewhere so as to maintain their support for necessary efforts to keep Southeast Asia from falling under Communist domination," was problematic.[67] To do so bridged legal and political boundaries that could further subject the US administration to political fallout if not handled well. The US Information and Educational Exchange Act of 1948, better known as the Smith–Mundt Act, officially restricted the Department of State and, unofficially, the Department of Defense from using propaganda against the American people.[68]

The minutes of the Honolulu Conference described a gloomy situation in which "there is no conceivable information program that will make defeats look like victories." Ambassador Lodge wanted "to press Premier Khanh vigorously to make fire-side chats on a regular basis," harkening back to Franklin Roosevelt and successful American political communications strategies. Ironically, Khanh had implemented this program the previous month. The conference noted that USIS film viewership in 1963 was 28 million. Using Khanh's voice in future films would expand his influence. The attendees urged the US Operations Mission (USOM—the American embassy in Saigon) to acquire 100,000 transistor radios to distribute outside the Saigon area and requested that the Japanese contribute 100,000 more transistor radios.

One unusual idea proffered was to find "a Vietnamese who can write a GVN imitation of 'God Bless America' to help rally people."[69]

The conference committee also recommended that at least one American propaganda adviser be assigned to each ARVN unit in the field with "a substantial number of American advisors" to focus on exploiting VC atrocities and GVN heroism. An example of this narrative would be to exploit the Vietcong's murder of a teacher, a common enough scenario. The American and GVN counterparts could "tape-record the widow's story of the atrocity" for use in broadcasts.[70]

The military training pipeline at the time had forty-two Americans programmed for service in the provinces. With the increase in American advisers needed for Vietnam, the Special Warfare Center at Fort Bragg began offering a six-week Military Assistance Training Advisor Course. This so-called long course for advisers was designed to "provide selected officers and enlisted men with a working knowledge of the duties of a military assistance training advisor in counterinsurgency operations, at the division and lower level." It included fifty-four hours of language training, four hours on the adviser's role in PSYOPs, and three hours on current PSYOP programs in Vietnam. Topics included the nature of psychological operations and the necessity of integrating PSYOPs with tactical operations and CA. This was geared toward ensuring that all advisers had at least a limited knowledge of PSYOPs, regardless of the military branch they were scheduled to assist in Vietnam. For those unable to attend the long course, MACV held an abbreviated version in South Vietnam. Additionally, a one-week PSYOP course was held in Vietnam specifically for newly assigned Sector PSYOP/S-5 advisers.[71]

The conference committee also expressed support for a CIA program to field two hundred People's Action Teams (PATs) to spread messages by word of mouth. PATs provided support to "armed drama teams, propaganda broadcasts over provincial radio, and 'black' teams posing as" Vietcong to help identify Communist cadres.[72] Nguyen Duy Be, in Binh Dinh Province and Quang Ngai Province, originated the program. He brought personal charisma combined with prior experience in the Vietminh. Teams were dressed in the black clothes normally associated with Vietcong, and the thirteen-week training course incorporated "successful techniques of the Communists, including self-criticism sessions, a communal style of living, and an intense focus on a simple, repetitive set of principles."[73] The teams were organized as an intelligence asset, but they also conducted "psychological warfare, propaganda, and information activities."[74] Nguyen Duy Be, a former member of the Vietminh, years earlier had "chafed at increased Communist Party control" and deserted in Laos. He developed a similar, small-unit political action group, training them first in Quang Ngai. Nguyen Duy Be sought "to identify with the people rather than forcing people to identify with the government."[75]

A memo from USIA director Rowan to the president concurred with many of these ideas. He noted how USIS had increased Voice of America broadcast to North Vietnam and reached an agreement with the GVN to install a 50-kilowatt Voice of Freedom transmitter in Hue to broadcast to the DRV. Additional help to the GVN film production and print capabilities had been promised as well. However, Rowan warned that the lack of coordination was the "gravest problem in terms of the information-psychological program." Since Barry Zorthian's arrival in Saigon as public affairs officer in January, Ambassador Lodge had refused to give him oversight on press relations. Lodge, in a telegram to Edward Murrow, emphasized that "I do all this work myself," according to Rowan. Given this attitude and the separate US military information program, the "result was that no one could be sure who was responsible for what," and he urged the president to give Zorthian overall responsibility in this area. This included "all activities involved in the orientation and indoctrination of military personnel."[76]

The Honolulu Conference also urged the administration to take steps to "restore some measure of credibility with American correspondents covering the war in Viet-Nam" who believed the military had "lied to them on several occasions." Conference attendees concluded that this hostility bred negative reporting. They urged more openness on the part of information officers and in-country travel assistance to reporters.[77] One aspect of war reporting that bothered Rowan was the tendency of American reporters "to emphasize American mistakes and acts that could be called 'brutal' and to give little coverage to Viet Cong atrocities." USIA recommended, and the military approved of, a plan to dispatch an army photographic team to South Vietnam to "cover combat actions and to make available to American and other newsmen the kind of photographs that put across the stories we want told."[78]

Despite this generally gloomy trend, innovation continued and some minor gains were made in the psychological war. Various Corps Tactical Zones began to experiment with a weapons turn-in rewards program, reporting positive results. Initiated by US Special Forces units, the program quickly spread and leaflets were produced in support. Additionally, despite an overall decrease in Chieu Hoi ralliers since the 1963 coup, military intelligence still reported gaining "valuable information from ralliers, such as unit designations, locations of arms caches, and indications of Viet Cong intentions."[79] By the first week of June 1964, utilization of aircraft for PSYOP support continued to grow. With this, the ability of the US Air Force to fill support requests increased. From previous lows of a 50 percent sortie rate, nearly 80 percent of the thirty-six requests for leaflet drops during the first week of June were flown. That same month, the ARVN 1st PSYWAR Battalion deployed its cultural/drama team outside of Saigon for the first time. Cultural/drama teams provided entertainment, such as music and plays, while supporting MEDCAPS or other PSYOP/Civic Action operations.[80]

In May 1964, as the GVN began to consider forming a POLWAR structure, the RVNAF's chief of the Psychological Warfare Division headed a group traveling to Taiwan to study the Nationalist Chinese POLWAR system. POLWAR differs from PSYOPs in that the former includes operations that could be termed "morale and welfare," designed to maintain public and especially military support for the government. Although political warfare was anathema to many Americans, fighting a counterinsurgency required such a capability. Without the support of the South Vietnamese people, or at least their acquiescence, winning the war would be impossible. Just as important was denying that support to the Vietcong. The military was another key audience that the POLWAR system was designed to target. The POLWAR system provided a positive lever to influence behaviors in the face of VC proselytizing. However, before POLWAR could achieve much, the ARVN PSYWAR commander changed in May—the fifth such change in six months. MACV lamented that the "continuous personnel changes in this key assignment seriously handicaps PSYOPS."[81]

Because of the continued instability, MACV noted a degradation of the overall pacification program. Despite the heroic stand of an ARVN ranger company in Duc Hoa, Hau Nghia Province, elsewhere in the country the tide was flowing against the GVN. Partly as a result of new assessment criteria, and partly due to VC successes, the number of hamlets considered to be under GVN control dropped from 7,344 in April to 4,905 in May. By July, only 30 of 219 strategic hamlets in Long An remained under government control, mostly along Highway 4.[82] In Kien Tuong Province, MACV reported that "there are not enough troops available to pacify and at the same time conduct operations outside the perimeter to keep the VC off balance"; other province officials mentioned similar problems. NLF propaganda seized on this instability. In Chuong Thien Province, the Front "published a letter listing the names of Mobile Action Cadre conducting pacification operations in Vinh Phuoc village in Kiem Hung district." It warned cadres to cease their operations or face punishment. Three allegedly kidnapped cadres were released after two days and warned to "cease their activities or be beheaded."[83]

As the situation continued to deteriorate in the summer of 1964, Johnson administration officials contemplated engaging in a wider war. The administration asked the US National Board of Estimates to analyze the question: "Would the loss of South Vietnam and Laos precipitate a 'Domino Effect' in the Far East?" Bureau chairman Sherman Kent replied that despite the lack of a domino effect, a loss would send a shock wave throughout East Asia. He predicted that the only immediate threat would be to Cambodia. However, the major problem would be the loss of American credibility, due to previous pronouncements of support for Laos and South Vietnam. "To some degree this will tend to encourage and strengthen the more activist revolutionary movements in various parts of the underdeveloped world," he warned.[84] Johnson therefore had to balance his decisions concerning Vietnam with a shifting set of Cold War domestic and regional goals. Jack Valenti, special assistant to the president, wrote on a cover sheet of Walt Rostow's memo discussing options for Vietnam that "the soft options in Southeast Asia have been used up by Eisenhower and Kennedy. Only hard ones lie ahead." According to Rostow, the psychological situation in Saigon was one of "disarray, hopelessness, and waiting for something to happen," because "Vietnamese leaders and people see no successful end to the war on present lines of action."[85]

In June, the Johnson administration also began to reexamine the information program for Vietnam. The issue over the secret nature of America's growing involvement, and the unpublicized deaths of American servicemen, helped to exacerbate a disconnect between the administration's rhetoric and media observations. Ambassador Lodge was a large part of the problem. After returning from a visit to Vietnam, Rowan told President Johnson that "Lodge's one-man rule over the U.S. Mission's public affairs program had harmed coordination of the overall public affairs effort." Barry Zorthian, Rowan said, "should take control of the entire program."[86] The president finally concurred, enlarging Zorthian's role to include responsibility for the US overt PSYOP program as well as public affairs. Zorthian immediately called for "maximum candor" with the press in an attempt to rebuild credibility for public affairs statements. William P. Bundy, deputy assistant secretary for international security affairs, hoped that the decision to put Zorthian in charge of all information activities in Vietnam would improve the press effort, which was burdened by Ambassador Lodge's actions. One possible concern with having Zorthian in charge of both public affairs and PSYOPs, however, was that it blurred the lines between information and influence operations. At the same time, Lodge resigned as ambassador to take care of his sick wife, and Maxwell Taylor took over.[87]

No sooner was Zorthian in place than he faced a public relations firestorm. Use of defoliants was growing in South Vietnam. MACV reported on details of the crop destruction program in May 1964, when Vietnamese Air Force H-34 helicopters sprayed approximately fifty-five hectares of VC crops in Mang Xim Valley. Prior to the operation, "PSYWAR leaflet drops and aerial broadcasts" advised people "of the reason for the crop destruction, harmlessness of the spray, and urging return to GVN control for resettlement with relief and assistance."[88] In June, the (North) Vietnam News Agency once again protested the use of "noxious chemicals," this time in the cities of Tan An, Cholon, Ca Mau, Ba Ria, and Rach Gia. The report stated that the "poisons affected many people, destroyed hundreds of hectares of rice fields and gardens, and killed many domestic animals."[89]

In July, Radio Hanoi charged the United States and South Vietnam with conducting biological warfare. The US Department of State advised the US embassy in Saigon, using the clipped language of diplomatic correspondence, to make a "straight simple denial of charges coupled with general expression of willingness to have impartial investigation."[90] The following day, the State Department delegated to Ambassador Lodge approval authority for further crop eradication operations, but it also required GVN approval for each operation. It admitted, however, that "crop destruction remains matter of serious political concern here and political aspects must be given careful consideration by Saigon before approval each operation."[91] Meanwhile, provincial officials began to settle claims with civilians related to the defoliation program. The first payment occurred in An Xuyen in June, compensating fifty-seven persons "claiming crop damage from defoliation operations." This is the first reported incident of indemnification payments being made for losses due to defoliation.[92]

In spite of the propaganda claims and negative press, the program continued. MACV viewed the twin goals of depriving the Vietcong of food and decreasing support for VC among the populace as critical to success. In III CTZ, aerial loudspeaker missions were flown in Binh Thuan Province to advise farmers that the GVN was destroying crops grown for the Vietcong. The Vietcong had promised to protect farmers growing food for VC benefit. Proving the inability of the Vietcong to do so was designed to convince farmers that it was "futile to stay in the Viet Cong controlled areas" and thus further deprive the Vietcong of support in rural areas. As these examples show, it was not about hearts and minds. The goal was to win. Though conducted for tactically sound reasons, the defoliation program continued to be a boon for North Vietnamese propaganda.[93]

As part of a National Day of Mourning, ARVN PSYWAR units exploited the massacre of forty family members of the Regional Forces (SVN local militia forces). An extensive national PSYOP campaign was begun to "further incite the people against the Viet Cong" for this and other atrocities.[94] Meanwhile, in July 1964, on the tenth anniversary of the signing of the Geneva Accords, GVN PSYWAR units began efforts to publicize a National Day of Shame "to promote loyalty to the government and militancy in the fight against the Viet Cong."[95] Professors and students of the Faculté des Lettres in

Saigon held an all-night vigil and march to pressure the government to strike northward.[96] The US embassy in Saigon advised Washington that General Nguyen Khanh, during his speech at the observances, echoed this goal, using the phrase "March to the North." That the GVN should attack to free the North from Communist domination became a popular theme among "refugee groups, political parties, religious groups, and student organizations" and was highlighted by several dailies. *Le Song Moi* (New Reason for Life) recommended that the "Geneva agreement must be swept away so [the] North can be won back and ten million compatriots liberated." One editorial wondered whether "Khanh dares unilaterally take [the] decision to march northward with or without US assistance."[97]

The CIA assessed that the reason for the March North agitation was Khanh's weak political position. A FLASH message (the highest-level message) from the embassy in Saigon noted that Khanh explained his rationale to Ambassador Lodge. Khanh "cited growing war weariness on part of Vietnamese people, slow pace of pacification efforts, continued high casualties," as well as "moves afoot to unseat him and replace him with one or another combination of generals," as the motivations for focusing the nation's attention northward. The upcoming US presidential election also influenced attitudes in the South. Republican senator Barry Goldwater's "nomination has led many . . . to believe that now was opportune moment to apply pressure on US to expand area of conflict," the embassy noted.[98] Despite VC threats to hinder the National Day of Shame observances, MACV reported that "Saigon remained quiet during the week." At the same time, Khanh was conducting secret talks with the NLF in hopes of drawing troops away from North Vietnamese control, and Buddhist political activists sought their own accommodation with the Front in hopes of neutralizing South Vietnam.[99]

The RVNAF's PSYWAR Training Center submitted a plan to the Ministry of Defense to increase the schools' capacity from 270 to 1,000 students per year to fill the critical shortage of PSYWAR/CA specialists. Additionally, that summer the training center continued to provide a series of PSYWAR orientation courses for ARVN commanders. During the second graduation exercise (for ninety-one officers), the chief of cabinet spoke on the importance of PSYWAR/CA to the overall counterinsurgency effort.[100]

Gulf of Tonkin Incident

On the ground, teams continued to expand face-to-face interactions in order to spread news of the Chieu Hoi, pacification, and recruiting programs. In addition, they continued to emphasize VC atrocities.[101] Communist activities exceeded weekly averages for the previous year. Company-sized attacks were "approximately double the monthly average for the first six months of 1964." Most attacks were in the IV Corps area south of Saigon. The Studies and Observations Group OP34A missions continued in June, when Hanoi reported intercepting South Vietnam Navy boats off Quang Binh Province. According to reports, the boats escaped and forced a Northern fishing boat to sail south with six fishermen.[102] The following week, Radio Hanoi provided details on commando training gleaned from a 6 June AP story. The story disclosed "that no success was recorded by the spy commandos sent by the US puppets for sabotage in North Vietnam," and it acknowledged that most of the agents parachuted into the North "have simply disappeared."[103] In the meantime, the army activated the US 5th Special Forces Group in Vietnam to oversee the now sprawling system of Special Forces camps and the army of Montagnard irregulars fighting an undeclared war.[104]

In June, the White House began drafting a resolution for approval by the US Congress authorizing Johnson to increase military activities in Vietnam. Except for the precise language in the "whereas" section, which pertained to the Gulf of Tonkin incident, this draft was nearly identical to the resolution that was eventually approved that August. The initial draft focused instead on Chinese support and stated that North Vietnam "has flouted its obligations under the Geneva Accords of 1954 and has engaged in aggression against the independence and territorial integrity" of South Vietnam by "furnishing direction, training, personnel and arms for the conduct of guerrilla warfare" there. The operative paragraph, like the Gulf of Tonkin Resolution two months later, would give the president the authorization for "all measures including the use of armed forces to assist that nation in the defense of its political independence and territorial integrity against aggression or subversion supported, controlled or directed from any Communist country." Initial planning suggested submitting the resolution to Congress after the fall presidential election.[105] With Johnson running for reelection as a peace candidate, he needed to put off a decision on escalating the war in Vietnam.

One series of SOG-initiated agent and sabotage missions at the end of July proved especially significant, however. The ongoing maritime attacks had heightened North Vietnamese vigilance, with the North Vietnam Navy augmenting its coastal patrols. On the night of 31 July, a patrol encountered the USS *Maddox*, which was conducting signals intelligence patrols in the Gulf of Tonkin as part of Operation Desoto. The US Navy conducted those operations to trigger North Vietnamese radar to assess reactions and capabilities. The clash between the North Vietnamese and American ships led to the Gulf of Tonkin incident and, subsequently, the expansion of American involvement in Vietnam. It also led to the congressional support Johnson had been seeking. Although the resolution was originally planned to be considered after the elections, the August encounter had forced Johnson's hand. The president introduced his previously written, reworded resolution to expand the US military role in Vietnam immediately after the incident. President Johnson initiated limited air attacks on coastal bases in southern North Vietnam.[106] The target list for the reprisal strikes was strictly limited, consisting almost exclusively of attacks on North Vietnamese patrol boats and limited attacks on oil storage facilities serving those boats. At the time, the Johnson administration repeated statements about the limited nature of the strikes to impress upon the North that this signaled no enlargement of the war.[107]

A PAVN press conference after the Gulf of Tonkin incident stated that the United States made a "hue and cry about what they call an unprovoked attack . . . to cover their acts of provocation sabotage, their violation of the airspace and territorial waters, and their encroachment on the territory and sovereignty of the DRV."[108] According to Radio Hanoi's domestic service, Ho praised the Vietnam People's Army antiaircraft and naval units for "the shooting down of U.S. aircraft, the capturing of a U.S. pilot, and the driving of U.S. destroyers out of our territorial waters, thus defending it" during the attacks.[109]

The psychological war continued even as the air war heated up. The week after the Gulf of Tonkin incident, ARVN 5th Division conducted the largest airmobile operation to date, lifting two battalions into assault positions near Ben Cat in Binh Duong Province. The RVN air force conducted one PSYWAR sortie that week using an L-20 light airplane equipped with loudspeakers. The US Air Force reported twelve PSYWAR sorties, and the US Army reported thirty-two sorties involving both fixed-wing and rotary aircraft. Also, I Corps units dropped 20,000 Katu-language leaflets around A Ro in the Central Highlands, using the theme "GVN help toward the Montagnards." Farther south, PSYWAR units dropped 25,000 leaflets in conjunction with loudspeaker operations along the IV Corps border with Cambodia. These used both Vietnamese and Cambodian languages regarding recent statements by President Johnson and General Khanh.[110]

After the Gulf of Tonkin incident, Radio Hanoi began broadcasting the lilting voice of the radio personality "Thu Huong" (Autumn Fragrance) across the airwaves: "This is your announcer Thu Huong calling American servicemen in South Vietnam." Thu Huong, better known to American soldiers as "Hanoi Hannah," was born Trinh Thi Ngo. She became the primary messenger of North Vietnamese propaganda directed at American soldiers for the remainder of the war. In December, she had this to offer:

> While you are worrying about your presence in South Vietnam, in Seattle, Washington, an American mother is asking herself: Why have (they?) been in

> South Vietnam? She is Mrs. James D. Hamlin, mother of American technician William R. Hamlin, who was listed by the Pentagon as missing during a fierce battle on the night of 8 December in (An Lao) valley, Binh Dinh Province, 300 miles northeast of Saigon. This sad news troubles her very much. Is William Hamlin still alive? If he is dead, why can his body not be found? Or he could have been taken prisoner? Mrs. James Hamlin told a UPI correspondent: "I just hope he is not lying out in some rice paddy."[111]

In an interview after the war, Ngo stated: "We wanted to make them a little bit sad." She stated that they used American news sources because "it would be more convincing" and remind US personnel of the war's unpopularity at home. Antiwar movement personalities visiting Hanoi from the United States and elsewhere brought music and magazines, which she used to entice soldiers to listen to her show.[112] US Marine Ken Watkins, who started listening to the broadcasts in 1966, later recalled that he "would tune in once or twice a week to hear her talk about the war, a war I was beginning to question and wanted to hear discussed." This put him squarely within the target audience that the station was trying to reach. "U.S. Armed Forces Vietnam Radio didn't talk about the war really, they ignored the issues or public attitudes at home," he said. Like Sweet Sister, which the South Vietnamese PSYWAR directorate later beamed at ARVN troops, perhaps the most appealing aspect of Hanoi Hannah's broadcast was that she "was female and had a nice soft voice."[113]

Ngo acknowledged the assistance of antiwar activists such as the journalist Wilfred Burchett, Cora Weiss from the group Women Strike for Peace, and Jane Fonda. She credited "broadcast tapes sent to us from Americans against the war" as among the most effective aspects. (This likely refers to the "Radio Stateside" program, discussed in Chapter 8.) The Hanoi Hannah broadcasts primarily supported the twin objectives of decreasing support for the war and dividing US troops from ARVN.[114]

Conclusion

During 1964, both sides expanded their psychological operations and targeted a wider array of groups. For the United States, the Chieu Hoi program became the major campaign, although rallier numbers dropped in the face of battlefield setbacks, unrest, and government instability. The South Vietnamese implemented corps-level PSYWAR coordination centers to help manage the psychological war. The GVN also began to consider building a POLWAR structure based on the Republic of China model (Chiang's nationalist movement). In February, the CIA had turned over control of the covert operations targeting the North to the Studies and Observations Group. While agent operations continued to fail, the increasing repression in the North indicated possible collateral effects of the covert program. Meanwhile, the North found a potent theme in attacking American use of "poison gas" in the war, and it expanded its international program with the *Vietnam Courier* newspaper. By targeting American troops, the Hanoi Hannah persona became a well-known and recognizable voice. Propaganda inconsistencies in her program likely made it ineffective on most troops. However, the messages may have acted as an insidious virus, feeding perceptions and beliefs held by many American soldiers about ARVN effectiveness and the US role in the war. All these activities increased markedly after the Gulf of Tonkin incident in August 1964.

Map 4. South Vietnam, 1964 (chapter 5 locations)

6 Organizational Problems, Fall 1964–Winter 1965

American involvement in Vietnam grew steadily after the Gulf of Tonkin incident. However, running as a peace candidate against Senator Barry Goldwater in the 1964 election, President Johnson continued to minimize the change in policy. In the face of repeated attacks on Americans, the president refused to act decisively until he had won a new term in office. Meanwhile, the organizational instability that marred psychological operations after the 1963 coup continued to harm the overall program. Public opinion surveys in Long An Province hinted at a minor improvement over the previous year but still showed how much work remained. The one bright spot was the creation of a South Vietnamese political warfare organization. The period ended with the Flaming Dart air attacks, which marked a shift from America as an adviser to a direct combat role.

ARVN's PSYWAR organization successfully tested a new concept in August. The plan involved creating a leaflet containing a letter from a recent rallier to his former comrades. Five Vietcong in the 7th Division Tactical Area reportedly rallied in the first ten days of this program, one of whom brought six weapons with him. According to the report, "He stated that he decided to come in because life among the VC was so miserable, and he knew that he would be well received and treated because he had seen Chieu Hoi leaflets."[1] By this point the Chieu Hoi program was back on track after the post-Diem collapse. According to the Binh Duong Province chief Colonel Nhieu, a Chieu Hoi leaflet drop offering money for mines and weapons proved successful. In sixty days, the effort had produced seven mines and eleven rifles. Even with this positive response, however, Nhieu "expressed concern for the continuance of this program since he said funds for the program were exhausted."[2]

New tactics for countering Front indoctrination at the hamlet level emerged as well. NLF propaganda teams normally entered disputed villages at night and conducted study sessions, held trials, and distributed propaganda. To counter this, the PSYWAR units in I CTZ began organizing night propaganda teams. These were designed to directly counter the VC teams and use loudspeakers to harass them.[3] For the civilians, later analysis showed that "prolonged exposure of the villagers to VC controls and policies has resulted in a growing disappointment in VC promises, while the intensification of GVN/US operations had led to a declining belief in VC victory. The VC increasingly appear to many villagers in the role of ruthless exploiters of the population and as a constant source of danger."[4]

Propaganda War at the Hamlet Level

To take advantage of this, VIS elements in III CTZ began broadcasting one-hour loudspeaker news programs in district marketplaces. Other corps remained active as well. The I CTZ conducted a PSYOP/CA course for two hundred village officials, and the II CTZ reported that PSYWAR teams attached to the ARVN 22nd Division visited area hamlets. These visits included films and gift distribution as well as information dissemination about the government pacification program. US Army aviation also continued to increase its number of PSYWAR missions. It was during this flurry of activity that the first American PSYOP casualty in Vietnam occurred. Captain Alan Harriman of the 14th PSYOP Battalion died on 15 August 1964 in a helicopter crash in Tay Ninh Province. He had been on temporary duty in Vietnam from Okinawa to support Special Forces PSYOP missions.[5]

After Tonkin, the North Vietnamese leadership decided to increase combat activities. President Johnson had meant for the pinprick bombing in response to signal American resolve to avoid a wider war. Operation Pierce Arrow consisted of sixty-four strikes on coastal targets related to the Gulf of Tonkin incident. Unfortunately, the North did not get the message. On the contrary, Northern leaders were convinced that Johnson's actions meant that the United States would escalate the war. Thus, the North had to strike first while it had the chance. Liberation Radio broadcasts called on the PLAF in South Vietnam to "step up the struggle in all fields and compete in scoring greater exploits and achievements in order to liberate South Vietnam and protect brother North Vietnam." Consequently, the Front increased its activities in the South, including a battalion-size operation in Phong Dinh Province. Attacks in Phu Yen resulted in the capture of a US embassy representative on 4 August 1964. Five days later, the prisoner was reportedly being used by the Vietcong "to support a propaganda lecture."[6]

That same month, USIS released the full Long An Province survey results. Five USIS Vietnamese locals and four cadres from Long An Province conducted interviews in contested areas of the province, under the supervision of Frank Scotton. Among the major findings was a lack of preparation by GVN ministries to support counterinsurgency operations. Ministries in Saigon were too slow and unresponsive to provincial needs and had not yet "decentralized control to allow provinces to function at best level." Nevertheless, the survey indicated a power shift "towards the GVN at the expense of the VC" since the January survey. Respondents reported few complaints against the government besides minor harassment. The Vietcong, by contrast, received many complaints "for taxes, recruitment by force, [and] harassment of hamlets." The inhabitants of one village reported being burdened with paying both VC and GVN taxes. The cost, as much as 1,000 piasters per hectare to the Front, was not unusual.[7] Operating farther north with the 4th PSYWAR Company, Frank Scotton found that "contributions" to the Front were more than the government taxes. Complaints had led the NLF to label the inhabitants as antirevolution. However, the cellular control structure that the Front imposed kept them in check. The villagers were caught in a double bind.[8]

A PSYWAR team estimated the loyalty in one contested village as 30 percent GVN supporters, 50 percent uncommitted, and 20 percent NLF. The report noted that the "VC paid more attention to the propaganda campaign which, with its attractive promises, have more effect on the fledgling minds of the youths." One problem the study highlighted was a lack of secondary student targeting. This key target audience for the Vietcong had no counterprogram by the government, leaving students open to Front influence.[9] As a result, youths tended to support the Vietcong. Front propaganda units operated boldly in the hamlet. They gathered inhabitants in the evening for meetings "to publicize their victories on different battle-fields and called upon the hamleters to pay their income-taxes to them, or urged the youths to join them." APTs visited at least every ten days. The ARVN's PSYWAR team noted that any government presence was minimal and suggested Civil Action projects such as bridge repair and agricultural aid to help swing people's perception of the government. However, without an active propaganda program, news of positive government actions failed to make inroads to the hamlet.[10]

In Khanh Hau village, containing 538 residents, the team reported twenty-one radios. While people liked to listen to Saigon, Long An, and Ba Xuyen radio stations, the team believed "that there are many people in this hamlet listening to the Viet Cong radio station." In fact, while the group broadcast the news from the Saigon station via loudspeaker, a youth requested that "the liberation radio station be turned on so that they can listen to it."[11] Another student requested the same and later claimed that "the people of Cau hamlet support the liberation." He claimed that inhabitants had removed all the team's posters after they left. This hamlet had an estimated 25 percent pro-GVN, 35 percent neutral, and 40 percent pro-VC split.[12]

By contrast, Phu Nhon Hamlet had "great sympathy for the government," though many residents sat on the fence hoping for government protection. Fear of VC impressments forced many youths into hiding nightly, after several had been kidnapped in July. According to the team, "families are willing to let their children join the Republican Army, rather than to let the VC force them to join VC ranks." In one hamlet, the team visited each family, distributing *Huong Que* magazines, posters, scrolls, and gifts to the people, as well as notebooks and pens to the pupils. On behalf of the chief of Long An Province, the team offered a small sum to console a sick indigent woman. Despite these good deeds, success required consistency of effort.[13]

PSYWAR and NLF teams fought a battle of slogans and posters in the villages of Long An, taking down and painting over each other's propaganda. This helped the PSYWAR teams track hamlet loyalty and control by observing how product was handled and whether it was still posted during the return visit. One team encountered five slogans near a village temple exhorting "religious sects to unite themselves, and oppose to the My-Khanh, to liberate the South and to join the Front for the Liberation of the South." These had remained up for a month, indicating NLF dominance there.[14]

However, the survey indicated that in Long An the Front's "anti-American campaign has not been very effective." Villagers judged Americans by the few they saw and how they behaved individually. Respondents with this experience regarded them as polite. The smiling and kind attitude presented "gave some a positive image." Having never seen an American, most had no opinion, and Front propaganda seemed to have little effect.[15]

In one village, respondents stated that during a patrol by two airborne companies "two US advisors often went to and fro in the hamlet, usually shook hands with the villagers with their broad smiles, especially with the old people, thus, they have got sympathy from the local residents." However, the populace was confused about the nature of American aid, and many "worried that it had to be paid back and wanted to know when and how much to pay." As for views toward ARVN troops, it depended on a particular unit's behavior. Some units came and behaved respectfully, prayed at the temple, and respected private property. Others took food and chickens, behaving unruly. Villagers in Long An tended to respect paratroopers and disrespect rangers based on these behaviors.[16]

As the war entered a new phase with increased American participation, Radio Hanoi heightened the targeting of American servicemen. It announced that "all hope of returning to your country in 1965 has gone up in smoke" and remarked on possible tour extensions. The station argued that "repatriation of all U.S. military personnel to let the Vietnamese people settle their affairs themselves is the only wise way out for you, isn't it?"[17] A special English-language broadcast to American soldiers that summer, meant to divide them from ARVN counterparts, carried tales of Americans betrayed after an artillery mission went awry. Radio Hanoi claimed that angry "security forces men beat a U.S. advisor to death and wounded another." In another story, the station claimed that twelve Americans were killed by South Vietnamese paratroopers in mid-May 1963. Urging the Americans to refuse to follow orders, it broadcast: "You American servicemen, as well as the South Vietnamese army men, will have the chance of living in peace, freedom, and happiness among your loved ones. If your government persist[s] in throwing you and the South Vietnamese army men into battle and intensifying attacks on the South Vietnamese people, the tragedy will drag on—of course, in higher proportions."[18] There is no indication that the events described ever happened.

Continued RVN Domestic Unrest

Meanwhile, the instability in South Vietnamese PSYWAR leadership continued into September. One bright spot was the return of Colonel Nguyen Ngoc Huyen, who held a doctorate in literature from Hanoi University, as the new director. Colonel Huyen had previously been the PSYWAR director and later the deputy minister of information.[19] Regardless, given the revolving door of leadership, it was difficult to complete long-term PSYWAR planning. Continued instability at the national level that fall exacerbated the turmoil, with yet another attempted coup against General Khanh, leader of the Military Revolutionary Council that ruled the nation. Postcoup unrest led to public-sector strikes, which Khanh eventually threatened to crush. Except for routine activities, most PSYWAR/CA units suspended operations during the turmoil. MACV reported that "the prevailing attitude of Vietnamese PSYWAR personnel was to wait and see what happens to the government." PSYWAR units concentrated on publicizing the Military Revolutionary Council's announcements.[20] Once again, propaganda efforts foundered on the inability to maintain consistent messaging.

MACV reported that new RVNAF printing plant equipment would finally arrive in August. To facilitate the startup, a Mobile Training Team based in Japan arrived on 24 August 1964. This team assisted in the installation and calibration of the new equipment and provided limited training to operators. MACV officials hoped that this plant would help alleviate the printing bottleneck. The RVNAF Printing and Publication Center became operational at the beginning of October. However, there was still a reported shortage of printing supplies. To overcome this temporarily, ink and paper orders were filled on an emergency basis from MACV stocks.[21]

In September 1964, McGeorge Bundy sent NSAM No. 314 to the secretaries of state and defense, in which President Johnson approved the resumption of operations in the Gulf of Tonkin, outside the North's twelve-mile limit, with air cover. Covert operations were to begin after a first Desoto patrol, but agent and resupply airdrops, as well as leaflet operations, were to be "secondary in importance."[22] Cyrus Vance, deputy secretary of defense, finally approved the first round of post–Tonkin Gulf OP34A operations on 1 October, including plans for upcoming PSYWAR missions. This plan outlined the potential timeline of operations from the date President Johnson granted approval, to be known as L-Day. Agent missions were to take place only with moonlight, while airdrops of propaganda leaflets could "be done at any time since accuracy is not critical."[23]

The plan authorized the capture of a junk on L+5, holding the "crew for intelligence purposes," booby-trapping the boat, and setting it adrift. The captives would be freed about L+13. That same day, SOG planned to capture an individual near Vinh, North Vietnam. "If the prisoner is military, retain as POW; if civilian, give him choice of remaining in SVN or being returned with Loki captives," according to the draft. Between D+15 and D+31, SOG planned to attempt "five single-purpose leaflet delivery missions, and on space-available basis deliver gift kits, leaflets, and deception media in conjunction with airborne missions." "Deception media" could refer to the practice of dropping ice blocks by parachute at night. The next morning, security patrols would find empty open chutes, indicating that agents had dropped in. The goal of this diversionary program, under Project Forae, was to "convince Hanoi that it had uncovered only the tip of the iceberg" and that enemy agents supported an internal conspiracy against the DRV. Dropping radios and sending radio deception messages ordering existing teams to join with other, nonexistent teams supported this fiction.[24] Lastly, Vance authorized two hundred poison-pen letters "delivered through third country channels to North Vietnam."[25]

To maintain the story line, per Vance's plan, broadcasting on the Red Flag covert station would continue the thirty-minute daily program, repeated once, and the Sacred Sword covert station would begin a weekly fifteen-minute broadcast, repeated once. Programs for that station were "recorded at makeshift studios in Saigon with poor acoustics and background noise to indicate 'shoe string' operation of a clandestine station struggling to get on the air." The overt Voice of Freedom radio from Hue would "continue 8½ hours of daily broadcast consisting of 4½ hours of original programs and 4 hours repeat programs."[26] SOG also initiated procurement of "7,000 transistor radios to be airdropped in DRV in order to increase potential radio audience."[27]

At the same time, South Vietnam began a much-needed change to its overall psychological program. On 26 June, now-President Khanh had indicated to Ambassador Lodge his intent to form a political warfare organization within the RVNAF. To facilitate this change, a Republic of China Advisory Group arrived in Saigon on 8 October 1964. The fourteen-man group, led by Lieutenant General Teng Ting Uyan, conducted a ten-day series of briefings and orientations before assuming a POLWAR advisory role. Teng had the mission of developing a new POLWAR department based on the Republic of China's model, including preparation of a POLWAR course for the coming year at the PSYWAR Training Center. The General Political Warfare Department was established by the prime minister's decree on 24 October 1964.[28]

To support training of PSYWAR officers, the RVNAF had operated the Political Warfare School at Fort Cay Mai, Saigon, since 1956. By 1961, its annual output was 212 individuals, rising to 940 by 1963. In 1964 the training center moved to Camp Le Van Duyet, Saigon, but the facilities at the new location were so inadequate that student capacity decreased and the 1965 output was still below previous years, with only 427 students graduating.[29] Eventually the GVN established the Political Warfare College in Dalat.

Domestic unrest spread throughout the Central Highlands during the fall of 1964, testing the ability of the government to respond. Though sparsely populated, the Highlands remained strategically important. The region contained supply routes branching off the Ho Chi Minh Trail, and the dense jungle offered ample hiding space for rest and recovery of units. US Special Forces operated a string of camps there to act as a tripwire for movement along these routes. The various Montagnard tribes were targeted by both sides in a separate propaganda war. Real issues of government fairness toward the tribes made them ripe for agitation. In September, possibly due in part to this agitation but mostly due to local issues, the region became a hotbed of rebellion. Quoting foreign news, Radio Hanoi reported that Montagnard soldiers deserted several Special Forces training camps to seize the Ban Me Thuot civilian radio station during an uprising on 20 September. It used the station to broadcast denouncements of "the oppression by the Nguyen Khanh administration" and demanded the "right of self-government for upland population in the high plateau."[30]

The same day, Radio Hanoi quoted a Radio Saigon report that "some 900 people in Qui Nhon city, Binh Dinh Province . . . last night occupied the Qui Nhon radio station and ran broadcast programs on orders from the provincial 'People's Council for National Salvation,' a secessionist organization of the province."[31] Radio Hanoi stoked the flames of unrest, using the Jarai language, calling on "soldiers in Pleiku and Ban Me Thuot who have risen up against the Khanh Clique" to refuse orders and join with the NLF.[32] These events indicated the reasonableness of President Diem's fear of establishing a decentralized radio system. Both stations were quickly retaken, however, and the PSYWAR and advisory efforts reacted quickly to the Montagnards' rebellion. A PSYWAR committee consisting of representatives of the II Corps G-5, the 23rd Division G-5, the ARVN's PSYWAR directorate, the Darlac Sector S-5, and the USIS coordinated the PSYOPs response. This committee "effectively coordinated psychological operations in support of the successful effort to quell the rebellion." With the ongoing unrest in the Highlands, one significant PSYOP activity included a leaflet drop announcing the attendance of Montagnard officer candidates at the Thu Duc training center. This was meant to show the positive relationship between the Highlanders and the government. Despite this, the simmering hostility between the Highlanders and the Vietnamese remained.[33]

Intelligence reports indicated the likelihood of demonstrations focused on Ban Me Thuot. A broadcast in Rhade from Montagnards regrouped in the North asked listeners, presumably Highland soldiers, "not to obey Khanh's call but be with the people fighting against Khanh." The speaker asserted that "the Saigon government is disintegrating" and urged continued unrest.[34] The broadcasts were instigating a divisive propaganda campaign to make the Montagnards distrust the government messages and "to separate the soldiers from the people."[35]

This sort of activity worked both ways. In October, following Operation Lam Son 129, the supporting PSYWAR team in Quang Tri Province transported nearby villagers to the battlefield. The operation had resulted in seventy-seven Vietcong killed. Supplemented with radio coverage and film crews, the team intended to show the dead Vietcong, along with captured weapons, as proof that ARVN forces had defeated the Vietcong, and thereby demonstrate the futility of supporting the NLF. This type of "inevitability of defeat" theme was commonly used by both sides to encourage one populace to reject the other.[36]

Expanded Propaganda Activities

At this point, radio was still the main form of mass communications in South Vietnam. In October, long-standing discussions concluded on developing a nationwide television system. USIA and USAID were the primary agencies involved with the South Vietnamese Ministry of Information on this issue. A subcommittee of the US Mission Council concluded that "TV could advance U.S. objectives, rural pacification, urban stability, national unity, Free World support, the U.S. presence in Vietnam, and allow the Vietnamese government to better project its image and that of the US to more of the people."[37] An additional factor in the decision to help construct such a system was the growing US military presence. The US Armed Forces Vietnam Network (AFVN) already operated a string of low-power AM and FM stations to provide US troops with news and entertainment. As the number of troops grew, MACV desired to install a television network as well. The GVN would allow the introduction of the AFVN television system only if the United States provided a system for South Vietnam civilians. The AFVN-built radio and television network became an informal, unanticipated information dissemination platform to many South Vietnamese.[38]

Meanwhile, a PSYWAR/CA course resumed at the PSYWAR Training Center on 15 October 1964. The overall plan called for Regional Forces to send one sergeant from each sector to this course. Company and battalion officer POLWAR courses were scheduled to start after the New Year. The PSYWAR Training Center later began a specialized PSYWAR techniques course for NCOs. Half of the seventy-four soldiers were from Regional Forces units. Also, a two-month basic PSYWAR course began for enlisted soldiers.[39]

A new group of American PSYWAR/CA advisers arrived in Vietnam by the end of October to assume duties as sector S-5 advisers. This positioned American PSYWAR/CA advisers across all levels and greatly expanded their influence. That same week, the first elements of the Republic of Korea Military Assistance Group arrived, eventually fielding its own PSYOP company in support of the war effort.[40]

The varied life of a sector PSYOP adviser is illustrated by an interview conducted with Marine Captain Joseph B. Knotts, PSYOP/CA adviser near Da Nang from September 1964 to May 1965. Shortly after his return from Vietnam, Knotts recalled that his responsibilities included "coordination through USOM, USIS, and other agencies that we had there, such as teams of American Sea Bees who were doing civic action work." He related that the MEDCAPs proved very popular with villagers. Teams were typically composed of Vietnamese medical personnel, American medical personnel, or a combination of both, along with psychological warfare teams. PSYOPs attracted people to the MEDCAPs using leaflets and loudspeakers, then talked to attendees to gather information, gave away product, and presented the drama teams. Among the products he distributed were "cooking utensils that had government slogans on the bottom of them." Knotts also worked with provincial radio stations and distributed small transistor radios so people could listen to the provincial broadcast. Remarking on the effectiveness of that program, he said, "There were a few instances where the Viet Cong conducted raids . . . to get the radios so that they could not listen to the government broadcast." The popularity of his activities attracted VC harassment fire as well, but Knotts believed success required persistence in such situations. The attempt to stifle information indicated at least a fear on the part of the Vietcong that the message would be effective.[41]

As the 1964 US presidential election heated up, Hanoi found fertile fields in American domestic political rhetoric. Radio Hanoi related a Barry Goldwater statement about the war in which the Republican presidential candidate had said: "The situation in South Vietnam seems to have deteriorated from confusion to chaos. Reports of collapse and crisis along the fighting front, in Saigon Government, in the cities and in the villages, have poured in." Goldwater continued, "The crisis of the week, which we have come to expect in South Vietnam, has now turned into a crisis of the day."[42] Hanoi timed these actions for maximum psychological effect.

As part of North Vietnam's post-Tonkin change of strategy, the Communist Party's Central Military Commission ordered the PLAF to prepare a major campaign in the South jointly with PAVN units to "annihilate a part of the enemy's main force units, destroy strategic hamlets, and expand liberated areas" along the borders. The commission planned to culminate operations with an attack on Bien Hoa Air Base.[43] Radio Hanoi later boasted proudly: "Let us acclaim the new resounding feat of the South Vietnamese People's Armed Forces." This rehashed straight news stories largely based on international press reports regarding the November attack on Bien Hoa Air Base.[44] The "brilliant exploit" destroyed twenty-seven US aircraft and included thirty-six US casualties. It "greatly frightened the enemy and has filled with joy the hearts of the peoples in the world who love peace, freedom, and national independence," enthused Radio Hanoi.[45]

Following the 2 November 1964 attack on Bien Hoa Air Base, President Johnson continued to vacillate between taking action and his desire to keep the war on the back burner. The struggle for balance crippled the Johnson administration's decision process to devise a believable public relations campaign. The deployment of the B-57s that had been destroyed in the attack were meant as a "signal" to the North. Now the president was hoisted on his own petard. In Vietnam, American military leaders were in a quandary over how to sell the plan to increase American action. With the election only days away, the pollster Louis Harris convinced the White House press secretary Bill Moyers that the United States need not respond to the air base attack. He assured Moyers that Johnson could delay a response with public approval.[46]

In a further attempt to divide American and ARVN soldiers, Radio Hanoi broadcast to South Vietnam an incident in Rach Gia. After an air strike in the area, which the station claimed resulted in "six Sky raiders and three jet helicopters" shot down, it noted that the Americans later "airdropped leaflets asking the people's forces to return the bodies of the four U.S. pilots," but it said nothing about the South Vietnamese men in the downed planes. Hanoi offered this as proof that Americans did not respect Vietnamese. In an October battle in Duc Hoa, it further claimed that US stray bombs killed or injured fifty-two South Vietnamese soldiers. Hanoi saw this as more evidence that "the Americans do not care about the lives of South Vietnamese" troops. Appealing to "dear friends in the South Vietnamese Air Force," the station asked: "With your patriotism, your desire for freedom, and the means at your disposal, you can perform brilliant achievements. How can you continue to kowtow before the American aggressors?" This propaganda understood the target audience members, effectively appealing to their self-interest while not belittling them.[47]

In the chaos and rapid government changes after the overthrow of Diem, the military situation on the ground had deteriorated considerably, despite progress in PSYOP training, innovation, and execution. Increasing infiltration from the North shifted the balance of power even more. The GVN steadily lost hamlets, and the number of Chieu Hoi ralliers dried up. Intelligence intake diminished. ARVN desertion increased. As a counter to the losses, PSYWAR took the lead on a national recruitment drive from 23 October to 3 November 1964. More than 170,000 posters and 7.5 million leaflets were produced in support of this program.[48] Initially focused on information about recruitment, the second phase presented the "punitive results for the youth who fail to register during the call-up phase." This indicated the keys to a successful PSYOP campaign: the carrot of product supporting attitudinal change, coupled with the stick of product clarifying punitive consequences. The initial results exceeded expectations, bringing in more than 11,000 new soldiers. However, failure to enforce strict punishments for failure to report for service "took the sting out of the program," and the numbers dropped in future months. Simultaneously, the government used propaganda to exploit the one-year anniversary of the Diem coup d'état as a national holiday. Military parades and celebrations were held and supplemented by PSYOP product. The badly shaken government attempted to build support by attacking the previous one.[49]

Showing another side of psychological operations, a government initiative demonstrated that its concern for the people of South Vietnam extended beyond the effects of the war. In September, PSYWAR teams distributed blankets, cloth, mosquito nets, roofing material, milk, and foodstuffs to victims of Typhoon Violet. Units once again aided devastated areas when two more typhoons struck in November 1964. Immediately, PSYOPs focus shifted to assisting humanitarian operations, supporting refugees, and providing relief assistance. The floods and high winds caused by Typhoons Iris and Joan ravaged the coast between Quang Nam and Khanh Hoa Provinces. USOM field representatives and MACV advisers assisted to ensure that the ARVN responded effectively. The USIS helped the ARVN PSYWAR directorate prepare leaflets to keep victims informed. Due to the isolation of the villages affected, the Vietnam Navy's PSYWAR section became involved in delivering tons of supplies by sea. The response to this disaster dominated PSYWAR activities for the remainder of 1964. The 3rd PSYWAR Battalion in I CTZ used its Cultural Platoon to give a "benefit performance in Quang Tri and Hue to raise money for flood relief," raising more than 122,000 piasters.[50] MACV reported that overall "the GVN both in Saigon and at the local level is doing an excellent job of psychologically exploiting the situation, the government's actions and the VC interference."[51]

Front interference abounded. To counter government success, Vietcong in Quang Tin Province claimed that the flood relief supplies came from North Vietnam. Meanwhile, Radio Hanoi professed anger "with the Americans and their henchmen, who are vilely taking advantage of natural calamities to drive the compatriots deeper into misfortune" rather than helping them. A related claim was that the government had failed to warn people of the typhoon. The accusation that the government also took "advantage of the southern people's misfortune to use troops and aircraft to carry on mopping up operations" had to be addressed.[52] The unusual amount of vitriol in the stories indicates that the Northern propagandists felt threatened by their enemy's ability to mobilize relief efforts in contrast to their own impotence. The GVN used psychological warfare to prevent these myths from taking hold. Countering such propaganda required analysis. To attack it directly risked spreading the Front's messaging. This case warranted an oblique approach, focusing on positive government actions. When the Vietcong's ongoing attacks failed, it tried the novel approach of ordering the "U.S. imperialists and their lackeys to bring relief at once to the compatriots," thereby taking credit for what was already happening.[53]

ARVN's dissemination technology improved as well. As the year drew to a close, heliborne loudspeaker use had been fully integrated with ARVN operations. PSYWAR teams used speakers to encourage surrender, remove civilians from the battle area, help refugees, and spread news about the Chieu Hoi program. The ARVN 7th Division implemented an innovative use of loudspeaker deception support to combat operations in November. During Operation Thang Long 27, aerial loudspeakers and leaflets warned a false target area of an upcoming operation. Civilians were told to evacuate along specific routes "where they would be safe and medical assistance would be available."[54] The actual target area received similar appeals, but once the attack began the "loudspeaker helicopter shifted to making surrender appeals." This led to the capture of fifty-four prisoners taken by the surprise change of location. The operation, in Dinh Tuong Province, wrested Ba Dua, six miles west of Ap Bac, from VC control and resulted in 136 enemy killed.[55]

Broadened Covert War against the DRV

As of 24 November 1964, SOG reported its Radio Red Flag broadcasting one hour daily from a 750-watt transmitter of the South Vietnam Ministry of Defense. SOG expected renovation of buildings for the overt and covert radio studios, including Radio Red Flag, to be completed by March. The new facility included constructing a 20-kilowatt station for the SSPL program. Meanwhile, the Voice of Freedom overt station expected to be on the air after receipt of a tuning crystal for the transmitter in Hue. The programs for this station were produced in a Saigon studio daily and "sent to Hue by courier and/or troposcatter link for retransmission."[56] SOG's other PSYWAR assets included an organic printing capability of 200,000 3-by-6-inch leaflets per eight-hour shift using a Harris high-speed press and a standby capacity of 500,000 per eight-hour shift, "provided by the USAB&VAPAC Vietnam Detachment w/mobile printing van." SOG delivered 31 million leaflets and more than 33,000 gift kits to the DRV in 1964. Despite this impressive production capability, in the period after 1 August 1964 dissemination was limited to 32,500 leaflets that fall "due to JCS restrictions and weather." Poison-pen operations dispatched 200 to 300 letters monthly "to specific addresses in North Vietnam through . . . third country mail channels." Often the letters were copied from originals by Paradise Island detainees, who also produced the gift kits.[57]

The Special Assistant for Counterinsurgency and Special Activities (SACSA) of the Programs and Review Division, Major General Rollin Henry Anthis, reported challenges in the Strategic Technical Directorate/SOG commando teams. They were "a long way from being volunteers," he wrote. The teams had become complacent, making them unreliable as covert agents and perhaps susceptible to the sort of propaganda aimed at them by the North. Agent operations since 1 February 1964 included eleven teams launched, consisting of seventy-six agents. Thirty-two of the agents were confirmed by Radio Hanoi to have been captured and nine more likely so. SOG still counted Team Eagle as viable. However, "all efforts to resupply/reinforce Eagle have been unsuccessful," per the report.[58] The memo stated that each agent team received 180 hours of PSYOP instruction at Long Thanh, indicating the priority given to this skill. A separate issue was the restriction on actually supporting a Northern resistance movement. As Anthis argued, "Until such time as this limitation is removed nothing in the way of sabotage operations can be expected."[59]

Regardless of these limitations, the program seems to have had greater impact on Northern civilians than is often believed. SOG had intelligence that people "attributed the actions of the SSPL not to the Southerners or Americans, but to the Soviet Union," specifically "two PAVN lieutenant colonels who had defected to the USSR in 1964." Project Forae also served the purpose of misleading the Strategic Technical Directorate that supplied the agents. Many in MACV-SOG believed that the directorate "had been penetrated by

Communist agents and that information concerning its agent team program was making its way to Hanoi." This operation maintained the fiction that "agent operations were doing well."[60]

Lieutenant Colonel Hale H. Knight reported that CIA operations against the North had been transferred to SOG by this point. This included PSYOP functions such as the overt and covert radio stations, the leaflet program, mail operations, and deception support. Considering the open secret that OP34A had become, SOG headquarters favored "surfacing" the programs. By this Knight meant creating an open, real resistance group based in the South. SOG HQ believed this could "give credence to the Sacred Sword Patriots League with the possibility of creating a resistance movement in [North Vietnam]." Knight's justification for surfacing the operations in the North was that the government could "acknowledge aiding the external support organizations of a resistance movement operation within [North Vietnam]." The US government could then acknowledge the activities and condone them even while not admitting involvement. This could enhance the overall credibility by acknowledging the truth.[61]

Exposing programs in this way would shift them from "the realm of random harassment" and put them in a "framework to accomplish the deterrent mission for which they were conceived." It would also exploit the already high level of paranoia and draconian internal security inside North Vietnam. SOG could attribute attacks to the SSPL, further legitimizing both actions, and the United States could say something besides "no comment." Colonel Clyde R. Russell, SOG's commander, argued that because "public opinion almost everywhere tends to favor a 'resistance movement,'" South Vietnam could also recruit motivated agents for a definite cause rather than the complacent ones Anthis encountered. Thus, the notional movement could grow to an actual movement and provide the GVN with a sound substitute for its "March to the North" program. He felt this would be less provocative to the Chinese and "more likely to enlist sympathy" in the North than would agent operations alone. Despite repeated requests to surface the program, as late as 1968, decision makers in Washington demurred.[62]

Effective deception operations tend to work along one of two tracks: ambiguity reducing and ambiguity enhancing. Ambiguity-reducing deceptions feed preconceived notions with confirmatory information so the target reacts in accordance with that. Ambiguity enhancing adds ambiguity and "white noise" to events so the enemy is unable to make a decision. Each tactic is designed to use the enemy's existing ideas against him. In the case of North Vietnam, the most effective choice was ambiguity reducing. The North "knew" and feared that a counterrevolutionary movement was operating in its midst. Feeding that paranoia with projects such as Project Forae offered a possibility of having a strong effect, even if the Northern security services detected some of the plan.[63]

The proposal called for setting up an official SSPL office in Saigon as a front group like the NLF. The SSPL would openly recruit agents, saboteurs, and pilots and provide aid to the operations. Its press office would also "make announcements to clarify slanderous remarks of the North Vietnam regime." Information to be used to support the deception narrative would come from mail intercepts. The SSPL would also sponsor deception kits dropped on the North containing information to confirm the false narrative of the existence of these operations. To further support that narrative, agents could decorate Loki captive debriefing sites, take credit for maritime operations, and plant encrypted one-way traffic on overt radio broadcasts. These needed to be real texts, so when the security services broke the code they would "have something plausible to worry about." Messages indicating SSPL negotiations with key DRV leaders might cause dissension, for instance. Attackers in the North would also perform psychological warfare tasks, and former POWs were to be used in speaking engagements to "appeal for reunification of a Vietnam not under Chinese domination."[64] The plan called for organizing the front agency in the fall of 1964. Despite the fact that this plan might have been more effective than other current operations, it was never implemented. In 1968, as SOG again moved in this direction, President Johnson ceased nearly all PSYWAR against North Vietnam.[65]

South Vietnamese POLWAR Grows

Throughout 1964, PSYOP mobile training teams assisted the South Vietnamese in developing capabilities in radio management, VC propaganda research, printing management and production, PSYOP instruction, and film production management. Additionally, during 1964, the USAB&VAPAC organized a detachment in Vietnam, which "contributed immeasurably to the PSYOPs effort in the GVN by filling in the technical advisory void" that existed in many areas. This detachment acted as the headquarters for teams of five personnel deployed on ninety-day rotations. MACV noted discernible progress "in all technical fields," resulting in a significant improvement in the RVNAF's PSYWAR capability. Besides Republic of China and US PSYWAR advisers, sixteen Filipino army officers arrived to help advise PSYWAR/CA companies in III Corps. They were assigned to ARVN PSYWAR companies.[66]

A ninth PSYWAR/CA conference was successfully conducted on 7 and 8 December. The goal was to provide fifty-three incoming PSYWAR/CA advisers "with more detailed information on policy procedures and problem areas." Corps, division, and sector PSYWAR/CA advisers attended, along with USIS field representatives. Observers came from the US Army Special Warfare School and USAB&VAPAC on Okinawa. Speakers included Major General Richard G. Stilwell, MACV's chief of staff, and Major General Xuan, the RVNAF's deputy chief of staff for military political warfare. Officials from the embassy, US Special Forces, and the Philippines' PSYWAR/CA contingent spoke as well.[67] The Philippines' PSYWAR advisers had proven so successful that sixteen more were requested in January 1965.[68]

A survey of activities at this time shows the breadth of psychological objectives being pursued. At year's end, the ARVN's PSYWAR directorate started a very successful radio program, featuring Da Lan (Sweet Sister). Like Thu Huong's broadcasts from the North, this program used a "warm, friendly, female personality" to talk to front-line soldiers, read the news, play music, and praise "victories over the VC" to boost morale. In just a few weeks the station received more than "one thousand appreciative letters" from armed forces members. These helped the PSYWAR directorate improve the program and "maximize its value."[69]

To decrease support for the Vietcong, the ARVN 1st Division conducted a major PSYWAR campaign to exploit the victory in Operation Lam Son 136 conducted between 27 November and 5 December. Later, it exhibited captured weapons in Hue and distributed leaflets publicizing the victory throughout I Corps. The 9th Division began to organize Chieu Hoi ralliers to enter contested villages to encourage further desertion. Throughout Vietnam, 513 rallied, including twenty-two political cadres. Farther south, in Binh Long Province, MEDCAPs proved very popular. Aimed at building support for the government, medical personnel of the 34th Ranger Battalion treated civilians in their area of operations. The US Air Force conducted seventeen PSYWAR missions that week across Vietnam. The US Army added thirty-five rotary and fixed-wing missions to this, with the Vietnam Air Force adding four.[70]

American authorities specifically assessed Chieu Hoi to be a valuable program to expand and so programmed sufficient funds to run it. MACV also considered expanding a bounty program that paid for killing Vietcong. This program, which presaged the Phoenix Program, was discussed at various staff levels, but no decision had been made by the end of the year. To a degree, this was a necessary part of a successful Chieu Hoi program. The carrot of leniency needed a stick. Regardless of how skillful the Chieu Hoi appeal was, however, desertion depended on other factors. Vietcong were unlikely to rally to a government perceived as losing. The precipitous drop in ralliers between 1963 and 1964 bore this out. Additionally, any bounty program for killing would naturally be open to potential abuse.

Whatever the level of progress or balance between the adversaries, PSYOPs could backfire. Lack of coordination among the various PSYOP organizations, in particular, often had tragic consequences. In November, the Ministry of Information conducted a national effort to induce ralliers. Because of successful military operations in IV Corps that fall, nearly 4,000 Vietcong reportedly heeded that call. Unfortunately, the province's administration was not aware of the propaganda campaign and was unprepared for such a large influx. With no food, housing, or medical care as promised, "most of the returnees decamped within a week," providing the Vietcong a "potent psychological victory."[71]

There were additional delays in expanding the psychological operations program, in part due to ongoing instability. At the end of the year much of the PSYOP budget remained unspent. MACV blamed the delays on a lack of classroom-trained advisers. It also urged that the RVNAF's entire Civic Action Program be revitalized. Attempts by USIA to spread information included posterboards that were often poorly maintained. According to a 1964 study, the boards were often only "empty symbols of community deference to external authority." While aspects such as "entertainment teams and mobile film programs were welcome and appreciated" in villages lacking such amusements, the political vacuum in rural Vietnam meant that tactical, rather than strategic, success was the order of the day.[72]

The MACV annual history closed with a cautionary note in 1964. It found that the Vietnamese concentrated too heavily on "gadgetry such as leaflets and loudspeakers while neglecting the advantages of the potentially more productive efforts of Chieu Hoi" and the weapons buyback program. MACV's *Command History* criticized the South Vietnamese "preference for the cheap, material gadget over the more difficult and demanding face-to-face approach with the people themselves." As a result, MACV reported that regardless of the tremendous potential, PSYOPs achieved little during the year. Although MACV blamed the disappointing results on the South Vietnamese, American advisers must share the blame, as they provided the gadgets and the training.[73]

Enemy strength in the South had grown steadily throughout 1964—a truly disastrous year for the Southern cause. MACV estimated that 90 percent of the 7,000 infiltrators that year were North Vietnamese draftees. Along with increased infiltration, accelerated recruitment and drafting in the South increased Communist main-force units by 50 percent to 33,000 in December. Ominously, PAVN/PLAF soldiers were increasingly equipped with new Soviet or Chinese automatic weapons using standardized ammunition, simplifying logistics and increasing firepower far beyond that of the typical ARVN soldier. As the intensity of combat increased, so did ARVN casualties, rising from 1,900 in January to 3,000 by December.[74]

President Johnson continued to fear repeating the mistakes of the Korean War and dreaded the thought of Chinese intervention, but events were forcing his hand. On Christmas Eve 1964, the Vietcong bombed the US officers' billets in the Brink Hotel in Saigon. The explosion killed two and injured sixty-three Americans and thirty-four Vietnamese. According to a typical Radio Hanoi exaggeration, the explosion killed seventy-one Americans and destroyed the nearby US AFVN radio station. Once again, Johnson did not respond after this attack. Rather, he sought reasons to do nothing. At a meeting at his ranch in Texas, he listened to various responses put forth by advisers and then responded to Ambassador Maxwell Taylor with "almost every conceivable objection, some of them bordering on the absurd." Once again, the president and the secretary of defense overrode a JCS recommendation to strike North Vietnam. Meanwhile, the Pentagon began contingency planning for Operation Flaming Dart, the expanded bombing campaign against an array of North Vietnamese targets.[75]

As 1965 opened, the American advisory system actively supported the Chieu Hoi program, built up South Vietnamese PSYOP capabilities, and conducted surveys to assess the program. US government agencies worked in conjunction with the Vietnamese to expand media facilities, and together the two allies established a system of corps-level combined PSYOPs coordination centers. In the coming year, however, this system faced unprecedented pressure.[76]

Barry Zorthian Manages Overt PSYOPs in Vietnam

On 26 January 1965, Barry Zorthian, chief of the USIS in Saigon, presented his plan for USIS activities in the coming year. From this report the outlines of the US psychological objectives began to emerge. Those objectives included increasing "national unity" and creating "confidence in the GVN." The goal of creating "dissension in the Viet Cong ranks and induc[ing] defection" was intended to drive a "wedge between the hardcore and non-hardcore elements." Among the intended supporting programs was "a chain letter designed to cause confusion in the VC ranks, and to spur the letters onward by means of periodic rewards along its chain."[77]

While the Zorthian plan announced a new robust approach to PSYOPs, the North was also intensifying its targeting of US troops. Comparing the service of Americans in Vietnam to that of their fathers during World War II, Thu Huong told them that the United States had fought that earlier war "to smash the common enemy of mankind." This generation, by contrast, was fighting an unjust war in contradiction to the "lofty ideals of the American people of Washington's, Jefferson's, and Lincoln's times," she said. Later in the broadcast, Huong remarked that Anna Louise Strong, "the noted American writer and journalist," had recently asked American mothers: "Did you raise your sons to be dynamiters of babies, rapers and torturers of women, poisoners of grain fields, devastators of land beyond the seas that never injured America?" Strong, active in the 1919 Seattle general strike, was later a correspondent in Moscow and founded the *Moscow Times* in 1930. She called on American mothers to end the war. Attacks such as Strong's were unlikely to endear her to the target audience: American servicemen.[78]

Huong later read an open letter, ostensibly to President Johnson: "You want the Americans to remain in South Vietnam, but public opinion in the states and even in the senate keeps on discussing animatedly the possibility of U.S. withdrawal." Those actually targeted in the broadcast, American soldiers in Vietnam, were meant to hear the dissension over the war, in hopes of leading them to question their own presence in Vietnam.[79] The objective of the messages was to create cognitive dissonance in the target audience. Introducing the nagging question of popular support in the audience's mind could open it to future propaganda themes.[80] According to Radio Hanoi, America was split by "the increasingly wide and deep struggle movement of progressive people in the United States including youths, students, intellectuals, mothers, personalities, and religious men." To diminish morale in the South, Radio Hanoi also hyped political divisions in the United States. Discussion about the war in American newspapers and political circles, as well as "struggle movements of peace-loving peoples opposing U.S. aggression in South Vietnam and supporting the [NLF]," indicated a lack of support for the Southern government, according to Hanoi.[81]

Zorthian and his staff had no choice but to counter charges such as this. Seeking to exploit the psychological vulnerabilities of a traditional peasant society, he employed several astrologers to write horoscopes and almanacs for PSYOP use, and he also interviewed "several magicians . . . for possible roving assignments countrywide, with a specific message worked into their performances."[82] He also announced the formation of smaller cultural/drama teams. These three- to five-member teams, rather than the normal twelve- to sixteen-person teams, were easier to transport. The plan envisaged reformatting drama teams to focus on "key themes of successful province achievements" rather than the more "popular, but ineffective, name calling anti-VC line which has for too long been a standard propaganda technique."[83]

Another way to drive a wedge between the Front and civilians was to have Hoi Chanh (Chieu Hoi ralliers) testify in village meetings about life under Communism, pointing out that "land reform is illusory—all land will revert to the state under communism and the peasants become slaves of the state."[84] Zorthian focused on creating an inevitability-of-defeat syndrome among the Vietcong by hyping VC defeats and ARVN victories. Publicizing reports of returned refugee numbers and aid from friendly nations to South Vietnam supported this stratagem. Ideologically, Zorthian sought to "reverse the VC indoctrination process by identifying psychological vulnerabilities in VC sense of 'historic mission,' 'nationalistic orientation' and 'upward social mobility,' then exploit these vulnerabilities to re-channel the process." Concentrating the Chieu Hoi effort on the "non-hardcore elements" seemed more useful. Like "Sweet Sister" or the North's "Autumn Fragrance," Zorthian wanted a similar female voice directed at the Vietcong.[85]

In a plan reminiscent of George Creel's World War I "four minute men," Zorthian envisaged creating a "Coffeehouse Lecturers" program, in which small groups would visit the countryside, lingering in "coffeehouses, bus stations, market places and other likely points of news dissemination." Using a "dialogue technique," the groups would spread messages informally.[86] In a similar program already underway, Vietnamese Information Service cadres assigned to checkpoints explained "resource control measures," placing the blame on the Vietcong. Another VIS program added information booths and loudspeakers at security checkpoints in Long An Province.[87] However, all this depended on a motivated VIS cadre to implement, which itself depended on government stability and focus. One additional dilemma was that the South Vietnamese military draft did not exclude VIS personnel. Drafting highly trained operatives for the infantry often left untrained personnel to run the field programs and critical face-to-face operations. They had to hire people not subject to the draft, so "often the results were personnel who performed marginally." The problem became particularly acute just as the war intensified and the RVNAF expanded.[88]

The USIS office in Saigon reported on a thriving campaign to inform the rural populace of successful government pacification activities in Phuoc Thanh Province in January. The report's author noted that prior to the start of this campaign, seventeen of twenty-one villages belonged to the Vietcong. Most residents believed that the government was incapable of securing or assisting the poor farmers. The USIS proposed a program based on graphically illustrating development results in Tan Ba village. A joint effort composed of the USIS, VIS, province chief, and other elements coordinated by the province and district PSYOP committees turned this around. Because of inadequate VIS or PSYWAR support in the area, other US and GVN agencies supported the program with printing, publications, films, radio broadcasts, and cultural teams to saturate the area. They publicized government activities using leaflets, signs, posters, movies, and metal signs "erected in front of new clinics, schools, wells, and market places saying: 'Another self-help project with the help of your local government.'" According to the memo, the USIS measured success by "the rising morale of local citizens, their increased willingness to provide information" on VC activities, and the number of returning farmers. The total cost of the campaign was about $279.[89]

Meanwhile, Radio Hanoi had turned to the issue of increased ARVN recruiting. In the North's view, the GVN was "unable to recruit enough soldiers to replace those lost." It claimed that more than 6,000 soldiers deserted in the central provinces in three months.[90] Radio Hanoi continued to persuade Montagnards to revolt, urging them to "join [the] people's side." Using a ranking Montagnard chief, Y Ngong, it played up the theme of South Vietnamese exploitation of natural resources. This was a long-standing fear among Montagnards. Y Ngong asked Highland soldiers: "How do you feel when your brothers are killed by your friends? Are you comfortable wearing heavy clothes and shoes and earning much money?"[91] This last question appears aimed at boosting grievances at having to forgo the traditional freedoms, associated with the loincloth lifestyle, for the constricted army life.

Flaming Dart Message Failure

On 7 February 1965, President Johnson authorized Operation Flaming Dart. This came after a morning attack on US facilities in Pleiku killed 9 Americans and wounded 108, in addition to destroying 21 fixed-wing aircraft and helicopters. Senior US officials recommended an immediate reprisal against the North, and this time the president agreed. Prior to the start of Flaming Dart, the United States had redoubled preparatory moves of equipment and supplies in a very open way as a signal to the North. The attacks were "not designed primarily for their potential military effect on the South Vietnamese military situation, but rather as calculated 'signals' to North Vietnam and, indirectly, to Communist China."[92] The president carefully graduated military escalation, each time communicating to Hanoi, China, the Soviet Union, and France the limited scope of its objectives. By considering military action to be an extension of diplomacy, alternate strategies based on military considerations were discouraged. Johnson's chief advisers fixated on the notion that "the use of force should aim to communicate with the enemy rather than to inflict destruction."[93]

The Johnson administration integrated into the planning for these and subsequent air strikes recommendations that "the U.S. be willing to pause to explore negotiated solutions." Using the bombing as a form of communication with North Vietnamese leaders, these policy makers believed it was imperative that "the prospect of greater pressures to come was at least as important as any damage actually inflicted." According to this view, the North would stop its aggression rather than suffer the greater risks of continuing. To make this work, however, it was important to ensure that there were sufficient numbers of high-priority targets left untouched that could be threatened.[94] The Barrel Roll bombing runs on the Ho Chi Minh Trail in Laos, begun in December, were likewise "considered more important for their political value as 'signals' to the North than as militarily useful interdiction operations." This policy ensured that the most important and influential targets were never hit.[95]

Despite this, President Johnson told the American people: "We have no choice now but to clear the decks, and make absolutely clear our determination to back South Vietnam in its fight to maintain its independence."[96] The administration linked the strikes to a "larger pattern of aggression" by North Vietnam, and it signaled a "change in the ground rules of the conflict in the South." However, President Johnson continued to thread the needle between war and peace. The administration's information policies, not the press, exacerbated the credibility gap. The media merely exposed the breach.[97] Meanwhile, Radio Hanoi ridiculed White House explanations for the air campaign against North Vietnam. It claimed that the response "by the U.S. imperialists was a prearranged and planned act" to expand the war and denied any connection to the recent attacks in the South. Flaming Dart remained a narrowly targeted affair. However, pressure increased on the administration after attacks at Qui Nhon killed twenty-three Americans.[98]

From the American perspective, the earlier Pierce Arrow attacks and Flaming Dart represented "restrained and discriminating use of force in pursuit of carefully limited objectives." Though the February strikes made the United States a cobelligerent due to the expanded target set, those targets were chosen with the idea of equivalence and were limited to the "extreme south of North Vietnam." However, the bombing campaign marked a change away from merely advising the South Vietnamese, toward expanded and direct US participation in the war. Following the 10 February 1965 bombing in Qui Nhon, the administration officials characterized a second set of strikes more broadly as a response to "continued acts of aggression" such as ambushes and assassinations. The 11 February strikes transformed the war and American messaging. They were a prelude to the Operation Rolling Thunder bombing campaign, which began the following month.[99]

Robert H. Shumaker was shot down near Vinh, North Vietnam, that day. His was the only plane shot down on 11 February, although Radio Hanoi claimed that "five U.S. aircraft were shot down." With this incident, Radio Hanoi began regular broadcasts of the names of Americans killed or captured to US troops in the South.[100] On 12 February, the North paraded Shumaker before cameras for propaganda purposes. The sight of Shumaker, according to the announcer, "aroused wrathful shouts of 'Down with the U.S. imperialists' at a recent press conference in Dong Hoi."[101]

Johnson's refusal to admit to the American people the significant change in US policy being contemplated gave a psychological advantage to the North. In a larger sense, the cycle of escalation was seen as a psychological action by President Johnson and Secretary of Defense McNamara—applying graduated military pressure in hopes of creating just the right amount of stress to encourage the North to back down. However, real human reactions are often irrational or driven by goals less prone to the sorts of bargaining that game theorists believed in. By initiating the graduated escalation policy, Johnson signaled his fear of widening the war and, perhaps, the limited American determination to win it.[102]

Theories such as these failed to consider the psychological tendency known as the "boiled frog" reaction. Raising the heat slowly, by degrees, gave the North the ability to adapt physically and psychologically to the pressures. More important, it gave time for antiaircraft weapons to arrive, along with thousands of Chinese troops. Additionally, the policy took no account of Le Duc Tho and Le Duan's single-minded goal of unifying the nation, regardless of the costs. Ho seemed to receive the signal, later warning colleagues at a politburo meeting "not to underestimate the United States." He desired to sustain the conflict as a "Special War" without American combat forces. However, Le Duc Tho and Le Duan had the whip hand, not Ho. As long as they remained in power and had the ability to send troops and equipment southward, the prospects of Johnson's messages achieving results were minimal.[103]

By February 1965, the Johnson administration released a long-awaited white paper detailing Communist infiltration of the South. Published in the *New York Times*, the document contained details gleaned from intelligence reports and prisoner or rallier interviews. The paper made a strong case for Northern domination of what the NLF claimed was a Southern insurgency.[104] However, many saw it simply as an effort to build support for the war. Procrastination had weakened the report's impact, and the United States had already lost the propaganda initiative. The Johnson administration fumbled with building the case of North Vietnamese influence over the insurgency. Trying to appear hesitant to escalate—even as escalation was visible to the press on the ground—mired the administration in credibility issues. Johnson allowed the North to control the narrative.[105]

Radio Hanoi responded to the white paper with mockery. Accusing the Johnson administration of "stepping up the special war," it called the paper "a typical example of the U.S. trick of calling for help while committing robbery." Radio Hanoi countered accusations of North Vietnamese aggression by asking how the "old slanderous arguments deceive anyone." It claimed that US public opinion opposed the war. However, that claim was based on intellectuals' letters and the perceived legitimacy of academics as a proxy for public opinion. No public opinion numbers appeared in the story, perhaps because high support would undercut the divisive theme. At this time, Gallup found that nearly two-thirds of the American public supported Johnson's moves. American public opinion did not flip until the fall of 1967.[106]

However, Hanoi's claim—that the administration had "uttered deceitful arguments in an effort to make public opinion" believe that the additional troops represented "only minor changes in tactics and weapons"—was in fact exactly what President Johnson had been doing. Although Northern broadcasts always contained considerable hyperbole, Johnson's pretense enhanced Radio Hanoi's legitimacy over time. Unfortunately for the president, his lack of candor played into the Northern propaganda line as the course of the war unfolded over the coming months.[107]

Conclusion

In the aftermath of the Gulf of Tonkin incident, North Vietnam under Le Duc Tho rushed forces into battle. The leaders there assumed that Operation Pierce Arrow heralded a larger American role and endeavored to win before US forces could be decisive. Although President Johnson avoided responding to repeated attacks targeting American forces (and causing considerable casualties), he eventually authorized the Flaming Dart bombings as a signal to North Vietnam. At each attempted signal, whether through bombing or pausing, the president failed. A combination of misreading and purposeful misinterpretation for psychological value marked the Northern response. After a pause in operations after the Gulf of Tonkin incident, SOG increased its deception operations against the North under Project Forae and expanded the SSPL deception strategy. For the South Vietnamese, radio opportunities opened up, and discussion began over building a television network. Organizational problems remained, but the creation of a political warfare system offered hope for improvement in that area as well. As spring arrived, both sides prepared for an expanded war.

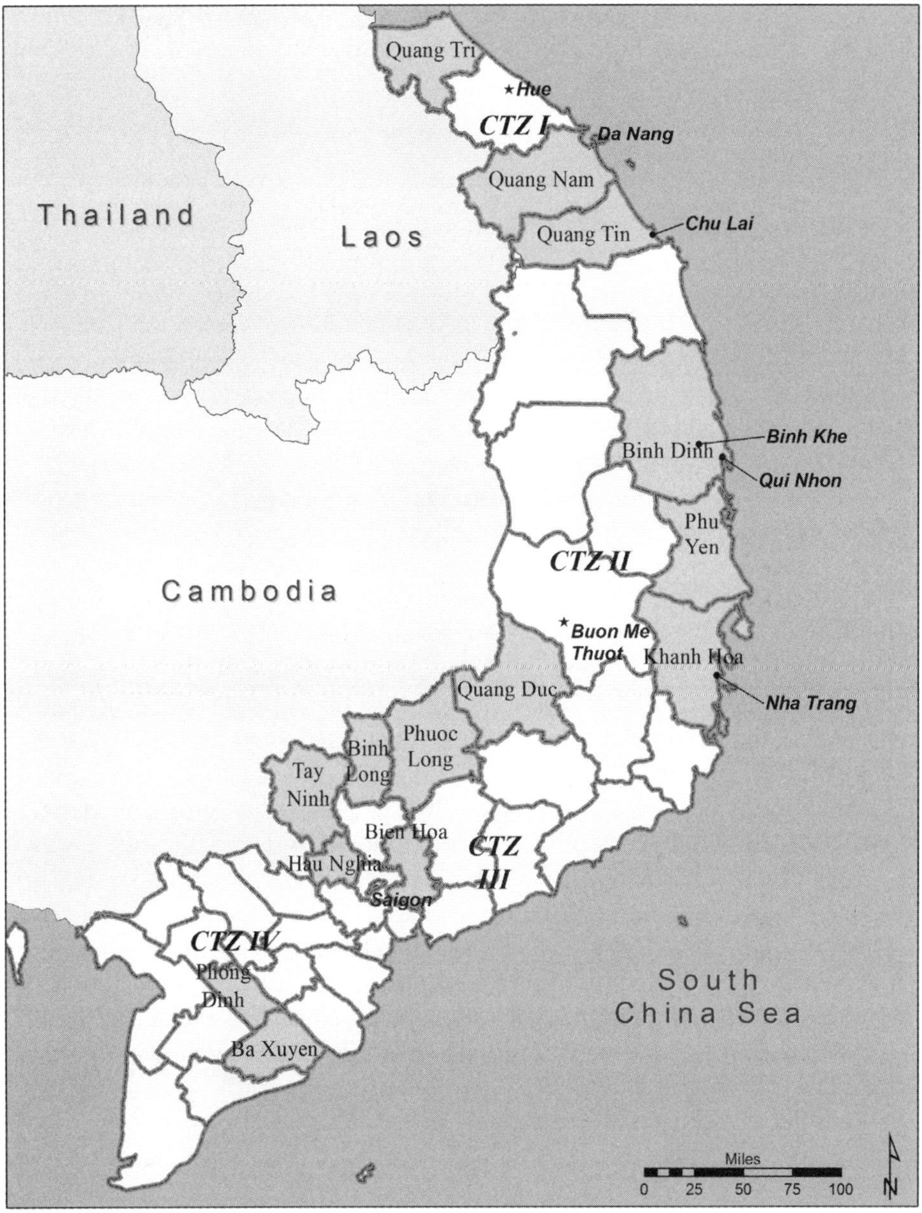

Map 5. South Vietnam, early 1965 (chapter 6 locations)

7 Expansion, Spring and Summer 1965

By early 1965, the American PSYOP force and its South Vietnamese counterpart were struggling to keep pace with the rapidly expanding war. The North increased infiltrations, hoping to win before the United States could bring the weight of its military force to bear. At the same time, the well-developed Northern propaganda force took the offensive, striking again at the sensitive gas-warfare theme. It also expanded targeting of Americans via the press, radio, and its deepening ties with the antiwar movement. The North's integrated propaganda system seemed strong, especially in comparison with their enemy's leaderless program. The entrance of foreign fighters, in both the North and the South, transformed the battlefield in 1965. However, in the face of increased US combat deployments and air support, the North's invasion stalled. By the summer, the United States had augmented its PSYOP forces, created a new structure to oversee operations, and contemplated further growth.

As the year opened, the American PSYOPs effort had grown modestly. The USIS Field Services Center was composed of thirteen USIS officers including Frank Scotton, two USOM and five MACV personnel, and a staff of seventy Vietnamese employees, twenty-three of whom were based in VIS provincial offices. MACV had an additional "seventy military American personnel engaged directly in psychological operations and civic action—working under general policy and strategy developed by USIA." Most of them were based at the province, division, corps, and Saigon levels. At the field level, USIS had an additional five Americans and fifty-seven Vietnamese producing "films, publications and leaflets designed for Vietnamese rural audiences." Fifty-one Filipino, Chinese, and Australian personnel were engaged in closely related operations. This number does not include the CIA, military PSYOPs, or South Vietnamese POLWAR units.[1]

As discussions over the need to escalate the war continued in the United States during the winter of 1965, the role of PSYOPs began to assume a higher precedence. Debate continued at USIA over how to manage the information program for an expanding war. Furthermore, the need for USIS to explain the American military presence and improve the nation's image in South Vietnam became vital. The uncoordinated nature of the developing MACV, USIS, and CIA propaganda programs clearly demonstrated the need for an organization to provide overall direction and to avoid competition and duplication of effort. More important, the different programs needed to avoid message fratricide. Rowan believed at the time that "the South Vietnam government's psychological efforts were floundering." From the American perspective, a drastic change was required.[2]

The USIS Research Reference Service published the results of a follow-up survey of Long An Province in February 1965. It revealed that security was among the most important measurable indicators and was "fundamental to any reversal of the present course of events" in Vietnam. To win, the government needed to "offer some credible hope for a better future." Saigon had failed to that point, at least as reflected in Long An. As a result, VC cadres had "the ears of the people." Front APTs operated in "all but the most secure hamlets, fanning the smoldering animosities between Buddhist and Catholic, and rich and poor; urging the relocatees to leave the fortified settlements and return to their ancestral homes; urging all to 'abandon the government which has abandoned them.'" Long An residents testified on the impact of VC terrorism, indicating that "judiciously employed" terror and assassination constantly reminded them of the "penalties attached to active support of the central government."[3]

Despite Front efforts to avoid offense to villagers, according to the study, "most villagers at first are not happy to see the VC, though in time they may be accepted."[4] Because the VC cadres regularly visited the contested villages, inhabitants had extensive interaction with them. Additionally, their performance was "quite effective," in contrast to the GVN, and reflected careful planning underlying the Front's campaign. In summary, the contested villages surveyed indicated that villagers leaned toward the Vietcong in sympathy. Their control of over two-thirds of Long An Province aided the VC propaganda effort there. In Front-controlled areas, posters, leaflets, and indoctrination lectures were constant reminders to the populace. Face-to-face contact spread insurgent messages in contested areas. Among the most important target audiences were ARVN troops and paramilitary forces. Proselytizing efforts were continuous.[5]

Targeting US troops in South Vietnam via radio, Thu Huong remarked on comments by Senator Wayne L. Morse of Oregon to President Lyndon Johnson regarding troop deployments: that the president was "ordering them to their death." Morse's consistent statements in opposition to the war became a regular feature of the programming. Thu Huong expanded on Morse's statements, urging listeners to understand that "President Johnson has no right to send you, as he has done recently, to your death in larger numbers."

Repatriation was the only way, she said, "to bring back happiness to your families who are desperately anxious over your safety day and night."[6] The next week, Huong implored American servicemen to understand that although it was honorable to sacrifice lives in defense of one's nation, dying "in South Vietnam to protect a regime which enjoys no support of the people is no honor either to yourself or to your country."[7]

Radio Hanoi, while maintaining a generally respectful tone in messages to American troops, in its broadcasts to the South Vietnamese soldiers attacked the "disdainful and overbearing attitude of the U.S. officers, who directly and openly command the units of the southern armed forces." The broadcaster claimed that American behavior aroused a "nationalistic spirit among Southern soldiers" and led to decreased morale and increased desertion. Amid an expanding war, the ARVN fought the Front for control of a critical resource—men of draft age. This was a zero-sum game: the side that acquired them would have an advantage. Both sides appealed to nationalism in this quest.[8]

DRV and Vietcong Propaganda System

One strong suit for the North Vietnamese propaganda effort was the understanding of the inherent political nature of the war. The North never suffered from the delusion that the conflict was actually two wars, one of pacification and the other of conventional forces. In theory, Hanoi integrated all its efforts and subordinated them to the politburo's political considerations. PAVN/PLAF units were organized under a dual command/political cadre system, which ensured political domination of the force. Unlike the ARVN's PSYWAR representative, who was simply a junior staff member, the Front and the North Vietnamese fully integrated indoctrination into the force. Front groups throughout the South and internationally allowed for varying degrees of control, and many coordinated their propaganda efforts toward a single goal. The NLF was only the most obvious of these fronts.[9]

In the PAVN and PLAF, "the political commissar led and the military commander commanded." The commissar's duty included motivating "the combat spirit of the soldiers through words and deeds." He acted as a surrogate parent to the troops, listened to their "complaints, cajoled and encouraged them, educated them, cared for them in difficult situations."[10] A MACV study of VC political indoctrination activities determined that they usually conducted troop political instruction for their units immediately prior to attacks, consisting of "political and ideological propaganda oriented" toward instilling courage for the impending battle.[11]

The propaganda program run by COSVN's Political Section consisted of four subsections, responsible for the Liberation Radio broadcasts, war-zone

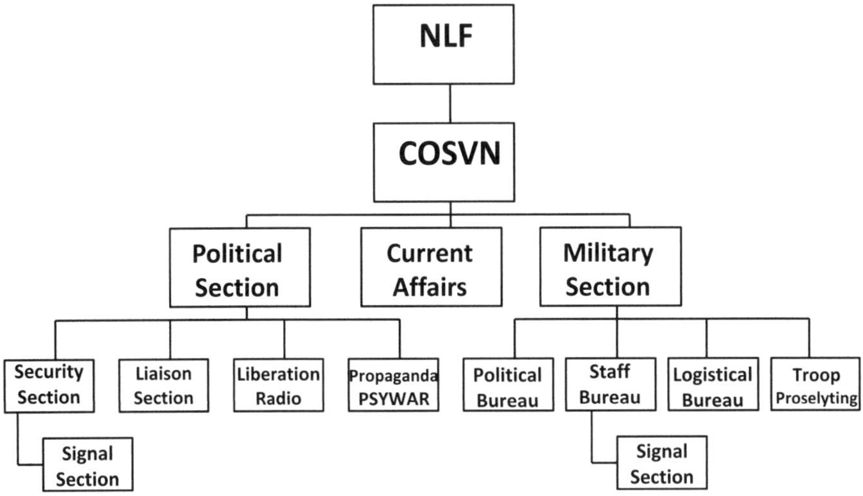

Figure 7.1. NLF Propaganda Organization

security, front organizations' liaison in South Vietnam, and the development of propaganda themes and directives for entertainment and political education directed outside the organization. A mirror organization existed at every command level down to the village, where mass organizations were added to aid control over the population. These last included groups for farmers, women, youths, and self-defense corps, aimed at ensnaring villagers in a web of NLF support activities "in one form or another, unwitting though it may be."[12] (See figure 7.1.)

A March 1965 Rand Corporation study titled "Effects of Military Operations on Viet Cong Behavior" found weak spots in this indoctrination that the United States could exploit. Discussing leaflets' effectiveness based on prisoner interviews, the study found that enemy personnel were "especially fearful of detection and attack from the air, and that intensified and continuous air surveillance and harassing activities by day and night are likely to have a major disruptive effect on Viet Cong living conditions, morale, movements, and operations." This presumably included loudspeaker and leaflet operations. Increased ground patrols also produced a disruptive effect. The study recommended that "protracted ground sweeps should be assessed primarily in terms of their harassing impact rather than according to the casualties they inflict on the Viet Cong."[13] (See figure 7.2.)

Keeping the guerrillas on the run and off balance, even without producing casualties, broke down unit cohesion. The Rand study found that VC units typically spent four months of inactivity between operations. This allowed

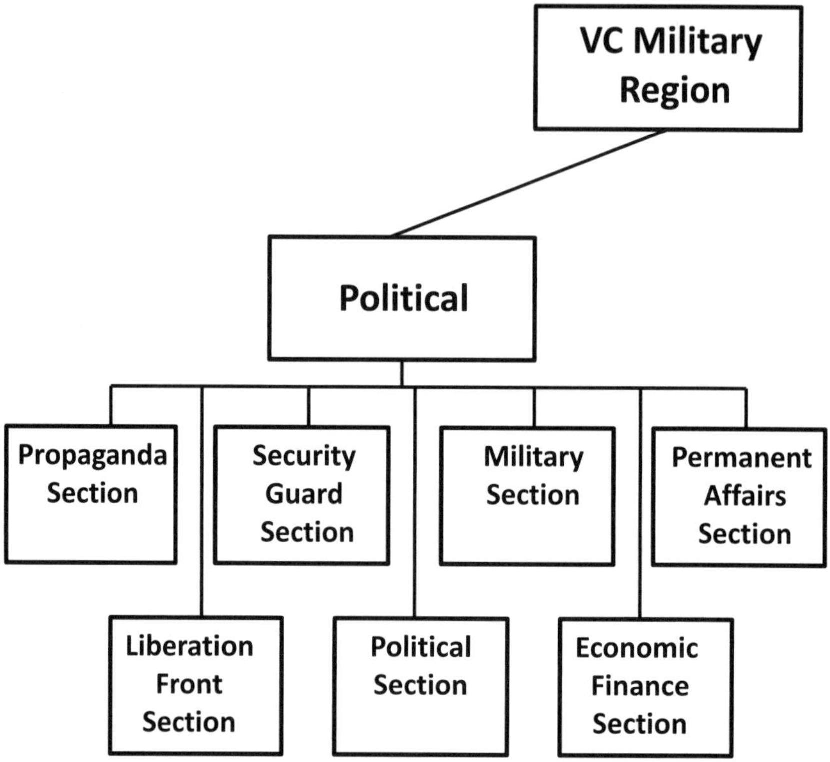

Figure 7.2. Vietcong Military Region Political Organization

the "cadres time and opportunity to propagandize and train their men, to deal with the adverse effects of losses, defeats, and bad living conditions, and to boost their confidence in the next operation." Indoctrination became especially important as the force filled with draftees to replace lost volunteers. These interviews revealed that many of the new recruits were "young boys of 16–17 forcibly drafted from hamlets."[14] Former VC who were taken prisoner said that "morale in these units is lower and that some were experiencing difficulty recruiting to offset a high desertion rate." However, the USIS in Saigon also noted that less than 45 percent of the interviewed subjects had seen or heard any Chieu Hoi messages. Clearly there was room to improve PSYOP targeting.[15]

But message targeting was not the only difficulty to overcome. During a visit to Saigon that month, the USIA's Director Rowan met with senior Vietnamese and American PSYOP personnel to discuss problems and potential solutions. Prime Minister Phan Huy Quat spoke frankly on the current psychological situation. Up to now, he acknowledged, "we have failed." The government was unable to present the Vietnamese people with a meaningful purpose in personal terms. Ambassador Taylor agreed, pointing out "that an understanding of the consequences of a Communist victory was an essential part of this information program." Rowan offered USIA "media and technical training" and requested that Zorthian coordinate this with the GVN's PSYWAR director. However, Zorthian soon had a larger issue to deal with.[16]

"Gas" Warfare Charges

The Associated Press reporter Peter Arnett wrote that same month that the ARVN had begun to experiment with the use of chemical agents to "temporarily disable opposition forces." During the winter, Radio Hanoi had frequently carried stories denouncing the American use of "noxious chemicals," and the coverage escalated in the wake of Arnett's story. Other than gratuitous allusions to German gas use in World War I, his brief article delineated the potential benefits of such gasses in Vietnam. Although it only discussed widely used tear gas, this story unleashed a worldwide backlash.[17]

The sensationalized, poorly sourced story by Arnett had set the administration on the defensive in trying to explain the nontoxic nature of tear gas to counter the North's ongoing propaganda narrative. That narrative produced results: restrictions quickly followed on tear gas use and crop eradication missions. A 24 March telegram from the US State Department to the embassy in Saigon suggested extending the currently planned crop destruction program start until the toxic gas story had run its course.[18] Ambassador Taylor responded that every day of delay would "reduce the effectiveness of the spraying since the crop was being harvested then." The spray aircraft at Nha Trang were supported by Qui Nhon–based PSYWAR teams with loudspeakers and leaflets. This preparation was visible and would likely come to the attention of correspondents.[19]

The Foreign Broadcast Information Service (FBIS) that week reviewed the widespread attacks on the use of gas in Vietnam and the tendency toward hyperbole. Moscow referred to it as a violation of the 1925 agreement to ban the use of "gas warfare," which "not even Hitler Germany" dared to flout. Later the East German newspaper *Deutsche Aussenpolitik* wrote "that the West German militarists are reemploying former Hitlerite experts in the use of poison and biological gases," citing supposed former Nazi chemists now employed and expected to be sent to South Vietnam. Mischaracterization of the tear gas used was prevalent in worldwide press reports. Radio Hanoi reportedly used up to one-third of its broadcast time in the days after Arnett's story on the issue of gas. One claim made by all Communist organs implied that the United States was bound by the 1925 treaty that it had not ratified.[20]

Claims to the contrary, the United States had not supplied or used treaty-covered chemical weapons in Vietnam. The embassy's postmortem on the incident concluded that "major damage was done by initial AP story which used emotional lead stating U.S. experimenting with gas warfare. While our handling after this initial story could have been better, fact is we were never able to catch up with impact of worldwide response to initial AP lead." Ambassador Taylor reminded the State Department that Peter Arnett had a history of "deliberately negative" stories about US involvement in Vietnam. Arnett had failed to even seek a comment from American sources prior to filing a story, "which even neophyte journalist would have known, would have worldwide anti-U.S. impact."[21] In response to White House queries, General Westmoreland answered that the first use of "riot control munitions" occurred on 23 December 1964 during an ARVN mission to extract American prisoners.[22]

Despite this propaganda setback, the CIA reported that ARVN forces had regained the offensive at last. "As for the political situation in Saigon," the director of Central Intelligence also wrote, "the South Vietnamese government, the military leadership, and a substantial part of the informed public have all received a morale boost" from recent US air strikes. The GVN welcomed an increased commitment on the part of the United States. However, the US Air Force was exhausting the list of meaningful approved targets in the North. The following week, both enemy casualties and weapons losses exceeded those of the ARVN in the South for the first time in months. The Vietcong lost two major engagements, and the majority of encounters in March were initiated by South Vietnamese forces. North Vietnamese propaganda in recent weeks had claimed to have shot down six times as many planes as were actually lost. A CIA report assessed that "this could indicate a growing concern on the part of the regime over the morale of the North Vietnamese population."[23]

However, the study also identified regular PAVN forces now operating in the northern provinces of South Vietnam. These forces seemed to be avoiding contact in preparation for an offensive. Also, VC units engaged were in larger formations and offered stiffer resistance.[24] Ominously, the NLF "announced that if the US continued to increase the commitment of its forces to South Vietnam, the Front might seek foreign assistance in the form of military personnel." The outgoing CIA director, John McCone, confirmed that thirty-six Chinese MiG-15s or MiG-17s were already operating out of North Vietnam. At least one engagement with American planes took place on 4 April 1965. Fear of an expanded Chinese deployment weighed heavily on President Johnson.[25]

Coincidently, Johnson authorized an expansion of USIA activities in Vietnam that same day. To fund this expansion the president tasked the USIA to "re-program funds" and "directed all agencies and departments to provide funds and resources as available to the Director of USIA." On 9 April, the president ordered the deployment of 18,000–20,000 logistics troops, along with two more Marine battalions and an air squadron. However, he ordered minimum publicity of all troop movements to downplay the significance of these increases. Despite Zorthian's professed policy of "maximum candor," this dissembling in a matter that all Vietnam-based reporters could see on the ground continued to undermine public information credibility. Johnson's efforts to mask the deployments worked for a short time, but they proved to be a disastrous error in the long run.[26]

DRV and the Antiwar Movement

With the onset of the Operation Rolling Thunder air campaign against the North that month and the lack of security for aircraft operating out of the South, Westmoreland requested additional Marines to guard the airfield at Da Nang. At the same time, Radio Hanoi added a new tool to influence US troop morale: Americans visiting the North. Robert Williams recorded a message for later broadcast on Radio Hanoi.[27] Williams had fled North Carolina due to problems related to civil rights activities, and he broadcast from Cuba on Radio Free Dixie. By this point, Williams had moved to China at Mao Tse-tung's invitation. In March, Radio Hanoi broadcast an appeal from Williams, who it claimed was a "former Marine officer" who had visited North Vietnam along with other Americans. Radio Hanoi replayed a quote from Williams recorded the previous autumn in which he told US troops: "Our enemy is the common enemy, the same man who blows the heads off blue-black [*sic*] girls in Sunday schools in Birmingham, the same man who blows the heads off blue [*sic*] girls in Vietnam in the paddy fields, the same man who rains death and destruction on the innocent people here who want freedom, who want to be friends to all of humanity."[28]

With the arrival of Marines, Radio Hanoi asked American troops to recall Williams's words "at a time when your Negro compatriots are being brutally suppressed by the racists in the state of Alabama."[29] Williams appeared in several other broadcasts in 1965. This racial wedge theme remained a staple of Northern propaganda throughout the war. The North tied the ongoing civil rights strife in America to the war, asserting that "the American ruling circles have strongly repressed American Negroes and schemed once again to launch a new campaign to repress the American communists. . . . The U.S. imperialists are the enemy of the American people; the U.S. imperialists are the enemy of the world's people."[30]

Foreign troops transformed the war at every level in 1965. Chinese arrived in the North while North Vietnamese, Australian, Korean, and American combat forces landed in South Vietnam. By the end of summer, the PAVN/PLAF reached an estimated combat strength of 221,000 fighters, "including 55 NVA battalions and 105 VC battalions."[31] What had been a low-level insurgency had now become a regional war. With the increased commitment of US forces, separating the American people from government policy became a major Northern psychological objective. To do this, Hanoi amplified antiwar activities in the United States in international, domestic, and US troop broadcasts. For instance, Radio Hanoi praised the Nobel Prize–winning chemist and peace activist Linus Pauling and "420 university professors and 122 renowned personalities" for declaring the war lost and urging that the "U.S. administration immediately put an end to this war." Announcers highlighted protests at colleges and universities across the United States.[32]

The Committee to Aid the NLF (CANLF) was a good example of this relationship. Walter D. Teague III founded the CANLF in New York City in April 1965 to "make it easier for Americans to turn against the war" and break down fears over North Vietnam.[33] However, the original overt purpose of the organization was to facilitate sending medical aid to the Front via the NLF mission in Czechoslovakia. Walter Teague, patronymous grandson of the noted American industrial designer, was a founder and spokesman for the organization until its dissolution in 1975.

The CANLF sold an extensive array of North Vietnamese propaganda through its offices on MacDougal Street. Much of the Hanoi Foreign Language Publishing House catalog—including *Ap Bac* and *Special War: An Outgrowth of Neo-Colonialism*, as well as the *Vietnam Courier*—were available. The committee advertised itself as having "the biggest supply of Vietnamese literature . . . in the U.S."[34] The organization sold individual copies of *Vietnam Courier* and offered a one-year subscription of forty-eight issues for five dollars. The CANLF also helped see that college libraries received these products.[35]

These activities and linkages allowed the North to spread inevitability-of-defeat themes and demoralization using unverifiable but conceivably true stories. In one case, Radio Hanoi reported an unsubstantiated case of twelve American NCOs in Rach Gia, 125 miles southwest of Saigon, refusing to take part in combat operations. With no names or confirming details, this appears to be an attempt to create an "emulation" campaign out of fiction. Supposedly, the soldiers "unanimously demanded their repatriation to the United States." Saying that the men had "not blackened the American people's love for freedom, reason, and justice," Radio Hanoi called on all American troops to "refuse to carry out their orders." It continued: "The real American is brave enough when he strives to defend his own country as he did in Washington's time when he defeated the British oppressors. The real American cannot tolerate fighting in the [face?] of aggression against another country, to enslave those nations who would like to live on friendly terms with the American people."[36]

US Forces' Expanded Combat Role

Conversely, the PSYWAR campaign against the North was conducted primarily through air-delivered means—leaflets as well as radio. For instance, on 9 April 1965, US officials approved overt leaflet drops prior to attacking targets in the North (the Fact Sheet/Frantic Goat campaign). The United States planned to warn inhabitants near potential targets that "people should evacuate all targets" considered to be military objectives. The first drops of this program began on 14 April, accompanied by reinforcing radio broadcasts.[37] The leaflet drops north of the 17th Parallel also stressed themes explaining the purpose of the air strikes as an "appropriate and necessary response to aggression by DRV against South Vietnam" and that the United States sought "no wider war." The operation's goals included increasing psychological pressure in the North and driving a wedge between civilians and the government.[38]

Most of the product was gray propaganda: produced by the United States but implied to come from the South Vietnam government. Themes included: the "Lao Dong party is needlessly extending the war," "destruction in the north will continue as long as the North attacks the South," and "the Ho regime is a puppet of the Chinese." Later the campaign informed people about peace talks to encourage war-weariness. The campaign, known as Fact Sheet, and later as Frantic Goat, ran through March 1968. After that, President Johnson limited drops to southern parts of North Vietnam.[39]

In June 1965, Washington authorities further delegated to the American ambassador in Saigon and to the Commander in Chief in the Pacific (CINCPAC), responsible for the air war, the authority to conduct leaflet drops as part of the overall air effort. These leaflet operations included two drops per week of about two million leaflets each. On 16 July, CINCPAC recommended that leaflet operations be conducted on the major North Vietnamese population centers, including Hanoi and Haiphong. Authorities approved this with the proviso that leaflet aircraft could not penetrate a 40-nautical-mile circle around either city. Leaflet drops for these cities utilized the wind-drift technique, calculating a high-altitude release point based on wind speed and direction along with the leaflet descent rate. Until the early part of September 1965 all these leaflet missions were executed by F-105 aircraft. This PSYOP onslaught forced the North to direct an extensive counterpropaganda campaign.[40]

Just as American deployments began to increase, the Defense Department issued a contract to American University, through the Special Operations Research Office, to produce a PSYOP study of Vietnam. Outlined in military operations order format, it compiled the most recent information of PSYOP relevance gleaned from the USIA, CIA, military intelligence, and open sources. This study intended to give incoming PSYOP personnel a basic understanding of the country and PSYOP infrastructure available. A second study compiled by SORO acted as a companion. It provided a political and historical overview of the nation and a breakdown of potential target audiences. The annex provided examples of psychologically exploitable situations, giving the background for each, how to exploit them, and the suggested target audiences. The annex listed the following PSYOP objectives for American PSYOP forces:

1. Create, build, and maintain confidence in the GVN.
2. Establish a climate of national unity behind the war effort.
3. Expose the true nature of VC objectives and induce disaffection among them.[41]

The SORO report noted that six 1,000-watt loudspeakers flown on UH-1 helicopters and four 250-watt loudspeakers flown by the USAF 1st Air Commando Squadron U-10s were available for support. By June, they expected ninety Tri-Lambrettas equipped with loudspeakers, movie projectors, and tape recorders to arrive in Vietnam. The systems were to be distributed two per sector. The report stated that in addition to the 44 newspapers available in major cities, 182 district-level newspapers, published two or three times a week on US-supplied equipment, existed. These each typically ran 150–300 copies per issue.[42]

By summer, the United States began a rapid growth of forces. Although PSYOP teams had rotated into the country for some time, they now began to multiply as well. The USAB&VAPAC ordered seven officers and seventeen enlisted men to report to Vietnam for ninety days as the 1st PSYOP Field Support Detachment (Provisional) on 22 July 1965. In addition to the US Army and US Air Force PSYOP elements arriving to support the expanded mission, the US Navy activated Task Force 115 on 30 July 1965 under Rear Admiral N. G. Ward, commander of the Coastal Surveillance Force. While its primary mission was conducting Market Time infiltration interdiction patrols by sea, the task force also performed pacification, advisory, and PSYWAR missions. (Market Time forces consisted of small US Navy boats operating along the South Vietnamese coast.) Because task force elements operated in remote locations, the personnel came "in close contact with a large segment of the population not touched by government influence." PSYOP product to support these missions came from the Naval Advisory Group's PSYWAR adviser. The task force's mission included PSYOP indoctrination of incoming US personnel.[43]

One of the major projects in adviser Marine Captain Joseph Knotts's sector that May was moving the Vietnamese population off property needed for the Marine base at Chu Lai. This sensitive subject had to be handled properly to avoid alienating families who were forced to move. Knotts stated, "We received prior notification to the landing at Chu Lai and that there was a certain area of real estate that they wanted for Marine Corps uses down there." He acted primarily as an adviser to a "totally Vietnamese operation; going in, clearing the area, getting the civilian populace out and paying the indemnification." This was important, according to Knotts, in order to maintain government legitimacy.[44]

South Vietnamese law provided for mandatory compensation for land requisitions. However, this was "not always satisfactory or timely because of the rigid procedures involved," Knotts said. Due to complexities of landownership, as well as missing owners or paperwork, some cases were not solvable. In other cases, such as the Phu Cat Air Base, officials milked the program for personal profit. In that case, "the province chief, Lieutenant Colonel Tran Dinh Vong, and the local military property chief, were both sentenced to life imprisonment."[45]

The stresses created by the removals and potential corruption opened exploitable opportunities for the NLF. "According to AP, the U.S. Marines based at Da Nang have also planned the grabbing of more than 100 square miles of additional territory west of Da Nang airbase," Radio Hanoi claimed.[46] With the looming deployments, Radio Hanoi urged South Vietnamese "workers, students, schoolchildren, youths, intellectuals, businessmen, and industrialists" to unite in opposing "the U.S. land grabbers and their henchmen."[47]

The first US Army combat formation sent to Vietnam was the Okinawa-based 173rd Airborne Brigade, which arrived in May 1965. Radio Hanoi carried a story to US troops in Vietnam stating that, before leaving, eighty soldiers of the brigade "threw stones, pulled down street markers, or broke the windowpanes of the inhabitants' houses to create incidents and give vent to their opposition to being sent to Vietnam." Furthermore, the broadcast warned incoming troops not to trust the Vietnamese, asking them to consider why the South Vietnamese "puppet troops" did not cooperate in combat.[48] Reflecting on the French defeat, it claimed: "Your authorities are treading the path of the French colonialists by relying on the puppet army for waging the special war in South Vietnam. You, servicemen, are fighting here for the interests neither of the American people, nor of your family, nor of yourselves. You are risking your lives uselessly in a very deadly war. Is it worth your sacrifice? Just think it over."[49] To further attack American morale, the broadcaster highlighted a growing antiwar movement "among American youths" that was "sweeping the campuses of universities and colleges in the United States."[50]

In the meantime, Radio Hanoi sent an opposite message to ARVN soldiers. It implored them to turn their weapons on US soldiers. It also made the accusation that US Marines were sent to "South Vietnam to massacre directly our compatriots." In light of the use of "chemical weapons such as napalm bombs and poisonous gas to barbarously kill children, old people, and women, including your family members," the peaceful Southerners had no choice but to take action.[51]

As the military PSYOP effort grew, so, too, did the civilian project. Rowan notified President Johnson of the increase of ten USIA personnel in Vietnam and the training of forty-two army officers to take over as provincial PSYOP/CA advisers at the district level. USIA had worked with the US Defense Department and the GVN to install a 50-kilowatt radio station in Hue to broadcast into North Vietnam, and it transferred three 50-kilowatt shortwave transmitters from Liberia to the Philippines to support additional broadcasts to Vietnam. The agency had extended Voice of America broadcasts to Vietnam to five and a half hours daily, recruiting Vietnamese personnel in support. This invigorated radio system offered "competitive coverage of North Viet-Nam unless the Communists resort to jamming." Although Rowan stated that a megawatt medium-wave station was needed to compete with Beijing, such a station could not be operational in less than eighteen months.[52]

One major problem Rowan noted was negative press stories derailing the mission, such as the earlier Arnett story. For instance, according to Rowan, *Washington Daily News* stories were "the products of reporters who go out into the field, gain the confidence of our soldiers and then pick up bits of information which they turn into stories that are not at all helpful to our overall mission." Rowan believed that these sorts of reporters were susceptible to little American influence, but he urged the president to allow Zorthian and the other USIA officers the freedom "to make whatever effort they can to inform and [give] guidance to American newsmen."[53]

However, the press did uncover real problems. A lack of synchronization between American words and deeds often hurt the information campaign. Upon the entrance of the American forces in 1965, "troops were to be briefed on the importance of saving civilian lives." This included clearly delineated free-fire zones that avoided populated areas, rules of engagement, and "joint operations with South Vietnamese troops . . . to avoid the appearance that U.S. forces were at war with all Vietnamese."[54] However, troops did not always honor these restrictions, and rowdy behavior by incoming Americans proved a lucrative propaganda opportunity. For instance, a "bloody brawl involving only the leathernecks" evolved into a street fight involving about fifty Americans, according to Radio Hanoi, "touched off by a fight between a white marine and a Negro marine."[55] Later it equated this incident with unruly French colonial troops, who "gathered for depraved activities and often engaged in bloody brawls, too." Coupled with reports of US Marines conducting "a scorched-earth operation in the rural areas south of Hue city," Radio Hanoi attempted to drive a wedge between US and ARVN troops, between Southern civilians and US troops, and among US troops themselves, noting, "They indulge in racial discrimination and look down on their Negro fellows." Addressing US troops, the station asked:

> Why did your mates, the U.S. Marines, use a strong hand to burn these peaceful South Vietnamese villages to ashes? What do you think about such things being done on orders of your government, in the name of your country with its tradition of freedom and justice, in a country half the globe away from yours, against a people who harbor no hatred towards the American people, who include your loved ones?[56]

This was a classic example of creating cognitive dissonance within a target audience. There is no record of the events described, but the charges had the capability to open nagging questions.

By late spring, direct allied support, including PSYOPs, expanded rapidly. Along with the deployment of US and Korean combat troops, 800 Australians arrived that summer. Operating first in the Bien Hoa area, the Australians later assumed responsibility for Phuoc Tuy Province and fielded their own 1st Psy Ops Company. Radio Hanoi attacked the Australian deployment announcement in strong terms. It accused the Australians of "conniving" with the United States in "the aggressive war in South Vietnam" against the "sovereignty and independence" of the nation.[57] In early June, the 1st Battalion, Royal Australian Regiment, which included 161st Battery, New Zealand Artillery, joined the 173rd Airborne Brigade as a third maneuver element. This was welcome news. Years of fighting in Malaya had taught the Australians "the importance of jungle warfare tactics." However, one battalion member was critical of the 173rd Airborne, observing that "our patrols do not fire off ammo or shoot up flares like the Yanks—they listen and move quietly." Major John Donohoe became the Task Force Civil Affairs and Psychological Operations officer.[58]

In May 1965, ARVN Major General Nguyen Ngoc Le returned from POLWAR study in Taipei and recommended establishing a special commissariat for PSYWAR. The next month, the RVNAF began to reorganize with "psywar, troop morale, civic affairs and social services."[59] To support this, MACV created the Political Warfare Advisory Directorate "to advise and support the General Political Warfare Department (GPWD) of the RVNAF."[60]

JUSPAO Formation

The following week, President Johnson approved the formation of the Joint US Public Affairs Office to oversee and coordinate all American psychological and information activities in Vietnam, as well as to manage the development of the South Vietnam government's programs. Zorthian and JUSPAO assumed responsibility for approximately 50 million leaflets per month targeting North Vietnam, 500 million leaflets per month targeted against the enemy in South Vietnam, and nearly 50 million toward friendly targets in the South. Through the USIS, four cultural centers and a book translation program were included, in addition to field support to the VIS.[61]

Upon the formation of JUSPAO, the first task Zorthian set for the organization was to develop a countrywide PSYOP plan with an effective monitoring system to ensure proper implementation. JUSPAO issued this in December. Zorthian also sought to create a handbook for psyoperators, identify "charismatic symbols and . . . slogans capable of serving as an alternative to the ideological vacuum," and suggest programs capable of inspiring the people of Vietnam. Among the publications JUSPAO prepared were 200,000 copies each of a weekly troop newspaper and monthly ARVN troop pictorial magazine, and 20,000 copies of a monthly magazine for officers. The JUSPAO Pictorial Center distributed commercial entertainment films and newsreels, ARVN training films, and Vietnamese-dubbed US training films. The PSYWAR Department also "produced the 'Voice of the Army' program for broadcasting over ARVN radio." In addition, Regional Forces had PSYWAR/CA teams in each province to conduct operations. JUSPAO requested fifty-seven PSYOP-trained advisers to support the expansion of the RVNAF political warfare structure and activities.[62]

Cracks in Vietcong Morale

Almost immediately after being formed, JUSPAO began issuing guidelines for conducting PSYOPs in Vietnam. In a decentralized operation such as this, these were essential to ensure message discipline and training for incoming soldiers. These directives bound all US military and civilian PSYOPs, but not the RVNAF's program. One directive, titled *Exploitation of VC Vulnerabilities*, discussed the negative effect the increasing intensity of US attacks was having on VC morale. It was based on prisoner interviews: "Many complained of being exhausted and discouraged by the frequent moves, the disruptions of their rest and cooking and by the constant fear of being detected and attacked." The interviewees indicated that their casualty rate had increased recently. The tremendous psychological effect of B-52 bombings was "widespread and not confined only to the areas which have experienced them."[63] These massive heavy bombers had a devastating effect on morale, with Front cadres noting "that in recent months the soldiers speak more often of dying in the next battle, of never seeing their families again or benefiting from a victory."[64]

By the end of the summer, it seemed that the tide had shifted, if only slightly. Through 23 August, 115 Vietcong had rallied to the other side. Defectors were offered food, shelter, and an easier life. After up to sixty days in a Chieu Hoi camp, they were resettled. On occasion, people would be counted twice or "defect" several times over. In addition, some agencies inflated the numbers to make themselves look better. However, by 1965 the sheer quantity of defecting VC was straining the ability of the Chieu Hoi centers to handle them. And with increased combat in 1965, peasants began to see the PLAF as bomb magnets, often blaming their presence for attracting air strikes. Additionally, the large numbers of Northern cadres present in the South caused cultural clashes within the NLF. The newcomers spoke a different dialect and "did not understand local customs," which antagonized Southerners. Under pressure to fill units that year, the Vietcong abandoned a program of having units serve in home areas, further contributing to the increasing Chieu Hoi rallier rates.[65]

A study of insurgents' motivations and morale found cognitive dissonance infecting regroupees as they came face to face with realities that were incompatible with the Northern propaganda they had been fed. Northerners in particular "found they were not universally welcomed as liberators in the South and that they were fighting other Vietnamese," contrary to what they had been told. Those who surrendered and rallied were particularly impressed with the military superiority of ARVN and US forces. Constant Front propaganda recounting VC victories with no casualties disillusioned them even further. They knew the reality was different. Regroupees also found the liberated zone much smaller than they had been led to expect. The better-than-expected living conditions among the Southerners was particularly shocking. For instance, one regroupee who rallied in March 1965 stated that after a visit to Ban Me Thuot on a propaganda mission he saw "that the living conditions of the people were not that bad, [and] I realized that the party had lied to us."[66]

People rallied for a host of personal and professional reasons, not all having to do with rejecting Front aims. One campaign specifically targeted midlevel cadres with access to radios. Messages stressed "obstacles to advancement in the Viet Cong based on merit," as well as the loss of popular support for the movement. Some left the NLF over broken promises of family support. One discouraged member stated that he was hurt most because VC "cadres did nothing to keep their promise that they would take care of my family. During my absence, they didn't give even one grain of rice to my family."[67]

Surprisingly, Front cadre counterpropaganda was another source of information for ralliers. A VC propaganda specialist learned about the harsh conditions in the North, especially the food rationing, by talking to returned regroupees. He stated that rather than the "fair" system the Northerners tried to portray, "this system is only designed to oppress the lives of the people and to impoverish them." He said he "understood very well the VC method of indoctrinating the people." As a result, support for the NLF attenuated. However, this did not automatically translate into support for the government.[68]

Incessant Front and North Vietnamese propaganda proclaiming imminent victory set fighters on an emotional roller coaster as the reality of protracted war settled in. They faced aggressive American troops who fought "even in a tight situation, usually inflicting considerable losses," which further depressed spirits.[69] To listeners in the South, Front propaganda persistently portrayed American forces as "brutal, sadistic killers who are indifferent to Vietnamese lives." Captured Vietcong, by contrast, found that US troops treated "captives well and that they generally behave well towards the population." To counter this, the Vietcong routinely mutilated and killed American captives. JUSPAO assessed this as an attempt to brutalize the war, spur US troops to respond in kind, and "provoke the U.S. forces into behaving in a manner which would give substance to VC propaganda claims so as to deter defection of VC soldiers to Americans and fraternization of Vietnamese civilians with them."[70]

Front social control was important in preventing desertion. The three-man cell was vital to this power, along with close supervision by cadres and political officers. Positive psychological support from cell members and higher cadres reinforced these negative approaches, inhibiting rallying. As for morale in the North, the long war and the mobilization of the economy necessary to support it opened exploitable opportunities for propaganda. An increasing percentage of the North's rural population departed for military service in the South. The diversion of productive capacity devastated the DRV's economy, and the nation avoided mass starvation only with generous rice shipments from China.[71]

The VC cadres most frequently cited leaflets and aerial broadcasts as the media through which they became aware of the Chieu Hoi program. However, they still had problems understanding the loudspeakers. Some also indicated the importance of radio and appeals from family members to their decision to rally. That last method had the greatest credibility and became an intermediate target audience for future Chieu Hoi campaigns. However, despite the increase in ralliers in 1965, the influx of forces from the North outpaced it, leading to a stalemate on the battlefield.[72]

DRV/VC Enhanced International Influence

In the summer of 1965, North Vietnam began a propaganda program to "mobilize manpower and resources" under the theme "Motivating the Peasants" to "stir up resentment in people" against "landlords, the Americans, or the GVN." Hatred helped to unleash an emotional reaction less susceptible to reason and thus to the American PSYOP program. For instance, in the My Tho district, a cadre described indoctrination sessions in which "we were told to write denunciations against the imperialists and the feudalists. . . . We had to remember all our ancestors' sufferings and make them known to other participants." Effective PSYOPs had to overcome these barriers.[73]

That summer, Thu Huong discussed Anna Louise Strong, then heading the delegation to the Second Conference of the International Trade Unions Committee for Solidarity with the Workers and People of Vietnam against U.S. Imperialist Aggressors. Huong told American troops that the "[eighty-year-old] American grandmother" was an "untiring peace fighter." Thu Huong said that the renowned writer "voiced the sympathy and support of progressive Americans for the Vietnamese people's struggle against foreign invasion." In the interview broadcast to American troops, Strong denied that Americans were "naturally imperialistic." She continued: "No American likes to believe that his country is imperialist. Yet in these years I have seen my country change from an anti-imperialist nation to the strongest imperialism on earth. In our land, which we love as other people love their native lands, lives a savage beast that threatens the human race." She further accused President Johnson of using "U.S. jets and pilots for massive bombings, adding poison gas and the fire storm to the murder weapons."[74]

Another technique used to attack morale among arriving troops was to highlight soldiers who refused to serve in the war. In one example, a Hanoi report on an AP story described how Lieutenant Richard B. Steinke of Wisconsin "refused to take command of a South Vietnam puppet Ranger unit as ordered by his superior." Steinke, a 1962 West Point graduate, "declared that he did not agree with the U.S. policy in Vietnam" and that it was "not worth an American life." According to Radio Hanoi, he was investigated, but "the hearing was later adjourned because the tribunal could not produce enough charges against him." This was meant to insinuate to the American soldiers listening that refusal would not draw a heavy penalty. (In fact, Steinke was convicted and sentenced to dismissal and forfeiture of pay and allowances.)[75]

Liberation Radio began broadcasting Front propaganda ninety minutes per day in February 1962. The broadcaster used mobile transmitters, but its main facilities were in Tay Ninh and Kien Hoa Provinces. Reportedly, "the antenna cable was buried underground and water to avoid aerial detection."[76] By late 1964, programming ran five and a half hours daily. The GVN jammed the signal in Saigon, but "a survey revealed that up to 60 percent of Liberation Broadcasts were still available there."[77] The NLF used radio to disseminate party doctrine, news, commentaries, and entertainment. One goal of Front broadcasting was to "link mass media channels to interpersonal [face-to-face] ones."[78] The NLF instructed village and cadre propagandists "to listen to broadcasts and follow up further dissemination of themes." One captured document ordered cadres to guide "people to listen to our broadcasts to be aware of current events and the general policy lines of the party." Listeners could then act as message disseminators. The NLF posted propaganda agents at places such as barbers, tailors, bicycle repair businesses, and coffee shops to encourage "broadcast listeners to [act upon the] propaganda." However, the NLF also used Liberation Radio as a means of sending operational messages to distant or underground groups.[79]

Broadcasting in South Vietnam was a noncommercial government monopoly under the South Vietnam Ministry of Information. Prior to 1965, the GVN had placed an emphasis on local and provincial broadcasting. However, as centralization continued after Diem's removal, more content originated in Saigon. The British Broadcasting Corporation, South Vietnamese radio and television, Voice of Freedom, US Armed Forces Vietnam Network, Liberation Radio and Radio Hanoi, and various covert stations were the primary competitors for listenership.[80]

RVN Forms General Political Warfare Department

With the advent of the POLWAR structure in the RVNAF, the training center was redesignated as the Political Warfare School, administered under the indoctrination directorate of the General Political Warfare Department (GPWD). During the coming year, the school graduated about 430 personnel and conducted shorter courses for enlisted men and one POLWAR instructors class. GPWD cadets were sent to the ARVN Infantry School to receive basic training and then began POLWAR training. The school assumed the mission of providing a two-year college-level training course to professional POLWAR officers after it moved from Camp Le Van Duyet in Saigon to Dalat the next year.[81]

Organized along Nationalist Chinese lines, the GPWD consisted of the POLWAR, Political Indoctrination, Military Security, Social Welfare, and Commissary Departments, plus directorates for Roman Catholics, Buddhists, and Protestants. It eventually controlled five POLWAR battalions, one for the South Vietnam Joint General Staff and one for each corps area. POLWAR staff members were incorporated into all levels of command. However, it was not until the nation regained a measure of stability in 1965 that the regime revived indoctrination in the RVNAF. This began with the Popular Forces militia under the Morale Rearmament Program, because their morale suffered the most during the 1963–1964 period. The program "sought to impart anti-communist ideology on [Popular Forces] troopers, teaching them the art and techniques of winning popular support and appealing for the enemy to rally."[82]

The PSYWAR directorate of the GPWD also administered the RVNAF's Civic Action Program. The program included sections for information and education, entertainment, gifts, repair and construction, and indemnification for damages and losses due to combat operations. Although the incoming US forces immediately began conducting Civic Action operations, South Vietnamese soldiers led most of the early missions. For instance, with US Navy help, the South Vietnam Navy converted a Landing Ship–Medium to use as a mobile hospital with a helicopter pad. The ship's crew treated thousands of civilians, and a "psywar team was often on board and gave cultural shows and other types of entertainment. Soap, food and clothing were also distributed."[83]

As these other forms of communication continued to develop, combat operations intensified during the summer. The US 173rd Airborne Brigade and the 1st Royal Australian Regiment conducted operations in Phuoc Tuy Province, attacking a suspected VC supply route into the heart of III CTZ.[84] The operation began on 28 July, supported by a PSYOP team. According to Captain Icenhower, of the brigade S-5, "We flew four loudspeaker missions in helicopters broadcasting to the individuals in the villages" on a self-protection theme, requesting that listeners help "us so that we can find the VC and not harm the villagers." The team also flew five leaflet missions distributing 15,000 "terror," 40,000 "surrender" and "show of force," 30,000 "protect yourself," and 2,000 "friendship" leaflets. The friendship leaflets were part of an ongoing series to convince readers that the Vietcong was being defeated. Icenhower said, "We are also helping the civilian population to rehabilitate their town and put it back into good working order again." Additionally, thousands of pounds of grain were distributed, along with 3,000 gallons of cooking oil. One tactic used was to distribute 20,000 packs of candy wrapped with leaflets to children. This technique used the kids as a distribution channel to remote families, informing them of Civic Action locations. Captain Icenhower stated that in the Ben Cat area people were starved for information because the Vietcong had cut them off for so long. He noted that they particularly enjoyed the 850 magazines distributed.[85]

Lieutenant Colonel George Dexter, commander of 2nd Battalion, 173rd Airborne, discussed unit operations in Phuoc Tuy Province that July. This was one of the first major American combat operations of the war. The unit found civilians dominated by, and given little choice but to cooperate with, the Vietcong. The Americans had entered the area, hurt the enemy, and conducted Civic Action, but Dexter had no doubt that "as soon as we walked out, the VC moved back in and continued to dominate them." Dexter asked the assembled brigade leadership: "Why do we run an operation when we don't have a government to move in behind us?" Without security, everything else was a Band-Aid.[86]

General Ellis W. Williamson, commander of the 173rd Airborne, added a word of caution in his operational summary:

> We do not burn villages. That is a terminology I would like never to hear again. We destroy some VC camps that we find out in the jungles. We do not burn villages. Psychologically that's a very poor term, and as far as we're concerned it runs counter to what we are trying to do with the local people. If it's an established village, we leave it to the RVN to make the decision whether it is to be maintained or destroyed. We destroy VC camps, we do not burn villages.[87]

Precise descriptions and nomenclature were important in a political war such as that being waged Vietnam.[88]

Shortly after the Phuoc Tuy mission, the 173rd Airborne deployed to Pleiku due to increasing VC pressure on the Special Forces CIDG Camp at Duc Co, five kilometers from the Cambodian border.[89] Peter Arnett reported on American casualties suffered in the 8 August operation. According to Radio Hanoi, he described the "bewilderment and weariness of the American troops at Duc Co."[90] However, the quarterly report from the 5th Special Forces Group noted a successful counterinsurgency effort to break the siege at Camp Duc Co in August. Units there "conducted an extensive program consisting of civic action, [PSYOPs], intelligence, and military operations." The results included resettling 2,300 civilians from the NLF to RVN-controlled areas "and 139 VC surrendering through the Chieu Hoi appeal."[91] The 173rd Airborne assisted, visiting fifty villages and treating more than 5,000 patients. In addition to the friendship and goodwill engendered, the operation produced "concrete results when a befriended Montagnard" led the unit "to a large arms cache of carbines, shotguns and ammunition." The brigade commander noted, "Civic Action is just plain good business for the military."[92]

However, propaganda produced by the North sought to undermine these good deeds. In a series of increasingly sensational claims made during that fall, Radio Hanoi accused the United States of sending a bacteriological warfare unit to South Vietnam. This unit, the 406th Medical General Laboratory, was actually the blood bank/preventative medicine unit coming from Japan to help with disease prevention among US troops as well as to oversee the increasing need for blood supplies.[93]

According to a New China News Agency report presented on Radio Hanoi, Japanese magazines "revealed that the Americans have a bacteriological warfare unit in Kanagawa Prefecture in the vicinity of Tokyo" that sent a detachment to South Vietnam. Over the next few weeks, Northern propaganda morphed what was referred to as the "406th Medical Experimental Unit of the U.S. Army Medical Corps" into a biological warfare research institute. According to the New China News Agency, the unit conducted "research on germs, biology, and chemical poisons." The report added: "The Americans recently spoke brazenly of the possibility of using bacteriological weapons."[94] It insinuated that the United States planned to conduct biological warfare in Vietnam. The unit became the "406th bacteriological and chemical warfare team" in later broadcasts, proving "that U.S. planned to use Vietnam as a testing ground for bacteriological war."[95]

By 10 September, the broadcaster referred to the unit as the "406th Bacteriological and Chemical Warfare Task Force" in a domestic report.[96] Later reports twisted the meaning of AP reports to make official statements sound sinister. Radio Hanoi stated that, at a press conference on the use of "toxic gas in South Vietnam, the U.S. military spokesman in Saigon stated overtly that Westmoreland, U.S. Armed Forces commander in South Vietnam, had the right to order the U.S. troops to use toxic gas to massacre our compatriots." Contrary to the facts, Radio Hanoi claimed that Westmoreland had admitted to using toxic poisons, and the station demanded that US troops "withdraw from South Vietnam the mobile institute belonging to the U.S. bacteriological and chemical warfare task force, which takes the pseudonym of unit 406."[97]

The next installment in this propaganda campaign came at the end of September. Radio Hanoi claimed that "the mobile research institute affiliated with the U.S. bacteriological and chemical warfare corps" came to Saigon in order "to intensify the use of poisonous chemicals as a war weapon in South Vietnam." The announcer said that US troops in South Vietnam were "equipped with cans of toxic gas, which are considered to be part of their ordinary equipment for carrying out raids." The North's propaganda system continued using the "toxic chemical" theme throughout the year in all its media organs. It relied heavily on reinforcement from Americans, Front groups, and friendly Communist news agencies to spread this story.[98]

American Psychological Objectives

As the US 1st Cavalry Division arrived in September, Radio Hanoi began an attack on American psychological operations, labeling them "specialists in committing unrighteous acts." It warned cadres of the importance that US PSYOP personnel placed on "psychological warfare designed to deceive public opinion in the world and the United States and to promote their bellicose and aggressive policy." PSYOPs were a key tool in the American strategy, per the announcer. The broadcaster also read an extensive *Nhan Dan* analysis of the American PSYOP effort in mid-September. This counterpropaganda study was meant to indoctrinate and inoculate cadres. The author had warned readers that the cunning "psychological warfare tricks of the U.S. aggressors" included "propaganda by means of broadcasting systems, dropping tons of leaflets and 'psywar boxes' containing children's clothes and toys to tempt them, ordering spies to stealthily launch false rumors to cause confusion, using aircraft and warships to carry out indiscriminate strafings and shellings, and launching flares to create tension and arouse fear." This litany captured most of the programs then in use. *Nhan Dan* listed the American psychological objectives as:

1. Increase fear of and respect for the United States.
2. Increase desire for peace among Northern civilians.
3. Increase dissension among civilians, the DRV and the PAVN.
4. Isolate the DRV internationally, especially from China.
5. To change white into black and blur the line between the just cause and the unjust one and between the warmongers and aggressors.[99]

Except for the last item, this was actually a more precise listing of objectives than the Americans had published for their own people.

To implement the US program, *Nhan Dan* wrote, American PSYWAR used techniques such as "distorting and sensationalizing the truth to attract attention" and creating "vain illusions to tempt and arouse personal desires." To fight this, cadres needed to "imbue everybody with a vigorous hatred of the U.S. aggressors." Only those in whom "the flame of hatred has not been kindled in his heart" would be affected by the PSYOPs. Emotions trump facts. Cadre had to "pay constant attention to propaganda among the people and permanently explain to them the party and government policy and lines and disclose the enemy plots and methods" to prevent the messages from taking hold. This task was more important than "combat and training."[100]

That same month, *Hoc Tap* (Studies), the North Vietnamese party ideological organ, published an article titled "Against the U.S. Psychological Warfare." According to the author, "thousands of old foxy politicians of the U.S. giant 'PSYWAR' complex in Washington" racked "their brains to find a way to 'undermine the morale' of the Vietnamese people." These people sought "to bribe, deceive, divide and intimidate our people, and to sow fear and confusion among our people in hopes of weakening our people's determination to fight the U.S. for national salvation." Reflecting on internal DRV politics, *Hoc Tap* also attacked Ho Chi Minh's supporters. The article argued that the United States sought to use PSYWAR to increase "the rightist and opportunistic tendency." Weak people become confused and are "easily conquered by the 'peace' theory, 'weapons' theory and other 'psywar' activities of the U.S. imperialists," according to the *Hoc Tap* writer.[101]

DRV/VC Propaganda Collapse in the South

One weakness of the North's propaganda effort was that villagers in the South exposed to VC control and policies grew disappointed. Repeated studies demonstrated this fact. At the same time, intensified GVN/US operations resulted in a belief that the Vietcong would lose and made them appear to "many villagers in the role of ruthless exploiters of the population and as a constant source of danger."[102] The Dinh Tuong Province Viet Cong Military Affairs Committee concurred, saying that "these psychological warfare [Chieu Hoi] activities of the enemy have affected the morale of the people and of our armed forces (especially the guerrillas)."[103] One Front message to propaganda cadres suggested increasing "control over the thoughts of the unit members" through the three-man cells and requiring fighters to report on family contacts.[104] A Rand Corporation study of VC morale determined several potentially useful themes for the allies to stress in the coming months. Themes like "the Viet Cong have been foiled in their objectives, that their timetable has been disrupted, and that their forces and political position have been weakened by their efforts to expand the struggle" now came to the fore.[105]

Similar morale problems arose among PAVN soldiers, who saw no way out of their predicament. Northerners did not think that the Chieu Hoi program applied to them. These soldiers also believed that defection "would preclude them from ever returning to their families and that it could result in reprisals against their families." At least 26,000 had infiltrated through 1965, nearly all of them ethnic North Vietnamese. This offered two psychological avenues for the United States to exploit: encourage surrender or increase intergroup suspicions within the Vietcong.[106]

Liberation forces executed two Americans, Master Sergeant Kenneth Roraback and Captain Humbert Versace, on 26 September 1965. The next day, Radio Hanoi reported that the NLF "has always implemented its policy of leniency toward the aggressors," but as a warning to the United States the NLF was determined to "punish the U.S. aggressors and their lackeys for having massacred our compatriots indiscriminately." The Front used the execution of three Vietcong in Da Nang on 22 September to justify this action.[107]

NLF propaganda had to thread a slender needle to market its message. Marxist ideology did not sell well among the bourgeois-oriented, largely undereducated rural Vietnamese. Agit-prop cadres narrowly focused each target audience on specific grievances to harness the social unrest and ongoing social changes as the entire region struggled to find its place in the technologically modern world. A JUSPAO study of Front propaganda revealed that it depended on "us versus them" and "bandwagon" techniques to arouse the population. Bandwagon themes insinuated that the NLF represented the future, so people should jump on board. Using a "loud, harsh, hortatory style of writing," propagandists dogmatically vilified the US and South Vietnam governments. Addressing these villagers, the NLF blamed "all 'your' troubles [on] the U.S.-Saigon clique." By 1965, a simplistic Marxist line began to emerge in messaging. It did not go beyond the traditional message: the inevitable collapse of capitalism and socialist triumph.[108]

To counter Front activity, the US propaganda program continued to expand. *Newsweek* in October carried a story about Barry Zorthian and the US PSYOP effort. The story noted the use of Air Commando loudspeaker broadcasts over remote jungles. "To their astonishment," the story said, those below heard "the eerie squeal of Buddhist funeral music followed by the thunderously amplified voice of a little Vietnamese girl pleading: 'Daddy! Daddy! Come home with me. Come home with Mamma. Daddy! Daddy!'" The author identified one man who turned himself in a few days later, holding a leaflet. *Newsweek* also referenced the supposedly black program to drop gift kits in the North. American planes flew above the 17th Parallel on 10 September, dropping "thousands of plastic bags each containing a piece of perfumed soap, pencils and three plastic toys along with the message: 'Happy child's festival from the children of South Vietnam to the children of North Vietnam,'" *Newsweek* revealed. Radio Hanoi reportedly ordered the gifts destroyed. This was, "in the view of Saigon's PSYWAR men, a sign of the project's success." However, with the secret out in the open, perhaps the SOG program needed to be brought out of the shadows as some in SOG wanted.[109]

The fall saw a great increase in available US PSYOP support. The I Corps PSYOP committee formed its Joint PSYWAR/Civic Action Center at the South Vietnamese Army 3rd PSYWAR Battalion Headquarters. Composed of ARVN, US Marine, and US Army officers, and receiving input from JUSPAO, the embassy, and the Vietnamese Information Service, this was a PSYWAR fusion cell to coordinate all PSYWAR/CA projects in I Corps. The arrival of more and more American forces greatly expanded propaganda capabilities. Most recently, the 24th PSYOP Detachment had formed at Fort Bragg and shipped to Vietnam with thirty troops in September. The detachment's mobile printing facilities "reduced the response time for large quantity production of leaflets from three weeks to only 36 hours" and improved their quality. The 25th PSYOP Detachment arrived in September, initially based in Nha Trang between Saigon and Da Nang.[110]

The US Air Force activated the 5th Air Commando Squadron at Nha Trang as well, to support the burgeoning PSYOP requirements. This action provided a squadron specifically to conduct PSYOPs with aircraft strategically located throughout Vietnam by the end of the year. The unit's U-10 aircraft mounted new Ling-Tempco-Vaught university loudspeaker systems, and the squadron's five C-47s each had Altec loudspeaker systems. The new speakers improved message understandability. These aircraft greatly augmented the "handful of Vietnamese light planes" that had been the mainstay for leaflet and loudspeaker operations up to that point in the war.[111]

Allied operations in the fall focused on securing base areas for incoming troops. These operations depended on PSYOPs to mitigate the implications of displacing and moving large numbers of civilians. To support this, the US 1st PSYOP Detachment, assigned to the PSYWAR/CA Advisory Team One, published a leaflet product catalog on 18 October 1965 for I Corps personnel, including the Province Civic Action Officers and the Marine Corps III Marine Amphibious Force (MAF). The 1st PSYOP Detachment also provided support to the incoming 1st Cavalry Division as it conducted Operation Happy Valley, twelve miles northwest of the division's An Khe base camp. Elements of the 2nd VC Regiment and the 95th PAVN Regiment had occupied Vinh Thanh Valley, forcing refugees out of the valley to the coastal plain. This search-and-clear operation was followed by pacification of the VC-controlled area. An air assault into the valley preceded saturation patrols and the seizing of VC rice caches. This was a joint/combined operation including US and South Vietnamese Special Forces, the US 101st Airborne and 1st Cavalry Divisions, and the ARVN 22nd Division.[112]

PSYOP elements used loudspeaker-equipped helicopters "to encourage the people to return to their villages, explain curfew and disseminate items of interest." Accompanying Civil Affairs teams, PSYOP teams conducted leaflet distribution and supported "medical treatment, distribution of CARE clothing and repair of buildings and roads in the hamlets." Leaflets used in the operation also urged civilians to remain in "the vicinity of homes, not to move on roads, and to observe the district curfew so that they would not be mistaken for VC and fired upon." The measure of success of these operations, per the report, was the increased flow of intelligence and volunteer informants pointing out local VC members. Additionally, two-thirds of the valley's 10,000 inhabitants had returned by the end of the operation. In the meantime, CIDG forces were recruited, equipped, and trained at Binh Khe to move into the valley to secure it.[113]

Displaced refugees were not happy but had no desire to return to Front control where conditions were worse. This dissatisfaction exposed grievances for PSYOP exploitation. In a Rand Corporation study of VC motivations, one cadre was quoted as saying that "those who left to go to GVN areas met GVN cadres who educated and explained to them what the opportunities of life in both areas were. The villagers could see for themselves that life in GVN areas bring[s] them freedom and comfort and so, they passed the word on to others." Whoever controlled the population had an advantage in the information war.[114]

Although Happy Valley was labeled a "complete success," the author warned that more cooperation from villagers would have been forthcoming "if tighter steps had been taken initially to preclude unnecessary property damage and personnel injury." The division subsequently allotted considerably more time to PSYOPs and CA programs because leadership understood that it "shortened the time for domination" of an area. After the clearance and eventual pacification, the Luc Luong Dac Biet (LLDB—South Vietnamese Special Forces) and CIDG took responsibility. A major factor in the operation's success was the Special Forces' focus on increasing the number of PSYOP-trained soldiers. With a small US military PSYOP contingent, the solution was to train locals. This was a potentially more effective method because of their intimate cultural understanding of the area.[115]

By December, US Special Forces were advising a force of 60,000 CIDG forces, often in conjunction with the LLDB, to secure the Highlands. Originally conceived by the CIA as a local defense force, the Special Forces had expanded CIDG to include a mobile strike force. The arrival of large numbers of US troops enabled the shifting of these camps to more remote areas. With the expansion of Special Forces activities, the number of personnel involved in PSYOPs increased as well. One novel use for these forces was to indoctrinate incoming US forces in Vietnamese traditions and customs.[116]

In Operation Clean House that month, the 1st Cavalry Division's 3rd Brigade received support from two PSYWAR teams of the ARVN 22nd Division. Throughout the region, the 1st Cavalry Division conducted CA projects such as the construction of schools, clinics, and refugee camps, as well as medical treatment programs "designed to win acceptance of US personnel and support for the GVN from Vietnamese citizens."[117] In the preceding three months, the division reported dissemination of 847,200 leaflets. The division commander, Major General Harry O. Kinnard, lauded the "close coordination with JUSPAO and the Vietnamese Information Service," saying that it provided "great assistance in the division's psychological warfare effort."[118]

Outside the 1st Cavalry Division base at An Khe, an American three-man audio-visual (HE) team assigned to the G-5 section sponsored movie showings in the area. The division's PSYOP plan included leaflets designed to induce VC defection and warned them "not to fire on helicopters." The team also explained that the US presence in Vietnam was to support the Republic of Vietnam and to appeal to "Vietnamese citizens to furnish information on VC activities." American forces captured, evacuated, and delivered to Binh Khe district headquarters 2,800 pounds of rice during the operation.[119] The 1st Cavalry Division, in an indication of the priority placed on PSYOPs, included a PSYOP annex in all of its operation orders. General Kinnard and his assistant division commanders endorsed this concept. Despite these successes, the JUSPAO representative complained of "a serious, sometimes stultifying, lack of equipment such as loudspeakers, microphones and tape recorders within the Division."[120]

Conclusion

Increased combat and psychological operations over the summer had blunted the North's plans, but this was hardly sufficient. The US PSYOP doctrine, developed in peacetime, was found lacking in specifics for counterinsurgency warfare. Additionally, equipment shortages and breakdowns plagued the Americans. Perhaps most glaringly, the units lacked cultural knowledge and had minimal PSYOP training. US forces could surmount these obstacles, but time was against them in the face of ongoing combat. Even as they overcame some of these problems in the fall and winter, the US forces expanded their target audiences. Likewise, the North built tighter relationships with antiwar movement leaders. Increasingly, degrading support for the war became a prime Northern psychological objective.

Map 6. South Vietnam, summer 1965 (chapter 7 locations)

8 Refined and Expanded Targeting, Fall 1965

By the autumn of 1965, both sides had augmented their psychological operations programs, and each targeted a wider array of groups. Throughout that fall, the North coordinated its propaganda with the American antiwar movement and a series of scheduled protests, culminating in a November demonstration in front of the Pentagon. This event, widely covered by North Vietnamese media organs, included the arrest of several protestors and a self-immolation below Defense Secretary Robert McNamara's office window. Back in Vietnam, direct combat between US and North Vietnamese troops increased the need for tactical PSYOP support. Among the American psychological programs, US Special Forces took the lead in utilizing PSYOPs to help secure the Central Highlands by mobilizing local tribes as a hedge against infiltration. By the end of the year, the US Army activated the 6th PSYOP Battalion to manage the mushrooming tactical PSYOP programs supporting combat operations. That same month, JUSPAO issued its first PSYOPs plan in an attempt to manage the hydra-like structure.

Emulation and Antiwar Propaganda

Radio Hanoi continued to use emulation as a motivator in its propaganda. FBIS editors noted that, beginning on 7 October, all North Vietnamese propaganda organs began to carry "an extraordinarily high volume of material relating to the [Front] appeal on the occasion of the first anniversary of the execution of Nguyen Van Troi" for his attempted assassination of Secretary of Defense McNamara in 1963.[1] Liberation Radio parroted the call, linking this attempt to the "glorious" Ap Bac victory. This praise illustrated the high status accorded him. Like the veneration of that battle, Troi became a propaganda device for the North. "The death of hero Nguyen Van Troi and his last words are appealing to the South Vietnamese soldiers and people to rise up and kill the enemy more vigorously," Radio Hanoi broadcast.[2] In programs to South Vietnam, Hanoi announced "a minute of silent respect for the memory of Nguyen Van Troi beginning at 1000 hours on 15 October and a general strike." FBIS editors noted this call, "repeated in approximately every second monitored Liberation Radio broadcast," contained demands to sabotage communications lines, roads, and depots. All target audiences received tailored appeals. Buddhists, Chinese, workers, and youth groups were each called upon to emulate Troi and "rise up and struggle against 'the enemy' and to respond to the [Front] appeal for a general strike on 15 October."[3]

The Front Peace Committee sent an open letter to the American people in connection with this campaign. The emotional appeal linked the fates of Vietnam and the peace movements in the United States. The committee warned that President Johnson "and his collaborators . . . relied on your name and on that of the world people to provoke bloodshed in our country." Tapping in to notions of "American exceptionalism," the NLF noted that "the American people, the descendants of Jefferson," would not allow US authorities to use "your money and the blood of your children to carry on their aggressive war in South Vietnam." According to Liberation Radio, the war tarnished America's image "throughout the world." This "City on a Hill" premise appealed to deep currents in American political discourse and therefore worked well as a manipulation tool. The broadcast praised friends in America and people such as Alice Herz, who immolated herself to protest the war.[4]

Radio Hanoi covered the American protest movement throughout the fall. This was part of an extensive campaign that linked internal and external propaganda with worldwide front groups and the mobilization of youths across America, timed for the anniversary of Troi's execution. Norman R. Morrison set himself on fire in front of McNamara's Pentagon office that November as the protests reached a crescendo. The thirty-one-year-old Baltimore Quaker and antiwar activist emulated Herz's actions earlier in the year. Radio Hanoi hailed him as "a freedom and peace-loving American," proclaiming that his "torch will blaze forever." The announcer said that Troi's wife wrote to Morrison's widow to profess her admiration.[5] The station proclaimed that by "all forms of struggle, including noble self-immolation . . . the American youth and peace-loving people have clearly shown that their hearts are with the Vietnamese people, not with Johnson." Troi became such an international celebrity as a result of this campaign that Jane Fonda even named her son after him.[6] Radio Hanoi also reported on 10,000 protest participants in Berkeley yelling slogans such as "To Hell with the Imperialist War!" and "U.S. Pull Out of Vietnam!" Additionally, student draft card burnings in New York City, New Haven, Boston, and Portland received wide play on Hanoi's domestic and Southern services.[7]

The DRV and Front groups praised the assistance of Americans from the Vietnam Day Committee, the National Coordinating Committee to End the War in Vietnam, and "American intellectuals, youths, students, and other American peace and justice-loving people who are taking part in the movement of protest against the aggressive war waged by the Johnson administration in Vietnam."[8] A month of protests peaked with reports on the "March on Washington for peace in Vietnam on 27 November" by the "progressive people of America." Among those leading the protests, Radio Hanoi singled out writer Arthur Miller, who "pointed out that the aim of the march is to mobilize the conscience of the American people to call upon their government to stop bombing North Vietnam and bring to an end the U.S. aggression in South Vietnam." Statements at the protests mimicked themes in North Vietnamese propaganda and helped reinforce and legitimize the messages through rebroadcasting.[9]

The CANLF and Walter Teague were prominent during the fall protests. Teague gave an 11 November speech at Columbia University titled "Why the American People Should Support the National Liberation Front of South Vietnam (Falsely Labeled the Viet Cong)." This was part of a lecture series sponsored by the Bertrand Russell Humanist Club.[10] Teague also advised the National Coordinating Committee on policy for protests scheduled later in the month. Seeking unity among the various factions constituting the committee, CANLF advocated a policy of actively supporting NLF victory, rather than simply pulling US support for the Republic of Vietnam. However, Teague's stance encountered pushback from some groups who did not support this extreme policy. In response, the CANLF defiantly wrote that at upcoming protests "we shall fly the NLF banner and make known their position through distribution of literature." Supporters handed this out, becoming a lightning rod for counterprotestors.[11]

Both China and the Soviet Union adopted the antiwar movement and attacked the claim that the North controlled the National Liberation Front. A Moscow broadcast in Mandarin to China called on all socialists to spur on the antiwar struggle in America. "We can see with our own eyes how imperialist forces in the capitalist world have been employing all possible means to halt the advance of history."[12] Being on the "right side of history" was a consistent propaganda theme designed to derail opposition to the socialist vision of the future. Any opposition was deemed to be on the wrong side of history, as if history moved in a predetermined course. The Soviet Union touted a broad-based front composed of the "Democratic, Social Radical, and People's Revolutionary parties," representing businessmen, landowners, intellectuals, and the working class.[13]

An example of such support was a letter that President Ho Chi Minh received from Nobel Prize–laureate Linus Pauling. In response to the letter, Ho scorned President Johnson's claim that America did "not intend to expand the war and is ready to negotiate." Ho attacked the "horrible massacres" and "devastation" the American military caused "in order to compel the Vietnamese people to lay down their arms and give up their legitimate aspirations." Broadcasting to the South, Radio Hanoi focused on the protests to show American inconsistency and internal divisions.[14]

After 1965, Ho Chi Minh became a propaganda tool for North Vietnam much like Ap Bac and Nguyen Van Troi. The politburo continued to defer to him on the world stage, but his primary role was to put a gentle face on the revolution. The kindly, old "Uncle Ho" visited "schools, factories, and collective farms to promote the cause of socialism and national reunification," all while standing strong in the face of aggressive imperialists. Internationally, Ho became a symbol of a weak nation set upon by an international bully. He represented a powerful tool in the Northern propaganda bag, but he no longer had much real authority. He had become trapped in a propaganda system of his own creation.[15]

As the war expanded, Radio Hanoi targeted propaganda at audiences beyond US and ARVN troops. In Korean-language broadcasts to South Vietnam, it appealed "to the officers and men of the South Korean Army" to "oppose the schemes and intrigues of the Americans to make Asians fight Asians!" It implored them to "take no part in the U.S. war of aggression." As in broadcasts to American troops, Radio Hanoi suggested that Koreans refuse to follow orders. Those who defected could be "sent to the DPRK" if they desired. Assessing the impact of these broadcasts is difficult. In the case of American troops, only two deserted to the enemy throughout the war. As seen below, however, overall American desertion numbers did rise worldwide; after 1968, so did indiscipline.[16]

Tipping the Balance

While the North made great strides in targeting international audiences, the war for control of the population in South Vietnam continued. US Special Forces were at the forefront in this regard, especially in the strategically important Central Highlands. In late October 1965, US Special Forces conducted a comprehensive PSYOP campaign in conjunction with conventional operations in the Plei Me area in the Highlands. The plan involved coordinated leaflet and loudspeaker missions, along with rapid intelligence exploitation, in order to harass the enemy and encourage surrender. The major units involved were the II Corps G-5, the 5th Special Forces Group G-5 and Detachment C-2, the US 25th PSYOP Detachment, and the USAF 1st Air Commando Squadron.[17]

A VC attack on Camp Plei Me on 19 October instigated this campaign. Three days later, intelligence sources reported Chinese advisers on the battlefield, as well as PAVN soldiers found chained to their machine gun. US

Special Forces quickly produced a leaflet showing an overbearing Chinese adviser yelling at a hapless, chained PAVN soldier. Further analysis of prisoner reports helped to refine and expand the message. PSYOP elements used this leaflet to thrust a divisive wedge between the Chinese advisers and the Vietnamese soldiers, something based on historically relevant fracture lines.[18]

On 24 October, U-10 aircraft from the 1st Air Commando Squadron arrived on station and dropped 70,000 leaflets over Chugo Mountain south of Pleiku and in the area north of Camp Plei Me. The aircraft broadcast loudspeaker messages until 2200 hours that night and resumed the following morning. At 1100 hours, a PAVN 32nd Regiment soldier rallied. Under interrogation, the soldier claimed that during the night his unit had listened to Radio Hanoi claims of a great victory in Plei Me. He knew this to be a lie. The rallier further stated that more soldiers would defect if they had clear instructions on how to do so and knew they would get medical care and good treatment. Based upon this, the PSYOP team developed a loudspeaker message:

> Vietnamese Brothers, You are now very hungry, tired, and sick, and you will have many more killed. It is not necessary. The government will welcome you, brothers, with food, medicines and a peaceful life. Leave your units, go north to highway 19 and follow it to the government post at [Thanh Binh]. Now is your chance to leave this miserable war.[19]

In conjunction with this message, the unit dropped 80,000 leaflets throughout the evening, using 1st Cavalry's recent victory in Bong Son as the theme. Additional messages were broadcast in local languages to urge listeners to provide intelligence on North Vietnamese troops. Appeals for information are not simply requests; they have a deeper psychological purpose. It is likely that PAVN troops would encounter such leaflets as well: knowing that every civilian they met represented a potential informer served to exacerbate already existing frictions between Northerners and Southerners. Whom could they trust? The cycle of drops and broadcasts continued until US forces lost contact when the PAVN 32nd Regiment fled to Cambodia on 27 October. Although this operation was generally considered to be a success, operatives also found that the loudspeaker broadcast coverage area was less than anticipated, requiring better planning for future operations.[20]

Throughout the operation, Americans seized "2,200 pounds of rice, 300 pounds of clothing, 1,000 pounds of medical supplies, and 400 surgical instruments." US forces destroyed most of the rice due to transportation shortages, with the remainder given to GVN officials for distribution to needy civilians. Approximately 2,700 refugees were transported from insecure areas to Le Thanh district headquarters. Medical personnel treated 777 Vietnamese patients during the operation.[21]

A USIS assessment of the operation noted confusion among staff regarding tactical PSYOP planning responsibilities. Regardless, close coordination between the US 25th PSYOP Detachment, the ARVN 2nd PSYWAR Battalion, and the JUSPAO representatives resulted in a skillfully implemented program. While the after-action report contained no specific number of ralliers, the USIS reported that the surrender of 100 PAVN soldiers holding safe-conduct passes (and many more passes found on bodies) indicated success. As one adviser noted, "The potentiality of PSYWAR is such that no American tactical unit can afford to ignore it."[22]

In addition, allied Special Forces operated beyond the Central Highlands. On 3 November 1965, guerrillas attacked the hamlet of Nhon Hoa, between Ap Bac and Saigon. Thanks to intelligence provided by locals, government forces laid an ambush in advance. The Vietcong withdrew after a short firefight, leaving two homes destroyed and several civilians injured or killed. In response, the province chief directed a tactical operation to the village as a planned psychological action. Three companies of infantry provided area security as a show of force on 5 November. Simultaneously, a team of medics, PSYWAR and Civil Affairs personnel moved into the village to provide medical treatment, pay death gratuities to families of those killed, and distribute food. At about 1100 hours the VIS representative assembled the hamlet's leaders and government welfare representatives and held an anti-VC rally, making speeches "discrediting the VC." Captain Howard Simister, the 5th Special Forces Group PSYOP/CA officer for the area, reported that the "rally was considered a great success."[23]

Product distribution by Special Forces PSYOP personnel that quarter alone included "over two million leaflets and more than 150,000 publications, such as bulletins, magazines, and local newspapers." Special Forces found that CIDG unit PSYOP/CA activities in hamlets and during military operations were so successful that they "initiated a program to organize and train CA/PSYOPs squads" at each Special Forces "A" detachment camp. In December, fifty-two Montagnards completed a ten-week course, and forty-two more began another at the Pleiku Montagnard Training Center. The course consisted of "practical training in motivation indoctrination, field expedient printing, loudspeaker broadcasts, agriculture, animal husbandry, carpentry, masonry, and blacksmithing." The influx of 21,444 refugees to the vicinity of US Special Forces camps was one measure of the effectiveness of these efforts.[24]

A US Special Forces PSYOP campaign to clear the enemy out of the Plei Me area near Camp Duc Co was considered to be a success. Goals of the campaign were "to enhance camp security" and to "gain and maintain contact with the population." The first phase consisted of indoctrinating CIDG troops "not to loot or steal from the people." Treating the population fairly helped to counter VC propaganda themes. Medical treatment provided to civilians, wounded VC prisoners, and Chieu Hoi ralliers also demonstrated "the good faith and intentions of the CIDG," per the report.[25]

As was typical, loudspeaker broadcasts and leaflets called on villagers to assemble, after which the senior LLDB officer explained the program. Teams contacted families of individual VC members to send relatives to "encourage them to surrender." Available Hoi Chanh (as those who surrendered were known) were "present at the formation where the LLDB or CIDG congratulate him and publically pay for his weapon if he had one and turned it in." Soldiers offered to resettle a family if it feared VC reprisals. The population in this area was split between Vietnamese and Jarai Montagnard tribesmen. The operation proved successful with both ethnicities, encompassing "over 6,000 people in different locations," with excellent potential for expansion. In just four months, more than 120 personnel had rallied, "many with weapons and grenades." Making the program work successfully required dedicated CIDG soldiers who were not distracted by any additional duties. An added requirement was close coordination among intelligence, PSYWAR, and Civic Action personnel. The Duc Co team warned, however, that for such a program to be successful "promises made in the leaflet, loudspeaker or person to person appeal must be honored."[26]

By the fall of 1965, the United States was actively involved in combat, attempting to drive the PAVN out of the Highlands to reduce the threat to coastal population centers. Among the first battles directly between the PAVN and US Army was the Battle of Ia Drang. The campaign resulted in significant casualties for both sides and ended with the PAVN troops escaping into Cambodia. Radio Hanoi termed the fighting in the Ia Drang Valley that November a resounding victory for the liberation forces. It maintained that the battle "demonstrated the irremediable collapse of the combat spirit of the U.S. troops fighting in South Vietnam." As proof, it claimed that General Westmoreland had sent 2,000 "puppet paratroops to save the U.S. troops" in order to limit American casualties. In reality, ARVN airborne troops simultaneously conducted a separate, related operation. As a consequence of the "victory," Radio Hanoi began to refer to the 1st Cavalry Division as the "Running Cavalry."[27]

The US 24th PSYOP Detachment, "blessed with resourceful and energetic personnel," attached teams with ARVN PSYWAR counterparts to each US battalion during the operation. The detachment's newly established production center ran three press shifts daily, working with ARVN peers to provide leaflets for the operations. Despite Northern claims to the contrary, the more than forty surrendered PAVN troops indicated effectiveness. Most of the soldiers confirmed hearing loudspeaker missions and carried surrender leaflets, using them as "safe conduct passes." The resulting ralliers were in turn used to record tapes and to write statements for further appeals to comrades who continued to fight. Soldiers also found leaflets on many PAVN bodies. Although the campaign resulted in many PAVN soldiers killed, the unit escaped to Cambodia. The inconclusive nature of the battle led to MACV's shift to reliance on body counts to measure progress. MACV collected a vast array of data and devised other metrics, but only body counts could indicate when the war of attrition General Westmoreland was fighting had reached the crossover point to victory.[28]

PSYWAR Counterattack: Revised Training and Doctrine

In response to the increasing military pressure throughout the South, Radio Hanoi began a counterpropaganda campaign. The campaign belittled the battlefield successes of the ARVN/US forces, saying that they represented "a few drops of tonic" to a dying patient. Attacking the "psychological warfare organs of both the Americans and puppet authorities," it claimed that any gains were a "product of their imagination." According to Radio Hanoi, three-fourths of South Vietnam was liberated, comprising a "population of over 10 million people." Despite such claims, the overall tone of the story belied an attempt to minimize setbacks for the North.[29]

The station also attacked "the depraved, obscene, U.S. cowboy culture" that poisoned the nation's youths and brought the "humiliating scourge" of prostitution. Charges such as these often contained enough observable truth for the average South Vietnamese civilian to believe. Just one American soldier causing a commotion was sufficient to lend credibility to the charge. The broadcast also attacked those who worked for or assisted the Americans by "bidding for construction contracts for roads, bridges, barracks, warehouses, bases, posts, and airbases and for the exploitation of forests so that the Americans might have the means to exterminate our compatriots." Those who refused to heed the warning would suffer once the Americans were defeated, per the Liberation Broadcasting Service.[30]

The American propaganda war also faced self-imposed setbacks. The PSYOP adviser in Quang Ngai and Quang Tin Provinces documented the frequent VC use of themes related to collateral damage caused by air and naval gun strikes. One adviser had "the clear feeling that little consideration if any . . . is given to the psychological consequences of air strikes and artillery and naval bombardments in many cases." This pointed to the problem of relying on attrition in a counterinsurgency fought within a populous district. Tight rules of engagement were necessary in order to maintain popular support. However, the widespread use of heavy weapons opened up the United States to credible propaganda charges, even in cases where the allegations were not, strictly speaking, true. Ironically, American firepower limitations through restrictive rules of engagement often gave guerrillas the chance to fight another day. The dilemma over when and when not to rely on firepower was never fully resolved.[31] However, to manage coordination between PSYOPs and combat formations, MACV issued Directive 525-3, emphasizing "discrimination in the application of firepower and the use of all available PSYOP resources during each operation."[32]

During the fall, the US Army authorized a study on improvements in the training of American PSYOPs personnel. A report issued in December pertained specifically to officers. The Special Warfare Center and School at Fort Bragg had responsibility for producing qualified special operations officers. It was scheduled to begin a six-week course (Military Assistance Training Adviser Psychological Operations Orientation) in January 1966. Some officers were trained at Fort Bragg, while others trained at civilian colleges. However, due to the specific multidisciplinary needs of PSYOP training, college training was not seen as an optimal solution. Regardless of the training received, once an officer completed an assignment in a PSYOP unit, he returned to his original career track in the army. Unless the officer volunteered for another assignment in PSYOPs, the army then lost his experience because there was no official PSYOP career path. The study recommended the creation of a specific PSYOP career path along with progressive training, concluding with a graduate degree for senior officers.[33]

The last quarter of 1965 saw an increase in the number of personnel assigned to PSYOP slots in Vietnam. The 5th Special Forces Group's table of organization contained a PSYOP section comprising two captains, three lieutenants, and a senior NCO to organize, plan, and oversee PSYOP programs throughout the Republic of Vietnam. Additionally, many Special Forces camps now had classroom-trained PSYOP officers. However, none of the forty-two recently assigned NCOs had PSYOP training. A four-day course in Vietnam conducted in January 1966 stressed the "practical 'nuts and bolts' of Civic Action and Psychological Operations at the district and hamlet level" for these soldiers. In order to overcome the shortage of PSYOP support at the camp level, the group expanded the CIDG agit-prop training program, creating Civic Action Squads to work at the grassroots level.[34]

The 5th Special Forces Group closed the year by praising the "integration of CA/Psy Ops into all aspects" of its operations. The operations exemplified the strong linkage between Special Forces, Civil Affairs, and PSYOPs in counterinsurgency operations. Among the important markers were 773 Civic Action projects begun, 79 percent of which were completed. This total represented nearly all the US Army Civic Action projects conducted in Vietnam at that time. Additionally, US Special Forces trained village health workers, who "treated 76,632 civilians at local dispensaries and on medical patrols," thereby gaining civilian support. The integration of local, LLDB, and CIDG officers as the face of these projects was the most important aspect. They planned and implemented the program, serving to buttress the RVN's legitimacy via propaganda of the deed.[35]

The USAB&VAPAC on Okinawa created the 7th Psychological Operations Group on 20 October and assumed responsibility for all American PSYOPs throughout East Asia. The group consisted of the 14th PSYOP Battalion and several detachments, including the Vietnam Detachment. With the additional PSYOP personnel in Vietnam, the entire system became more nimble. Additionally, a new 55-kilowatt radio station in Ban Me Thuot was nearing completion. That power would make it "the most important Vietnamese radio station outside of Saigon."[36]

The South Vietnamese program was improving as well. In Quang Tri Province, the VIS joined with cadres from the Rural Construction, Social Welfare, and Provincial Health services, police, the office of education, Chieu Hoi, and province officials to conduct a concerted campaign to provide food, blankets, and information to refugees. One aspect of this program was labeled Operation Harelip. The program brought together American and Vietnamese doctors to conduct a series of cleft lip restoration operations, publicized by VIS. With JUSPAO's assistance, a poster on the successful program was developed.[37]

One event indicating the increased capability to exploit VC atrocities occurred on 12 December 1965. The VC forces massacred twenty-five unarmed canal workers at Tan Huong, near My Tho. The VIS and JUSPAO channeled the resulting popular anger and indignation, "resulting in a large and impressive" anti-VC demonstration. According to JUSPAO, the VIS district chief and the American subsector adviser photographed the massacre scene. Robert Dickerman, the JUSPAO representative, "reported that their photos showed not only the tragic dead lined up in the pagoda yard, but the shocked grief of the survivors." One photo showed a nine-year-old boy orphaned by the "massacre being comforted by a GVN official," as well as "a young woman and her baby who had also lost a loved one."[38]

Rapid work by the VIS and JUSPAO personnel included a special helicopter to transport the prints to Saigon. This quick response enabled the photos to be released to the international press the next morning, "resulting in worldwide coverage of the incident." A follow-up rally organized by VIS cadres included an anti-VC banner. Government officials provided funerals, "obtained the services of bonzes to pray for the dead, and obtained gifts and emergency funds for the survivors." They also placed a memorial plaque to remind inhabitants that the NLF's objective—stopping the canal construction and progress—had been thwarted. In another incident, 7 persons, including 5 children, were killed and 41 wounded when NLF personnel tossed a grenade into a crowd of 600 people at a VIS film presentation on 17 December 1965 in Vinh Long Province. The VIS film operator, the target of the attack, was killed as well. Villagers rallied the next day to protest the attack.[39] Elsewhere, the Phu Yen Province adviser was "impressed with VIS activity," noting that "two bulletin boards were observed in Tuy Hoa with up to date news written in chalk upon them." The PSYOP staffs in Long An and An Xuyen Provinces organized a "Letters-to-the-North" campaign. Schoolchildren wrote more than 30,000 letters to be air-dropped in the North.[40]

Despite these successes, the VIS never achieved nationwide operational consistency. Attracting and maintaining quality personnel in the face of violence and South Vietnam's bureaucratic inefficiency and counterproductive policies remained a problem. As one adviser remarked, VIS representatives in Binh Thuan continued "to have good ideas very poorly executed by incompetent personnel." The cultural team there appeared to the advisers as "just another body of cadre slouching arrogantly and ineffectually about the hamlets." The cultural team in Ninh Thuan Province was "endowed with skill and musical talent," as witnessed at a "farewell party for the departing Sector Advisor." However, in the hamlets, the team lapsed "into its dull repetition of poker-faced singing."[41] In the wake of the Plei Me fighting, some PSYOP advisers sounded an exasperated tone regarding VIS activities. Passive VIS cadres had to be needled into "using loudspeakers to announce the news," one wrote. In Binh Dinh the adviser referred to VIS "inactivities."[42] The service remained an unreliable force.

DRV Targeting Deserters and Prisoners

Notwithstanding Northern advances and successes in targeting international audiences, messages targeting civilians in the South began to depend more on fear than on a positive, encouraging agenda. In IV Corps, USIS advisers found Front propaganda to be "shrilly anti-American and heavily doctrinaire." One rumor campaign in Ba Xuyen Province passed "word that American and ARVN medicine was poisonous or had fishhook embedded in it." This campaign reduced participation in MEDCAPs until "corpsmen started breaking up pills to show people that our medicine was completely unadulterated." With the lie exposed, attendance resumed. Civilians in that area were also instructed by the Vietcong to destroy their government identity cards or face punishment. Another campaign sought to portray district officials as "totally corrupt," which an adviser reported in this case was false. He remarked, "In both Ba Xuyen and Chuong Thien [Provinces] they appear to direct most of their abuse towards honest and effective officials and leave the rest alone." Good administrators were a great threat to the NLF, so they had to be marginalized or destroyed.[43]

By this time, US troops began to appear in Northern propaganda. On 28 September 1965, the Vietcong captured Marine Private Robert Garwood. The precise events leading to this—whether he defected or only subsequently became a supporter of the NLF—are still in question. A US Navy study later found that Garwood had "drifted steadily from collusion to defection, beginning with the making of propaganda tapes in exchange for preferred treatment and eventually wearing a Viet Cong uniform, interrogating and guarding his own countrymen, and, according to some reports, fighting alongside the VC." Garwood was eventually declared a deserter.[44]

On 5 December, the first reported leaflet containing Garwood's signature appeared, targeting American troops. The leaflet said, "We hope that you, too, find yourself, as a human being, unable to tolerate this nightmare war, and we hope that you will oppose it. A growing number of GIs have already refused to fight in Vietnam and have been court-martialed." The leaflet urged soldiers to talk to one another about the war and referred to their responsibilities in light of the Nuremberg trials. This was an attempt to urge soldiers to spread the propaganda further and to refuse orders. Although Garwood later admitted to signing the leaflet, the writing and content indicate that it was not his composition. As a Marine, he would never refer to himself as a "soldier," and the text included several misspellings and incorrect dates. Additionally, the leaflet listed him as a chaplain's aide, rather than his correct position as a motor pool driver. The following year, Radio Hanoi used the leaflet and the text of the letter in its broadcast propaganda.[45]

Claude McClure and George Smith, who had been captured in the 1963 Hiep Hoa attack described above, were released in December 1965. After the execution of fellow captive Roraback, the prisoners agreed to sign statements against the United States. *Nhan Dan* put it this way: "The statement of these two U.S. servicemen also expressed the aspirations of the American people who are resolutely opposing the aggressive war of the U.S. imperialists and respecting the independence and sovereignty of the South Vietnamese people."[46] Radio Hanoi claimed that Smith said, "I believe that the American people will understand the real situation in South Vietnam and that the U.S. Government can by no means gloss over the realities I have met." The soldiers were later charged with treason, but not convicted.[47]

American operations in December extended the reach of US forces in III Corps. Between 21 November and 17 December 1965 the 173rd Airborne Brigade launched Operation New Life. The effort had three objectives: to destroy local VC units, to ensure that the farmers could harvest and sell their rice, and to help restore government to La Nga Valley. Since the mid-1950s,

Long Khanh and Binh Tuy Provinces had been major rice-producing areas. Until the early 1960s, the hamlets in the valley had supported the central government, but in 1963 the Vietcong arrived in force, seizing about half of the rice crop that year and the entire harvest the next year. In addition to road-clearing, "air assaults, night patrols, and other actions to secure the sector and provide protection for the harvest," US Army and Philippines medical, dental, PSYOPs, and Civil Affairs teams worked to break the villagers' ties to the Vietcong. "In the end, the farmers harvested their crops without interference" and reinstalled the district-level government.[48]

Operations such as this were fine, but gaining control of contested villages required more effort. The CIA advisers saw People's Action Teams as a potential counter to the Vietcong's APTs in the hamlet-level battle. By 1965, Brigadier General William DePuy, Westmoreland's operations officer, recognized the government's failure to "capture the psychological initiative in the competition for peasant loyalty" in the hamlets. Though he was initially hesitant to accept CIA leadership in pacification, observing the "performance of PATs in II Corps and Quang Ngai Province" overcame this. That fall, DePuy proposed that General Westmoreland "recognize the PAT as the psychological and political adjunct to the U.S. military operations now enjoying considerable success."[49] One technique to attack the Vietcong's infrastructure in the hamlets was to disseminate "false information in order to discredit them in [VC] eyes." Rumors that an individual spied for the GVN put lives at risk from Front justice. An alternate method was to shower the cadre and their families "with gifts to encourage neighbors to suspect that they were working for the Allies."[50]

Southern peasants were becoming war-weary by 1965, even as attacks thwarted the PAVN in areas like Ia Drang. Ground combat intensified near populated areas, reinforcing a "climate in which popular sentiment seems to have been shifting away from the insurgents." One study found that almost all of the 200 Vietcong interviewed agreed with this statement, "which they attributed not only to allied sweep operations and B-52 bombings but also to an easing of popular hostility toward the GVN."[51] Even VC land reform programs proved less effective over time. Besides being "unwilling to accept land which had belonged to fellow villagers," the high property taxes on top of NLF taxes and perceptions of favoritism to "Viet Cong cadres in the distribution of land" lessened the impact.[52]

JUSPAO Planning

The JUSPAO National Psychological Operations Plan issued in December noted that PSYOPs "cannot live up to their full potential unless committed to the offensive." Technical capabilities were desirable but not sufficient or essential to success. By contrast, endlessly repeated "clear and simple approaches to elemental emotions and reasoning" by key communicators could "spread like wildfire." "Key communicators" referred to respected local officials, such as teachers, who could be used to spread information. Overly sophisticated appeals could "doom the effort."[53] The plan stated that by the end of 1965, despite a Maoist doctrine based on voluntary popular support for the revolution, the allies had boxed in the Vietcong. Due to increased US/ARVN military pressure, the Vietcong could no longer avoid taking actions that alienated the populace. Consequently, the message to the Vietcong that "all hopes for victory are doomed" had credibility.[54] Likewise, the plan sought to convince South Vietnamese citizens that the nation would "win its struggle against aggression and subversion." However, this theme depended on credible statements that the Americans would continue to support South Vietnam.[55]

Priority target audiences under the JUSPAO plan included "opinion makers and key communicators: village chiefs, members of village councils, [and] officers of village social organizations," all persons with widespread influence. Additional target audiences included "civil servants, mass media personnel, educators," students, military members, and businessmen. Parents of those serving with the Vietcong and VC sympathizers were the key target audiences for the Chieu Hoi program. The preferred technique was face-to-face communication.[56]

Although the military had a difficult time assessing the value of the overall PSYOPs effort, the cost-benefit ratio of the Chieu Hoi program was becoming clear. Between 1963 and 1965, 27,789 had rallied, at an estimated 1967 per capita operational cost of $250. In 1965 alone, 11,123 Vietcong defected, an estimated 5 per 1,000 soldier desertion rate for 1965. However, this is only an estimate, as the NLF's total strength was not precisely known. Another way of looking at it was the then-current ratio of combat deaths: the neutralization of the 17,671 guerrillas that rallied in 1967 would likely have cost more than 3,000 friendly lives. A later study found that "the military effectiveness of the GVN" was among the key reasons that fighters rallied. The operational tempo put pressure on the target and provided opportunities to escape NLF control procedures. Any program that could remove this number of enemies from the battlefield was worth pursuing. American officials decided to give Chieu Hoi a higher priority and made a greater commitment of resources. By 1967, the government had created the Ministry of Chieu Hoi and employed 1,615 personnel and 132 advisers to manage the program.[57]

In contrast, per MACV's *Command History*, the ARVN desertion rate for 1967 was estimated at 14 per 1,000 soldiers and 28 per 1,000 for the Popular

Forces. However, many of the ARVN deserters were soldiers who would be classified as AWOL under American standards. Nevertheless, raising morale and loyalty among the troops fighting on behalf of the government was another important task. MACV worried that US reporters gave Americans the "false impression that the total RVNAF was incompetent, corrupt, and ineffective."[58]

By comparison, a congressional study found that 40,000 US military personnel throughout the world deserted in 1967, most of them for personal reasons. The report noted that a few may have had political reasons, but contained no solid numbers. Interestingly, courts-martial and US soldier unruliness in Vietnam dropped through 1967. In light of this, there seems to have been minimal recognition that American soldiers might be susceptible to enemy propaganda, and consequently little troop indoctrination was conducted. One example of indoctrination was the film *The Unique War*, produced by the US Army Pictorial Service and narrated by the actor Glenn Ford. This film, shown to incoming American soldiers, clearly articulated the PSYOP and Civic Action responsibilities of all soldiers. The army originally produced this as an episode of the *Big Picture* television show (discussed below).[59]

Radio and TV Expansion

Radio was another important medium through which to spread news and PSYOP messages into remote hamlets. By 1966, the earlier North Vietnamese broadcast advantages had been overtaken thanks to American aid. Radio Vietnam (using the VTVN call sign) consisted of three channels based on target audiences: A for civilians; B for ARVN; C for minority groups. Additionally, the US-operated Armed Forces Vietnam Network began broadcasting in August 1962 to US troops. AFVN covered 91 percent of the Southern population by 1970. Often, these were low-power stations of 1–25 kilowatts, except for the 50-kilowatt AM station, 100-kilowatt FM station, and later the TV station. Broadcasts typically interspersed music, news, and command information for American troops. Examples included health and safety alerts, such as: "Take malaria pills, practice fire safety or beware of venereal disease." The Korean Army also operated radio stations.[60]

Discussions between Vietnamese and American personnel had been ongoing regarding development of a South Vietnamese television system as well. In mid-October 1964, the US Mission Council received a briefing on possible proposals, putting off the final decision. Premier Nguyen Cao Ky recalled later that at the fall 1965 Honolulu planning meeting he had asked President Johnson for television capability to help "build a national consensus and maintain political stability." According to Ky, Johnson turned to his assistant, Jack Valenti, and said: “Don’t we have a flying TV studio that we could send?”[61]

President Johnson was referring to Project Jenny, which deployed US Navy NC-121J Super Constellation aircraft under the code name Blue Eagle. The United States developed the project to introduce television broadcasting in Vietnam. Project Jenny was a stopgap until ground stations came on line. Three aircraft arrived in time to provide airborne radio broadcast support for the 1966 Tet campaign. Each aircraft contained all the necessary broadcast equipment and a small studio for live production. Two Tan Son Nhut–based aircraft broadcast AFVN and the South Vietnamese THVN television station four hours per night. AFVN prescreened its programming content to remove material that might be culturally objectionable for Vietnamese. However, the American and Vietnamese systems had competing goals. As the former USIA official William Hoffer wrote, while “AFVN television was geared to help the GIs forget the war, THVN aimed to help Vietnamese remember it.” A third aircraft based in Da Nang flew SOG PSYWAR missions over the Gulf of Tonkin beginning on 1 June 1967 as part of Project Humidor. This airplane broadcast PSYOP radio to the North. This included CIA Project Treat broadcasts such as Radio Red Star and Mimic. Red Star purported to represent the views of dissident Vietcong members, along the line of the SSPL. Mimic was a covert program to copy legitimate Liberation Radio broadcasts but changed the message slightly to undermine Front propaganda.[62]

Television broadcasting began on 7 February 1966, eventually encompassing eight ground stations. As the ground stations began to come on line in October 1967, the Tan Son Nhut aircraft shifted to broadcasting over the Mekong Delta region, expanding the coverage. Entertainment represented 81 percent of the programming. To support the system, the USIA distributed community TV sets to villages; however, viewership was concentrated in urban areas. Television proved very popular for those within broadcasting range and to some degree acted as an effective and broader counter to the Vietcong’s face-to-face communications.[63]

Not a formal part of the PSYOP effort, the American radio and television stations nonetheless became part of the broader information environment in Vietnam. Although only a small percentage of civilians eavesdropped on AFVN radio, a substantially higher percentage watched American TV shows, which were popular among urban Vietnamese. Voice of America also broadcast to Vietnam. Vietnamese listeners rated the BBC among the most respected radio stations.[64]

JUSPAO conducted numerous public opinion surveys to determine Vietnamese viewing and listening habits. The results indicated that the most popular program types on South Vietnamese television were so-called Cai Luong dramas telecast on Friday and Saturday evenings. These melded Vietnamese and French drama styles with Southern folk music. Surveys indicated that they "attracted the largest Vietnamese audiences even when up against the American television program *Wild, Wild West* on the AFVN channel." Occasionally, political messages were conveyed through the Cai Luong shows.[65]

In the information battle, both sides sought to prevent the spread of the other's messages, often with dubious results. The South Vietnamese government jammed Liberation Radio to limit the signal in Saigon. Likewise, listenership of South Vietnamese broadcasts in Front-controlled areas was high enough to require a steady stream of directives banning such activity. To counter black radio broadcasts, Liberation Radio notified its listeners to beware: "The enemy recently set up a false self-styled Liberation Radio using our frequency and sign on music . . . and it has been broadcasting false news and distorting arguments about the struggle of the South Vietnamese people."[66]

The Oliver Quayle Company conducted another public opinion survey in late 1965 under a JUSPAO contract. Over an eight-week period, Vietnamese trained by Dr. Robert Sullivan of the USIA completed 974 interviews with citizens. The survey authors first set the attitudinal baseline by analyzing studies done since the fall of 1964. Among the most striking findings of the review was a pervasive apathy in 1964, "caused in part by a lack of education and communication and in part by the sense of hopelessness that inevitably grows out of a quarter of a century of armed conflict." Additionally, the absence of security led many citizens to sit on the fence while awaiting a clearer outcome. This hesitancy was exacerbated by a perception that the ARVN was guilty of petty crimes. A final aspect revealed by the examination was a sense of government failure at all levels.[67]

Front Morale Problems Grow

Although it was difficult to achieve a good survey sample during a war, the authors wrote that "the final sample represented about 75 percent of the population of South Viet Nam, excluding the Montagnards." They believed that the final report was an accurate representation and that the respondents were being honest. By 1966, the apathy level had waned. Southerners believed by 83 percent that the war was important and that they could influence events. In a series of questions designed to indirectly elicit opinions on Communism, 81 percent expressed negative views. A clear majority of those surveyed viewed the NLF as a "communist effort to rule South Vietnam," and 54 percent expressed a positive view of Americans. Less than 21 percent expressed strongly anti-American views. Three-quarters or more of those with an opinion disagreed that this was an American war or that Americans should leave.

However, the lesser majority that disagreed with the statement "The Americans should stop their bombing" indicated that work needed to be done in this area. In addition, the 30 percent without an opinion on whether America should stay involved in the war represented a pool of undecideds susceptible to propaganda from both sides. Given the results of this survey, the Republic of Vietnam had its work cut out for it in transferring the lost Front support from apathy to active government support.[68]

The NLF recapped its military achievements for the previous year in a January 1966 broadcast. The story claimed that it fought nearly 40,000 battles and "put out of action 227,500 enemy troops, including 19,200 American aggressors killed, wounded or captured" during the previous year.[69] However, VC documents captured in 1966 revealed developing moral problems in that organization as well. The Vietcong noted that the Chieu Hoi program "has brought us a lot of difficulties and great losses."[70] The Communists credited enemy PSYOPs with deepening "schools of thought of enjoying life, balking at making sacrifices and enduring hardships." They blamed this on "inadequate ideological indoctrination and lax political activity which make some people vulnerable to enemy propaganda." Calls for intensified indoctrination grew throughout the year.[71]

Morale problems permeated the Northern populace as well. Occasionally, documents stressed the ineffectiveness of US/GVN propaganda. For instance, an editorial in *Quan Doi Nhan Dan* (Hanoi's *People's Army* newspaper) proclaimed PSYOPs to be a complete failure. "The army and people in North Vietnam who, under the leadership of the Vietnam Workers Party and President Ho Chi Minh, have long been tempered in revolutionary struggle, have not only displayed matchless bravery in the armed struggle but also high vigilance in the political struggle with the enemy."[72] However, the efforts spent on counterpropaganda belie such claims. The claims seem more like exhortations of expected action and an attempt to "condition" North Vietnamese civilians on how to respond to enemy propaganda. For instance, another document stated that "many children have torn to pieces the leaflets dropped on fields by enemy aircraft. Whenever the enemy dropped toys or clothing, the people immediately collected them and brought them to the local administration or fighters of the People's Security Armed Forces nearest them." This again illustrates the proposed actions that the target audience should take.[73]

At the same time, North Vietnam expanded its links to overseas groups opposing the war. It then used those connections to reinforce messages targeting US troops and the South Vietnamese. Themes supported the objectives of demoralizing the troops and proving the lack of American commitment to the South Vietnamese. For instance, Herbert Aptheker, a Marxist American historian, visited Hanoi and broadcast to American troops via Radio Hanoi, hailing the self-immolation of Norman Morrison the previous fall.[74] Aptheker said he "read with great interest and deep feeling" Ho Chi Minh's New Year's message to America and was impressed with his "references to his country's struggle for national independence and freedom."[75] The professor traveled to Hanoi along with the activists Staughton Lynd and Tom Hayden. Lynd, a history professor, was an antiwar protestor and a colleague of Howard Zinn. According to Radio Hanoi, after Lynd returned to America he "gave a talk on his recent visit to North Vietnam together with two other Americans at a mass meeting organized by the Women Strike for Peace Movement." Lynd told the group "about the war escalation, the use of gas, the destruction of crops and . . . the scorched earth tactics by the U.S. forces in Vietnam." In Hanoi, they met with an unnamed POW whom Aptheker later described in terms that made the POW seem to accept the injustice of the war. However, Hayden later wrote that "decency compelled us to leave the man in peace and not humiliate him further," understanding that he was just another propaganda prop like much of what they saw on the tour.[76]

In coordination with groups in the United States, Radio Hanoi broadcast "Radio Stateside" to American servicemen for the first time in January. The thirty-minute program, recorded in Los Angeles, was "an effort of a group of students and young working people to bear testimony to a radical militant friendship in America's youths." The self-identified far-left hosts, Steve Fisher and Joe Epstein, played rock and folk music. "Hi, guys, this is Radio Stateside coming to you from Watts," they opened, as they played the popular Donovan song "The Universal Soldier," written by the Native American folk singer and political activist Buffy Sainte-Marie. They claimed it was number one on the charts; however, this was a lie. By asserting that the antiwar song was number one (it really only reached number fifty-three on the charts), they meant to imply that the movement was mainstream. Regardless, the speaker claimed that this song "is appealing to you, our GIs in Vietnam, to lay down your arms, to refuse to fight, to go on hunger strikes, to become reclassified as conscientious objectors, and to return here to liberate your own country, to bring about a little bit more civil rights and civil liberty here in America." Fisher and Epstein referenced the Nuremberg trials and an individual's responsibility to refrain from committing war crimes.[77]

Epstein said, "One of the reasons why this broadcast comes to you from Watts, the ultra-left Negro ghetto in Los Angeles[,] is because this is where we show to the world that we don't have liberty . . . when we are sending hundreds of thousands of American troops to Vietnam and billions of dollars to deny another nation its liberty." Steve Fisher then read an "interesting article" about tear gas, which he called "inhumane," and exclaimed: "The United States should never do anything like this." Overall, the initial show reflected amateur propaganda techniques and a poor understanding of the target audience. Still, it was an improvement over Radio Hanoi broadcasts to US troops. Additionally, it was consistent with DRV propaganda objectives and themes. The show appeared several times in the coming year, with improved quality.[78]

1966 PSYOPs Campaign Plan

Most of the US military operations in 1966—and by extension tactical PSYOPs—centered on clearing routes and bases for new units arriving in-country. The US Marines had incorporated Civic Action into their operations since their entry, providing medical treatment, supporting orphanages, and building schools. However, they quickly discovered that without PSYOP support the "immediate assistance to the local populace would not provide the necessary essentials to give the peasant a positive identification with his own government."[79] The measure of success they used in these activities was the "willingness of the peasantry to give information about VC activity."[80]

On 8 January 1966, the US 173rd Airborne Brigade initiated Operation Crimp, designed to destroy the political-military headquarters of the Vietcong in Military Region 4. It was a region long held by the NLF; the inhabitants were thoroughly indoctrinated and willingly supported the Vietcong. Operation Crimp involved removing the inhabitants to clear the region for military operations. Removal was similar to the conventional war concept of clearing the battlefield, but it made far less sense in the context of a counterinsurgency. Simply removing people was not preferable in a war that was focused on the population.[81]

PSYOP teams flew seven loudspeaker and leaflet missions over the area, "emphasizing the overwhelming strength of the Allied Forces and the ultimate destruction of the VC if they continued to resist." After evacuation, all refugees were briefly interned at the brigade's POW collection point, where they were fed and given medical aid; US personnel distributed 150 pounds of clothing, including 394 T-shirts bearing the slogan 173D AIRBORNE BRIGADE, AIRBORNE ALL THE WAY to the children. After interrogations, the forced refugees were sent to the processing center located at Trung Lap. At the processing center "it became evident that adequate foodstuffs, housing and means of control did not exist at the District and Province level to continue their evacuation," leading Vietnamese officials to cease refugee evacuations.[82]

MACV began the largest coordinated PSYOP campaign to date in support of the 1966 Lunar New Year (Tet). Using a "nation-wide, all-media campaign by all GVN, RVNAF, and US civilian and military PSYOP agencies" to encourage ralliers, the campaign's goal was to use Chieu Hoi themes extensively during a period when fighters would be homesick.[83] The effort consisted of 92 million leaflets printed locally and 69 million printed and shipped from Okinawa. The USIS assisted the South Vietnamese government by printing additional materials in the Philippines and conducted "an extensive multi-media command information program" to indoctrinate commanders and troops about Chieu Hoi and the Tet campaign.[84]

During the Tet holiday, Marine Corps CA/PSYOP teams worked closely with "Vietnamese revolutionary development in an attempt to convince the people of the importance of allying themselves with the government." The Marines integrated Civic Action, PSYWAR, and cordon-and-search operations into "County Fairs" in an effort to destroy VC influence in hamlets and restore government control. Typically, Marine Corps units assumed blocking positions on the outer cordon of a selected hamlet in the early morning. At daylight, ARVN troops and government political cadres "would enter the cordoned area and move all of the civilians to a pre-designated collection point where they would be fed, counted, and identified, given propaganda lectures, drama presentations, and shown movies." The Marine Corps deemed these operations so successful that by the end of March 1966 they became the model for their operations. The "County Fair" became known as "Hamlet Search" and "Go Team" operations as the idea spread to other units, leading to the Chanh Luu operation described in the opening passages of this book.[85]

PSYOP support of US Marine Corps operations was often provided by ARVN PSYWAR battalions. The February 1966 arrival of the 3rd Marine Division Band and Marine Drum and Bugle Corps provided another tool for entertainment during these operations. The band used a drumhead's image depicting the US and Vietnamese flags, "with a hand shake symbol and the words 'Friendship through Music,' written in Vietnamese." They reportedly received a "warm response" from inhabitants. Band members also taught music appreciation and English classes at local schools "in an attempt to appeal to the Vietnamese interest in music and drama."[86]

Despite this positive view that MACV described, the USIS officer Frank Scotton found them to be culturally insensitive. He may have been right in a strict sense: the performances could have been conducted in a more culturally appropriate way. Quoting one Vietnamese observer: "How would you feel if a bunch of burly foreigners invaded your hamlet, took away your men, and played weird foreign music to 'entertain you'?" Others described the operations as "ham-handed and clumsy attempts to win over the people." However, the overriding military goal was to eradicate the Vietcong. The music was simply a way to mitigate the discomfort for civilians in the midst of conflict, and the medical aid provided a positive counter to the VC program. There appear to be no surveys of how the villagers actually felt about these events, and it is possible that both MACV and Scotton were at least partially correct.[87]

Regardless, the Marine Corps found that the combination of increased security, PSYOPs, and Civic Action were "a potent force in combating the guerrilla[s] and destroying the VC infrastructure within the village and hamlet." This was especially important given the population density during the operations southwest of Da Nang and south of Chu Lai, where CA/PSYOP teams contributed to success. The units hoped to shift the inhabitants from VC support to government acquiescence and, later, even to government support. This culminated in the development of Marine Combined Action Platoons, stationed in villages to work with local forces to promote security.[88]

Throughout fall 1965, the US Army had rapidly expanded its PSYOP forces in Vietnam to meet tactical and advisory needs. The estimated US military PSYOP presence in South Vietnam by then was about 500 personnel, not including other agencies and local hires. PSYOP detachments fell under "the operational control of senior CTZ advisors" to work with and to professionalize ARVN PSYWAR units while helping Americans overcome the shortage of experience with the Vietnamese language and culture. The Joint Chiefs of Staff approved the establishment of the US 6th PSYOP Battalion in Vietnam in mid-December. The unit was formed from existing PSYOP detachments in Vietnam. This completed the gradual buildup of US Army PSYOP detachments, and provided command and control for US Army PSYOP units in the South. On 10 February 1966, the 6th PSYOP Battalion was activated in Saigon, comprising three tactical companies. The battalion headquarters was formed from elements of the 7th PSYOP Group's Vietnam Detachment formed out of the USAB&VAPAC on 20 October.[89]

The 1st PSYOP Field Support Detachment (Provisional) at Da Nang and the 27th PSYWAR Detachment at Quang Ngai merged to form the 244th PSYOP Company, with initial headquarters at Da Nang. The company's mission was to assist the ARVN 3rd PSYWAR Battalion and combat units in I CTZ. The Headquarters Detachment was collocated with Provisional Detachment 1, which consisted of two each loudspeaker, audio-visual, press, and mobile propaganda and research teams at Da Nang. The similarly structured Provisional Detachment 2 was based at Quang Ngai.[90]

The 24th PSYWAR Detachment at Qui Nhon and Nha Trang and the 25th PSYWAR Detachment at Pleiku merged to form the 245th PSYOP Company. The company had a small headquarters element based at Nha Trang and a current intelligence team (FC), loudspeaker, and audio-visual elements at Qui Nhon. The remainder of the company constituted Detachment B based in Pleiku, to provide print and field support to American combat units, and the ARVN 2nd PSYWAR Battalion, and a radio station. The 245th PSYOP Company also supported the Republic of Korea's PSYOP Company as well as ARVN combat units. Farther south, the 20th PSYOP Field Support

Detachment (Provisional) and the 26th PSYWAR Detachment merged to form the 246th PSYOP Company, located at Bien Hoa, near the Saigon-based 6th PSYOP Battalion headquarters.[91]

It must be noted that US Army doctrine at that time had no specific organization for a PSYOP battalion or company. PSYOP units were based on the cellular structure described earlier and tailor-fit to the circumstances. Among the early problems identified were a shortage of intelligence and analysis personnel in the table of organization and a critical requirement for more interpreters. Additionally, since the detachments that formed the companies were on temporary duty in Vietnam, many experienced personnel were due to leave soon.[92]

Each PSYOP company was authorized with two press teams, two light mobile teams, three-man HB loudspeaker teams and three-man HE audio-visual teams (four each), as well as broadcast radio advisory and propaganda analysis teams. The loudspeaker teams used the AN/UIH-5 loudspeaker with a range of up to two kilometers. Additionally, each US infantry brigade had an ARS-4 public address system as a supplement.[93] In response to a request from 2nd PSYOP Group at Fort Bragg for information on the performance of the UIH-5 loudspeaker system, Captain Howard Holiday, the 6th Battalion adjutant, responded: "In general, the loudspeaker sets performed very satisfactorily in field operation." A set of eight cones mounted on a helicopter could be heard clearly up to 1,350 feet during daytime and 1,600 feet during nighttime. A set of twelve cones could be heard plainly up to 1,750 feet during the day and up to 2,000 feet at night; one level of jungle canopy did not appreciably affect range. Local innovations had to be devised to mount speakers on helicopters, however.[94]

Supporting the entire 6th PSYOP Battalion, the USAF 5th Air Commando Squadron used detachments at Nha Trang and Pleiku to provide leaflet drop and aerial loudspeaker support, using four U-10 and two C-47 aircraft. During the previous three months, this unit alone dropped more than 112 million leaflets and conducted 463 hours of broadcasts. Among the operations they supported was Operation Windy, targeting North Vietnamese soldiers along the border. This campaign reinforced the messaging from the Ho Chi Minh Trail and North Vietnam campaigns by using safe-conduct passes and dropped 16 million leaflets in nine aerial missions, targeting a depth of fourteen miles into Cambodia.[95]

Effective leaflet themes directed at infiltrating PAVN troops were often straightforward. For instance, one (with a B-52 bomber image) stated:

> As you leave home to kill the peaceable people in South Vietnam, you leave behind burdens which the women, the old ones, and the children will

> HAVE TO BEAR. THEY DO NOT COMPLAIN, BUT THEY KNOW THAT THEY MUST PREPARE THEMSELVES TO MOURN YOUR DEATH AND SHAMEFUL BURIAL IN AN UNMARKED GRAVE. DURING YOUR THREE-MONTH MARCH TO SOUTH VIETNAM, EITHER YOU OR ONE OF YOUR TWO NEAREST COMRADES WILL BE STRICKEN WITH MALARIA. IF YOU TAKE YOUR MALARIA PILLS REGULARLY, YOU MAY LIVE TO DIE IN SOUTH VIETNAM. YOU WILL NEVER SEE ONE OF THESE AND PROBABLY WON'T HEAR IT. IT IS A B-52 BOMBER WHICH CARRIES 29,700 KILOS OF BOMBS AND CAN DROP THEM WITH PIN-POINT ACCURACY.[96]

During this period, the US military command element for II CTZ, 1 Field Force Vietnam, began advocating PSYOP support to reach remote villages via the US Navy Swift Boat and Vietnamese Navy Junk Fleet interdiction programs. US Navy Beach Jumper Unit 1 (BJU-1), headquartered at White Beach, Okinawa, began rotating teams through Vietnam in 1966. The first BJU-1 element to deploy to Vietnam was Detachment Alpha, consisting of two officers and ten enlisted men, and was assigned to the operational control under US Navy Amphibious Ready Group Bravo in support of Marine Special Landing Force operations. Detachment Alpha conducted PSYOPs as one of its primary missions, coordinating leaflet drops and loudspeaker broadcasts, as well as face-to-face communications. The I Field Force operational report from July 1966 revealed plans to provide loudspeaker capabilities to both the Swift and Junk fleets and to coordinate the arrival of boats with leaflet and loudspeaker missions. These PSYOP missions later provided an unclassified cover for covert and deception operations that BJU-1 undertook.[97]

Meanwhile, in the Rung Sat Special Zone southeast of Saigon, the US Navy Advisory Group coordinated with the RVNAF PSYWAR Department to initiate "an all-media PSYOP campaign" to explain security measures within the zone. Farther north, after a series of 1st ARVN Division and combined Marine/ARVN victories in operations such as Lien Ket 26 in early March, I CTZ PSYOP personnel "launched an all-media PSYOP campaign" to publicize these "important and decisive victories."[98] PSYOP personnel were present during the operations and, to some degree, the planning of all US military services in Vietnam.

As an example of how to provide advice and assistance to a host nation fighting a counterinsurgency, the period 1960–1965 in Vietnam offers many useful lessons. The innovative use of aerial loudspeakers was one positive example. US forces and the ARVN used them to assist stranded refugees, take part in humanitarian actions, encourage surrender, spread national-level themes and messages, and harass the enemy. Printed matter was ubiquitous, with US and allied forces disseminating millions of leaflets, magazines, posters, and other products. But the development of South Vietnamese PSYOP capability was the most important achievement. Training in PSYOP techniques was widespread, from short classes for Montagnard tribesmen to months-long training for RVNAF officers at schools in the United States. Training included development of the all-important face-to-face component to break the VC monopoly in this crucial area. Development of television and radio also blunted Front propaganda.

An important lesson: black propaganda actions had a high potential payoff, even though they were difficult to conduct. SOG dropped more than 209 million leaflets and tens of thousands of gift kits over North Vietnam. Even as agent infiltration operations continued in the North, the inherent difficulties in running these missions in the tightly closed North meant that most teams were unsuccessful. However, US officials judged these missions to be successful based on the North Vietnam government's reaction. Hanoi had to admit that in recent months "some 'backward elements' had been enticed by the gifts and duped by the broadcasts" and that "some inquisitive people had become propaganda disseminators." DRV officials were concerned over countering "deceitful propaganda arguments" aimed at the North. Increased repression there demonstrated how the counterpropaganda reflected a real fear that the messages would be effective. MACV analysts also surmised that the selective nature of the allied bombing campaign against the North helped to lower morale there. Bombing was not so widespread as to cause the "rally effect." Coupled with "the hardships and dislocations caused by such bombings . . . and the factual data contained in the leaflets," MACV believed that PSYWAR was effective. For instance, by late 1965 Northern fishermen were required to drop anchor 100 meters offshore and send a swimmer to contact shore patrols prior to landing or risk being sunk. Prisoner interrogations indicated that Radio Red Flag effectively confused midlevel cadre.[99]

Another lesson began to emerge by this time. Effective PSYOPs were not simply an extension of the marketing and advertising techniques familiar to Americans. Information was useful. However, motivating action and short-circuiting enemy propaganda required more. Leading someone to make a call to a tipline to turn in an enemy—an action that could potentially result in the caller's death—required a deeper understanding of various behavioral motivators and levers. While sharing many of the same techniques and technologies of typical advertising, PSYOPs were not exactly the same thing. PSYOPs required detailed knowledge of the target audience and culture. Hatred was a powerful motivator. This was an area in which the North and its agit-prop tactics excelled. Effective PSYOPs also required the willingness to embrace an idea that no advertiser ever could: sending men to certain death in the North to achieve a psychological goal.

Conclusion

Before his ouster, President Diem had been able to hold the country together in a way that his immediate successors could not. This control allowed the pacification campaign to move forward—however haltingly—and allowed the persuasive effect of PSYOPs to transform actions. The coup ended that. After two years of political instability that distracted the nation and prevented focused PSYOPs, the South Vietnam government made a move in the right direction through a series of elections. Nguyen Van Thieu became head of state, with Nguyen Cao Ky serving as premier. Although imperfect, this change stunted the negative drift at last. Looking at the trend of Chieu Hoi ralliers during this period underscores the dramatic change. Between Tet 1963 and the end of that year, roughly 11,200 people took advantage of the program. Throughout the entire year of 1964, less than half that number rallied. Despite scoring tremendous successes in building South Vietnam's PSYOP capability throughout 1964, the program could not overcome the inherent instability of the nation after November 1963. If the coup had never happened, it remains an open question whether the situation would have degraded so quickly. Not until 1965 did rallier numbers approach those of 1963. (See figure 12.2.)

As the 1966 Tet celebrations faded into the past, the war continued to expand. JUSPAO began to develop longer-term plans, conducted additional detailed studies, and provided message guidance on a wider scale. US Special Forces and CIA programs looked promising and utilized APT concepts that the Vietcong formerly dominated. The creation of the 6th PSYOP Battalion gave the expanding tactical PSYOP program a clear organization for command and control. An especially promising new front in the psychological war had opened as well: television and expanded radio coverage. For the North and for the NLF operating in South Vietnam, the international antiwar campaign looked favorable. All sides had expanded their target sets and capabilities to prepare for a monumental struggle for control.

Map 7. South Vietnam, fall 1965 (chapter 8 locations)

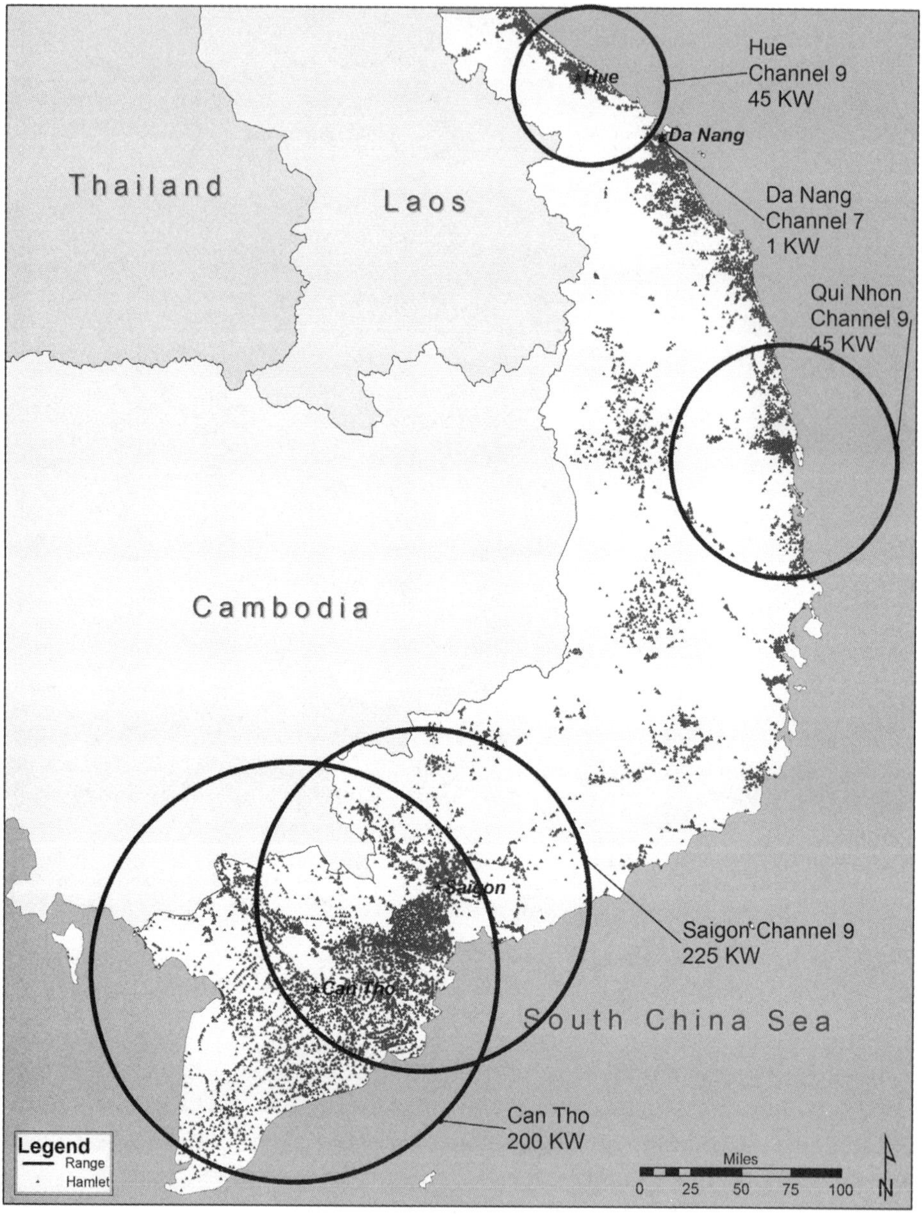

Map 8. South Vietnam television broadcast coverage

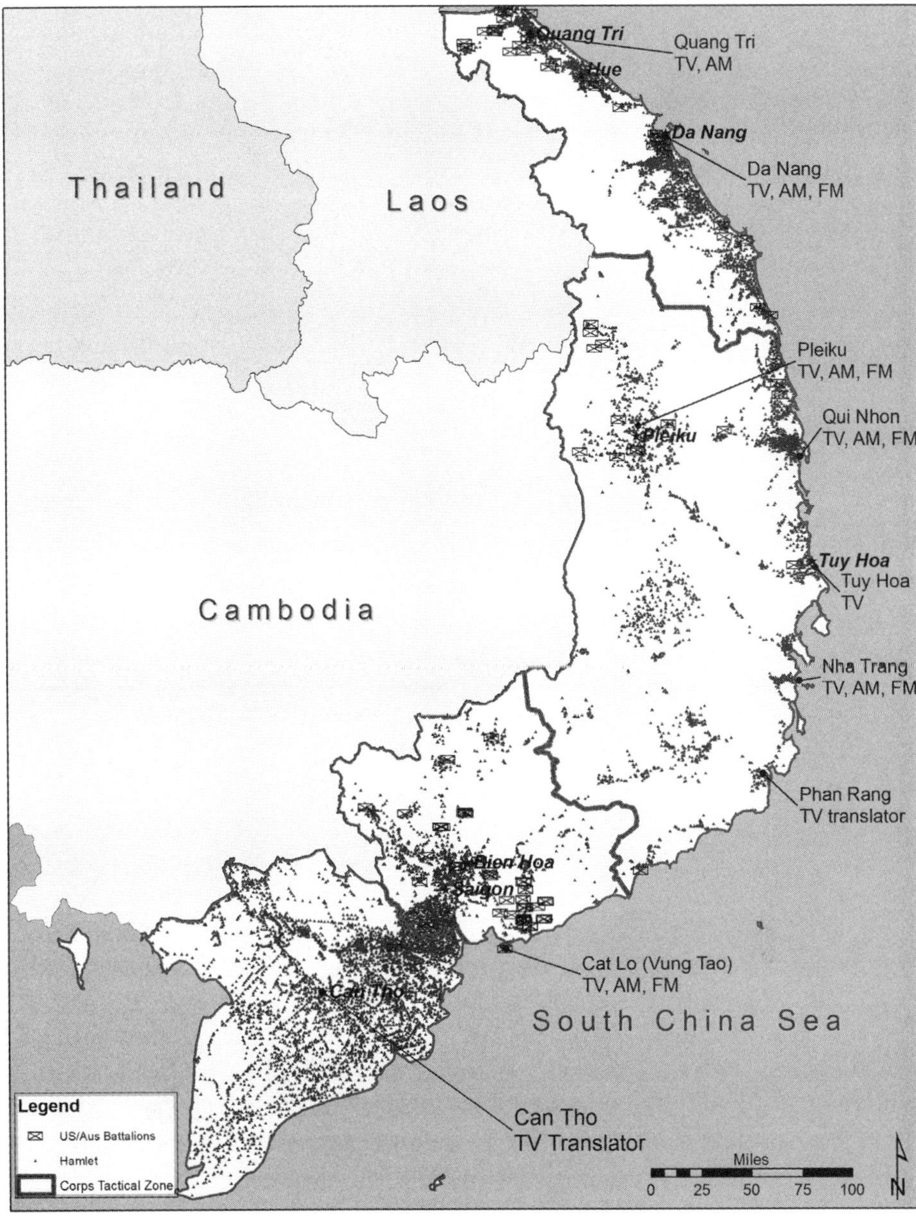

Map 9. Armed Forces Vietnam Network broadcast facilities

7. South Vietnamese PSYWAR team disseminating products, Dinh Tuong Province, November 1964. VA009238, Edward P. Metzner Collection, The Vietnam Center and Archive, Texas Tech University.

8. Members of the 246th PSYOP Company and Vietnamese soldiers load leaflet bombs at the Bien Hoa Airfield, undated. NARA 111-CCV-443-CC36725. Courtesy of Fold3.com by Ancestry.

9. First Lieutenant James R. Paris, on temporary duty status from the USAB&VAPAC, Okinawa, drops leaflets from a helicopter over Binh Duong Province, north of Saigon, 28 September 1965. NARA 111-CCV-443-CC32166. Courtesy of Fold3.com by Ancestry.

10. Private First Class Carl J. Schroder (*left*) and Private First Class Leonard A. Ryan with 3rd Marine Division hold up a captured Vietcong propaganda banner: When Americans and Soldiers of Republic of Vietnam Come, They Shoot Villagers, 7 May 1965. NARA 127-GVB-65-A184135. Courtesy of Fold3.com by Ancestry.

11. The 3rd Marine Division Drum and Bugle Corps with Major General Lewis W. Walt, commander of III MAF. The drumhead says: "Friendship through Music," 28 February 1966. NARA 127-GVB-189-A186757. Courtesy of Fold3.com by Ancestry.

12. Second Lieutenant Robert J. Serry, 6th PSYOP Battalion, TV director for ARVN, and a Vietnamese cameraman watch the monitors during a television program, September 1966. NARA 111-CCV-443-CC36738. Courtesy of Fold3.com by Ancestry.

13. A view of the loudspeaker used to talk to the Vietcong from a C-47 aircraft, September 1966. NARA 111-CCV-443-CC36733. Courtesy of Fold3.com by Ancestry.

9 Second Republic Surge, Spring and Summer 1966

By the beginning of 1966, the US military went on the offensive, seeking to clear a region long held by the NLF. That spring, the United States was also poised to conduct the most intense PSYOP campaign in its history, and the new 6th PSYOP Battalion stepped off quickly to support the expanded combat operations. The formation of American PSYOP companies and a PSYOP battalion proved the benefits of the cellular doctrine in adapting to local circumstances. However, the organization was unable to match demand, despite augmentation. By summer, it was clear that one battalion was not sufficient to meet the needs.[1]

As fall approached, both sides could claim improvements and some successes. The North and the NLF had tightened their links to the antiwar movement. They were successful with that audience, but questions lingered over the degree to which that campaign would influence decision makers in the United States. Likewise, the effects of the American program targeting the North were still in doubt. President Lyndon Johnson's misplaced attempts to send political signals via bombings continued to backfire, and the North was clearly losing the battle to attract Southern civilians to its cause. By the end of the period, indications of decline in VC morale had become more commonplace. However, the government lacked an ideological attractor until the September Constituent Assembly elections. The effect of the election portended further collapse of the Front position in the South.

Development of Propaganda Organization and Objectives

To manage the American PSYOP program, the United States had created the Joint US Public Affairs Office the previous May. By the opening of 1966, the organization expanded and refined the PSYOP program. The Johnson administration's decision to create JUSPAO was novel, but it was not an unprecedented organization. It contained many functional similarities with the Office of War Information from World War II, but with greater integration and message discipline at all levels. Civilian control and guidance over semi-independent military PSYOP organizations was a hallmark of the American system. Perhaps the most controversial aspect—the inclusion of the Saigon embassy's Public Affairs Office in the system—created the perception on the part of some that it was spreading propaganda into the news cycle in the United States. However, it must be remembered that the USIS had been in control of the news output from the embassy for some time and that MACV's Public Affairs Office remained independent.[2]

Organizationally, JUSPAO consisted of several branches. The Policy and Planning Branch issued PSYOP guidance to coordinate messaging among the American psychological operators. The Survey and Research Branch produced target audience analysis studies and managed "systematic surveys of Vietnamese attitudes and reactions" to improve program effectiveness. The Evaluation and Analysis Branch evaluated PSYOP products and conducted post-tests to determine product effectiveness, using an "evaluation panel composed of ordinary Vietnamese citizens representing various segments of Vietnamese society."[3] The Field Development Division produced products and developed, coordinated, and supported campaigns such as the Chieu Hoi, Refugee, Public Safety, and Revolutionary Development, as well as economic and agricultural propaganda programs. Among the publications JUSPAO prepared, *Free World* targeted adults with a digest of US magazines. *Quest,* targeting students with popular science articles, ran 100,000 copies per month and was so popular that this free magazine was reportedly resold on the Saigon black market for about sixty cents American each.[4]

JUSPAO also had responsibility for cultural exchange programs, providing "scholarships, libraries, English instruction, visiting lecturers and professors in Vietnamese universities," as well as informing the Vietnamese people about American society and policy. Perhaps in an overextension of its mission, JUSPAO also embarked on the task of explaining South Vietnam to the world, a problematic endeavor since it undercut the theme of an independent, self-reliant South Vietnam. For instance, JUSPAO's psychological objectives in the South included increasing the "Vietnamese people's identification with and participation in their government in the war against communist subversion and aggression" and promoting the development of political institutions at all levels.[5]

JUSPAO's North Vietnamese Affairs Division, in close coordination with MACV's PSYOP directorate and other Republic of Vietnam and US government agencies, planned and directed PSYWAR aimed at North Vietnam as well as PAVN units anywhere. Primarily, Operation Fact Sheet battered North Vietnamese citizenry with leaflets. The operation's goal was to cause dissension and to pressure the Lao Dong Communist Party "to cease its aggressions against the RVN." The program intended to drive a wedge between the Northern population and the government and party by informing them "of the true nature of the aggressive war in South Vietnam and relating their hardships and privations to the continuation of the aggression." Fact Sheet stressed American efforts to find a "peaceful settlement." A supporting objective was to convince the people of the futility of the "liberation" war in South Vietnam and that Northern "defeat in the South is inevitable and that unification, a common aspiration of both North and South Vietnamese, can only be achieved through peaceful means."[6]

In this goal, the Fact Sheet campaign was a failure. To be effective, the target audience needed to be able to influence events. The people of North Vietnam were an inappropriate target because they could not affect government policy. The campaign served another purpose, however. The program forced the DRV to expend resources in the fear that it would increase internal discontent. Indicators of a government clampdown quickly emerged. However, due to the extension of President Johnson's Christmas bombing pause, Fact Sheet drops stopped as well. In response to a 7 January 1966 query from the State Department, the US Mission Council in Saigon recommended that operations should restart only after the bombing pause ended, apart from "a special Tet Greetings drop." However, Secretary of State Dean Rusk did not approve the embassy's plan due to hopes of a "favorable response from Hanoi to US peace initiatives" and recommended that leaflet printing cease as well. The Johnson administration continued to place hopes in its messaging/action policy, to which North Vietnam clearly did not respond.[7]

In January, President Johnson devoted a considerable portion of his 1966 State of the Union speech to discussing the Vietnam War. He again highlighted his quest for a peaceful resolution of the conflict as long as it considered the desires of South Vietnamese. He asserted that the bombing pause was a symbol of that desire. Radio Hanoi commented on the speech, saying, "Johnson's deceitful peace argument cannot conceal his dark intention of pursuing the escalation and expansion of the war of aggression in Vietnam." It began a full-court press to attack Johnson's peace moves across all media. The level of vitriol in the broadcasts indicated a fear that the message would affect Northern morale and required a hard response.[8]

Specialists on both sides struggled to refine their own psychological objectives and themes while at the same time attempting to understand and counteract the other side's program. Examination of North Vietnamese and Front propaganda indicates the development of the following psychological objectives by this point in the war:

1. Divide United States and South Vietnamese military personnel.
2. Decrease the U.S. public's support for South Vietnam.
3. Reduce the effectiveness of the Government of Vietnam.

4. Increase worldwide support for the Democratic Republic of Vietnam as the legitimate representative of the Vietnamese people.
5. Gain international support for the NLF as an independent movement representing the true Southern aspirations.[9]

To South Vietnam, Hanoi exposed the high cost of war to insinuate that America would not continue to fight. "Surely the U.S. people will not stand idly by and let Johnson strangle them and extort money from them and send their children to South Vietnam to serve as cannon fodder," Radio Hanoi said.[10] Meanwhile, the Liberation Broadcast Service took the opportunity of the New Year's cease-fire to state, "Those [ARVN] enlisted men and officers who abandon their ranks to return to their families and to the revolution will be guaranteed their rights to freedom and will be helped to find jobs."[11]

Expanded Tactical Psychological Operations

Meanwhile, the ground war relentlessly expanded. Operation Masher began near the end of January 1966 when elements of the US 1st Cavalry Division, an ARVN division, and a South Korean battalion "engaged in the first division-size search and destroy operation of the war." This six-week operation cleared the enemy around the village of Bong Son, located between Qui Nhon and Quang Ngai. PSYOP soldiers coordinated with the 5th Air Commando Squadron for U-10s to drop leaflets and broadcast messages, with ground and air strikes. After each battle, the planes "broadcast funeral dirges and wailing sounds to play on the enemy's superstitions." However, due to command concern over the aggressive-sounding name, the operation was renamed White Wing during the second phase.[12]

A good example of successful Special Forces "propaganda of the deed" occurred during the Plei Me campaign that had begun in the fall. In conjunction with the 245th PSYOP Company, JUSPAO, and the Vietnamese Information Service, the operation sought to return the area around the Duc Co Special Forces camp in Pleiku Province back to the government from its current "contested" category. An extension of the program discussed earlier, this unified effort targeted all hamlets within three miles of the camp, supported by MEDCAPs, face-to-face interactions, USAID, and CARE (then known as the Cooperative for American Remittances Everywhere). The VIS team held internal training sessions prior to the events, emphasizing local and national goals for its members to spread. The primary theme was that the government was working for peace, whereas the NLF prevented this. The teams held sick-call and met with local officials while delivering PSYOP materials to the people and schools. These included distribution of *Huong Que* magazine and posters showing ARVN victories.[13] Teams found exploitable "resentment toward Viet Cong methods of recruitment and forcing attendance at interminable propaganda sessions," as well as anger at increasing Front rice confiscation from inhabitants.[14]

At the same time, Special Forces continued to improve the quality of its PSYOP personnel. In January, the 5th Special Forces Group held a US Special Forces NCO CA/PSYOP course. It also coordinated with the Vietnamese Political Warfare School for Vietnamese special forces to attend a six-week CA/PSYOP course. The 5th Special Forces Group further endeavored to ensure that American NCOs assigned to PSYOP positions in the future were trained at the Fort Bragg PSYOP course prior to deploying to Vietnam. The lack of a Military Occupational Specialty for enlisted personnel in the PSYOP field was a limiting factor. Any intelligence NCO could be sent to fill a PSYOP slot, regardless of prior training and experience.[15]

The 1st Cavalry Division continued operations to systematically clear the area around its base at An Khe, often in conjunction with Special Forces. Operation Jim Bowie sought to clear the longtime VC-held valleys of Binh Dinh Province. Not many civilians were expected in the area. Beyond a few "slash and burn" tribe populations, the area was nearly devoid of villages. As such, the American messages focused on VC surrender. The 245th PSYOP Company provided leaflets and HB loudspeaker teams, and the 5th Air Commando Squadron provided U-10 aircraft for leaflet and loudspeaker missions in support of the operation. PSYOP elements for the operation included the 1st Cavalry JUSPAO representative, who "provided professional advisory services and interpreter capability." Special Forces also provided interpreters and personnel to test PSYOP appeals on tribal groups in the area.[16]

As part of this operation, the 1st Cavalry Division's 1st Battalion, 9th Cavalry Regiment "conducted deception operations in the Suoi Ca Valley with the objective of misleading the enemy as to the division's area of operations." The deception consisted of phony rifle platoon insertions by helicopter and reconnaissance operations supported by aerial rocket artillery along with PSYWAR activities. One loudspeaker mission on the theme "Don't Fire at Aircraft," coupled with six leaflet missions, supported the deception. The operations kept the guerrillas guessing as to the point of attack. This showed a successful way that PSYOPs could support deception missions by integrating them into the tactical plan and helping to cover the actual assault point. Additionally, the 1st Cavalry conducted thirty-nine loudspeaker missions totaling fifty-seven hours during the operation. Primarily using Chieu Hoi themes in North and South Vietnamese dialects and tribal languages, PSYOP personnel conducted forty-three separate leaflet drops totaling 1,675,000 leaflets during Operation Jim Bowie.[17]

Interrogators during the operation found it best to emphasize good treatment, food, and medical aid and to avoid the term *surrender* in future broadcasts. To diminish the stigma of surrender, fighters were told not to raise their hands but rather to wave their shirts to get the attention of helicopter personnel. Hoi Chanh (people who rallied to the government) reported that the loudspeakers were more effective than leaflets. Loudspeaker surrender instructions were very clear, telling listeners where to go, how to carry their weapons, and so on. During the morning of 23 March 1966, the 3rd Brigade captured a VC cadre who "had read the instructions on a safe conduct pass and was observed standing on a bald hill, weapon slung muzzle down, waving a towel at passing aircraft."[18]

Captured VC Party Province Committee documents indicated morale and recruitment problems because of these operations. A March document captured in Binh Dinh stated that the "fierce sweep operations conducted by U.S. and [Korean] troops" seriously affected cadre morale. Many left for government-controlled areas "or surrendered to the enemy." Another document, from this same committee, noted the need to "stabilize the morale of the population in the areas already cleared by enemy operations and where pacification activities were in progress." Continuing VC recruitment difficulties reflected low youth morale, per the committee.[19] Higher-ranking VC ralliers included a PAVN first lieutenant and a VC Regional Force platoon leader. At this time, the Liberation Press Agency issued a quarterly review of the war. It attacked the ongoing pacification effort. Per the agency, the United States "put into effect the Lodge-Lansdale plan of indiscriminate raids and bombings to raze to the ground one region after another and of large-scale sprays of chemical poison along with intensified psychological warfare to realize their rural pacification scheme." Front assaults on Chieu Hoi centers also spiked, indicating their perception of the defection threat. In response, South Vietnam increased construction of Chieu Hoi centers.[20]

During the spring, domestic unrest struck South Vietnam again. Along with the Buddhist protests, another Vietnamese propaganda source arose. Groups of Buddhist supporters took over radio stations in Da Nang and Hue, using them to rally supporters against government policy. By the end of March the situation became critical. In Hue, a radio broadcast announced protests against Marines for tearing down an English-language banner denouncing the United States. The station later broadcast an apology from "the major general commanding the U.S. Marines" for this action.[21]

Saigon's domestic service broadcast a message from President Thieu denouncing the takeover of stations that had spread to Dalat, Nha Trang, and Ban Me Thuot in the following weeks. After one station was burned, Thieu said that the situation could not continue. Meanwhile, the North stoked the flames: "What is noteworthy is that day by day the movement has manifested more and more clearly its anti-U.S. character, and become more active."[22] In May, a crowd torched the USIS Library and Cultural Center in Hue.

At the same time as this unrest, Liberation Radio opportunistically encouraged anger at monetary inflation in South Vietnam. "Let everybody rise up and ask for salary increase, living cost reduction, and a solution to the food shortage problem," the broadcast said in a story that blamed the problems on the United States. "The 200,000 U.S. expeditionary troops who have invaded South Vietnam for a year" escalated prices, according to the Liberation Broadcast Service, seeking to exploit economic issues to divide the Vietnamese from the Americans.[23] In orders sent to cadres broadcast over LBS, the NLF praised the successful "revolutionary proselytizing among the soldiers of the puppet armed forces."[24]

Meanwhile, Radio Hanoi once again denounced the American use of gas. Returning to the biological-warfare theme, it accused "U.S. ruling circles" of bringing the "Mobile Research Institute of the U.S. Bacteriological and Warfare Corps No. 406," to South Vietnam. This institute was helped by "West German militarists" in producing "combat gases," the broadcast alleged. This story seems to have been a response to the pending deployment of a West German hospital ship to Saigon to aid civilians.[25]

During the first three months of 1966, the 5th Special Forces Group continued to build a tactical PSYOP capability to counter the enemy. The group enhanced the development of CIDG CA/PSYOP teams to provide them with "additional and effective means of conducting civic action and psychological operations." Success with the test led to the authorization of one team per camp. By April, 140 more troops had graduated the ten-week course at the Montagnard Training Center in Pleiku, and the Political Warfare School planned to present a six-week course in July.[26] Organized into twelve five-man units, Motivation Teams were "capable of training 680 men per month."[27] The teams included a team leader, an assistant, and two each medical, CA/PSYOP, intelligence, and survey members. Civic Action was essential to team success. By the end of April 1966, US Special Forces had completed 4,261 CA projects, mostly in the Central Highlands. Among these, they constructed or repaired 47 schools, 50 dispensaries, 118 wells, 97 bridges, and 46 roads. They had also provided relief supplies for more than 18,000 refugees and conducted numerous medical patrols and other projects "designed to gain the support of local civilians." The numbers represent a significant percentage of the inhabitants of the sparsely populated Highlands, where most Special Forces camps were located.[28]

The 5th Special Forces Group increased the intensity of PSYOPs along with CA, dropping over 23 million leaflets and distributing "over 185,000 publications such as bulletins, magazines, posters and newspapers." Furthermore, the group issued "portable loudspeakers, tape recorders and mimeograph machines" to all detachments to "carry out more timely and effective propaganda in the local villages and hamlets." For instance, Camp Tien Bien's CA/PSYOP NCO produced a bimonthly newspaper named *Truth* that targeted the Vietcong, and he reported "five VC returned to GVN control carrying a copy of the newspaper."[29]

Captain Frank J. Leach, assigned to Team A-234 in An Lac, Darlac Province, used his CA/PSYOP team to transform villagers' behavior. Leach noted that, upon arriving at An Lac in April, he "found the local inhabitants completely hostile toward the American and Vietnamese soldiers." The people openly supported the Vietcong, and the inhabitants avoided all contact with Americans. "Whenever an American entered a village, the people would run to their homes and close the doors," and the "children were even afraid of the Americans." Leach devised a four-step program to overcome this hostility. The plan began with a dinner for the village chiefs at which the team explained that "we were here to help in any way we could," promising "protection from the Viet Cong."[30]

Leach's activities were a solid example of the importance of developing relationships for effective PSYOPs to occur. The team proposed to elders a plan to send medical patrols to the area and to distribute "clothes, toys, soap, salt and food to the needy families." Next, Leach contacted USAID in Dalat and Ban Me Thuot and asked for support and CARE packages. During each medical patrol, the camp commander "talked with the people explaining that we wanted to help." Because the people refused to exit their homes, the team left supplies with the village chief. According to Leach, "The next time we visited the villages we found a few people who would come out to see us but they were very suspicious." After repeated visits, the ice melted, and "soon the children were accepting candy and clothes from us" and "villagers were looking to our next visit."[31]

The second phase entailed securing each village with an American-advised squad to help build houses, dig wells, and plant gardens during the day and provide protection at night. "Soon villagers were inviting us to their homes for meals and local celebrations," Leach wrote. This program resulted in the active and willing support of the villagers, and "smiles and laughter" greeted the team's visits. Sincerity and perseverance were required, according to Leach, as well as a willingness to "accept and honor their customs and habits."[32]

The management of unintended consequence often fell to PSYOP teams. The Ben Soi Civic Action Team dealt with a refugee problem resulting from American air strikes and ground operations conducted in the area. The team initiated mitigation to address the consequences of February air strikes. Combat operations in Thanh Dien had left approximately 300 people temporarily homeless and about 30 civilians killed or wounded. The team coordinated with a VIS loudspeaker team "to assemble the people to a designated temporary refugee center" and used money supplied by USAID for immediate relief. The province and district Civic Action staff supplied cooking utensils.[33]

Special Forces Detachment A-321 helped move 218 families to GVN-controlled hamlets during this mission. Despite the negative circumstances, many of these refugees provided valuable information on the enemy and aided the war effort. This camp was in a formerly VC-controlled region, but the team increasingly found people voluntarily seeking the protection of the camp. The report's author stated that "civic action projects such as this show that winning the people's hearts and minds is not just a useful cliché but a meaningful phrase." Working at the grassroots level and showing government interest in locals' welfare paid dividends.[34]

PROVN Study and Troop Indoctrination

In March 1966, the Office of the Deputy Chief of Staff for Military Operations produced a report titled *A Program for the Pacification and Long-Term Development of South Vietnam* (normally abbreviated as the PROVN study). This reevaluated the American strategy and operations in Vietnam. It urged a single-manager concept for pacification, which was largely implemented that year. Although MACV incorporated some of the ideas, it opposed recommendations to take over internal South Vietnamese affairs. MACV believed this would backfire and undercut RVN legitimacy. Per the PROVN study, emphasizing that villagers' hardships were the fault of the Vietcong and focusing on the "light at the end of the tunnel" would be effective themes.[35]

The study noted the great increase in PSYOP capabilities in South Vietnam and the quality of the overall plan. However, the authors warned that "the whispered word from an NLF cadre, plying enlightened self-interest, is more effective than a bank of airborne loudspeakers blasting the merits of greater sacrifice from 3,000 feet. Unkept promises and empty clichés constitute the core problem."[36] A coherent "national ideology" alone was not sufficient, according to the authors. An effective pursuit of needed social reforms "can motivate and polarize the peasants. Nothing less will succeed."[37]

By spring, the South Vietnamese decision to create a political warfare system required a change in the American advisory effort. MACV issued Directive No. 515-1 regarding American support to the ARVN's POLWAR system on 26 April 1966. This document established POLWAR advisers at all command levels down to company-size units and delineated staff responsibilities. Among the objectives of the POLWAR system were establishing and maintaining a "high degree of loyalty, discipline, and morale within RVNAF" and gaining the "support of the civilian populace by improving the civilian-military relationship." The system consisted of five POLWAR battalions composed of four POLWAR companies and one Civic Action company each. The 10th POLWAR Battalion was located in Da Nang, the 20th in Pleiku, the 30th in Bien Hoa, the 40th in Can Tho, and the 50th in Saigon. Within each battalion, POLWAR companies were numbered sequentially from one to four. For example, 10th Battalion had companies 101 through 104. Each company had six audio-visual teams. The Republic of China was responsible for the organization and initial training of the POLWAR units.[38]

While JUSPAO focused on the strategic level and MACV Psychological Operations Directorate (MACPD) the tactical level, there was some overlap. The director of the Psychological Operations Directorate reported to the commander of MACV and served as the senior adviser to the RVNAF's General Political Warfare Department. This included all aspects: training and doctrine, equipment procurement, and message dissemination. The MACPD had operational control of the 6th PSYOP Battalion and conducted its own research and target audience analysis.

Poor political indoctrination of US troops, however, often hampered attempts to build the positive relations needed to succeed in pacification operations. AFVN radio did offer some programming to overcome this shortage. However, US troops "received virtually no political guidance," in contrast to Front and PAVN soldiers' continual indoctrination from "the political cadres who accompanied every unit."[39] One VC cadre captured in Cu Chi told interrogators that although US operations hurt Front recruitment, this was mitigated to some degree by "nearly daily incidents concerning bad road discipline on the part of American vehicles, resulting in injuries or damaged crops." Overall guidance to US soldiers stressed the need to maintain positive relations with the Vietnamese, but more emphasis was needed.[40] The ideas found in the film *The Unique War* required wider diffusion. Australian soldiers faced the same issues. Each deployed soldier received an "Australian Nine Rules" card to set expected behavior standards. Among the most important directions: "The Communists will use any weapon to discredit the Government and countries, like ours, in the eyes of the Vietnamese people. Don't let your behaviour be a propaganda weapon which helps in any way to destroy Vietnam." Even with this, some Australians "simply ignored their training" and committed acts that brought discredit.[41]

Additionally, soldiers received mixed signals from counterproductive incentives. For instance, the US 25th Infantry Division awarded platoons "ten points for each 'possible body count,' one hundred points for each enemy crew served weapon captured, and two hundred points for each tactical radio captured." For every American soldier killed in action, the platoon lost five hundred points. Using those metrics, the division identified "productive platoons." As such, soldiers carried out actions necessary to boost their ratings. Without claiming that any soldier specifically targeted civilians to earn points, the measurement system may well have led some soldiers to feel pressure to achieve a higher kill ratio. Americans were not the only ones using body counts as a measurement of success, though.[42] The NLF also rated units by the number of enemies killed, and a soldier "who killed five Americans or the individual guerrilla who killed three was awarded the prestigious title of 'Heroic American killer.'"[43]

American troop awareness of PSYOPs also needed to be raised. One lesson learned: "Friendly troop units must be aware of the purpose of the broadcast and of any part they may be expected to play." One solution was to broadcast a translation of the message in the "language of the friendly troops, English or Korean, explaining at the same time what is expected of ground units." This was especially useful when giving specific surrender instructions, and it ensured that friendly ground troops acted accordingly. Doing this also offered the "benefit of providing an on-the-ground check of loudspeaker effectiveness" to determine the best flight path.[44]

Meanwhile, North Vietnamese targeting of American troops became more varied, and the themes used were more likely to hit home. Shortly after the US-conducted Operation Birmingham in War Zone D, *Nhan Dan* announced the "mutiny" of an American battalion. The story said that officers and men of a unit in Lai Khe had refused orders on 24 April to go into the rubber plantation. *Nhan Dan* claimed that these soldiers had sustained "stinging blows" previously and "on that day made quite a din outside the battalion headquarters, scattered their guns on the ground, and refused to board the planes. Many of them ran back to their barracks and sat there weeping bitterly; others took refuge in the houses of the local people or hid in trenches." *Nhan Dan* claimed that this mutiny was a sign of collapsing American morale. The LPA carried the same story ("A new phase of development of the antiwar movement among U.S. servicemen") in a broadcast to the South.[45] No record of the incident could be found.

One broadcast replayed a November antiwar speech by Robert Scheer, *Ramparts* reporter and North Vietnam visitor, in Berkeley. Another read a statement, attributed to the deserter Robert Garwood, urging "troops to stop fighting." "I am calling on you to stop terrorizing and massacring the people of South Vietnam. Don't burn their villages, bombard their rice fields and homes or run tanks into the fields," said Garwood. He said he was well treated and that soldiers who refused to fight would be "sent back to the United States where you belong."[46]

In May, Radio Hanoi broadcast stories of American pilots shot down over the North, including one in which Charles Boyd recounted what happened to him. "Although hatred was reflected from the eyes of my captors, to my surprise the people and local authorities immediately gave me decent food and medical attention for my right arm, which had been seriously hurt on ground contact," he said. Boyd admitted, per the broadcast, to committing a crime by attacking the North and begged for forgiveness.[47] The station also read from a story in the Japanese paper *Akahata* (Red Flag) containing a statement from another captured US pilot. Per *Akahata*, Colonel Robinson Risner said,

> I now realize that the air strikes against North Vietnam are illegal and that they are only helping to steadily heighten the morale of the Vietnamese people. I have learned from my own experience how high is the morale of the Vietnamese people and how firm in their determination, and I believe that these are the main factors that have enabled you to carry on your resistance war for more than 10 years.[48]

As usual, Radio Hanoi laundered one of its own stories through an overseas group to provide reinforcement and legitimacy for its propaganda.

Excerpts from an article by General Vuong Thua Vu, a high-ranking officer of the Vietnam People's Army, read by Hanoi Hannah, urged US servicemen to consider why they fought. "At this point, some of you would shrug your shoulders and laugh: 'Hanoi Hannah again talks fiction and gives fishy facts,'" she joked. She urged her listeners to use their own minds, arguing that "military technology is no match for PLAF morale." Hanoi Hannah closed by saying that although listeners might not agree at the moment, they should "discuss his ideas and see whether they are correct or not." All these stories and reinforcements were designed to sap American troops' morale and to make them less likely to risk their lives. Although designed to produce cognitive dissonance, as exemplified by Hanoi Hannah's questioning, it is difficult to determine the degree of success.[49]

Combat Offensives Overload Propaganda Operations

Binh Dinh Province had long been a Front stronghold. In fact, Ho Chi Minh had lived there for a period before leaving for France. Pacification of the province fell predominantly to the 1st Cavalry from its base in An Khe. In an after-action review for Operation Crazy Horse in the Vinh Thanh Valley, the 1st Brigade's commander stated that "the 'talking' helicopter is a must . . . in connection with PSYOPS." Despite this praise, technical and coordination issues hampered operations. He urged that a loudspeaker helicopter and PSYOP team be stationed at the division's forward headquarters. After the next major 1st Cavalry operation, Davy Crockett, Colonel Harold Moore, commander of 3rd Brigade, recommended that each brigade receive its own loudspeaker set. "Often the requirement for psychological operations is so immediate that normal support units cannot react," he said.[50] However, this high demand did not consider the scarcity of men and equipment for support. The Pleiku Detachment of the 245th PSYOP Company that supported the operation, conducting sixty loudspeaker missions and dropping nearly six million leaflets, was at its limits.[51]

A psychological warfare team attached to the 1st Cavalry Division conducted a leaflet drop shortly after Operation Davy Crockett began. This covered the entire sector, emphasizing the Chieu Hoi program and VC hardships. During the operation, 5th Air Commando Squadron aircraft dropped leaflets within hours of B-52 strikes, covering the "radius of the bomb-blast noise envelope to exploit the fear and disorganization resulting from the raid."[52] Loudspeaker broadcasts and another leaflet drop followed. "Two specific missions were conducted using an airborne interpreter" to urge entrenched Vietcong to give up, resulting in twelve fighters surrendering. One lesson reinforced during the operation was that the Vietcong preferred to rally to ARVN forces due to the language barrier with Americans. Another lesson was that some ralliers felt that the VC cadres did not take the same risks they forced the soldiers to take, indicating another potential fracture line within the NLF.[53]

As US commanders increasingly requested PSYOP support, MACV ordered a comprehensive analysis "to determine whether the current MACV PSYWAR organization was adequate" to satisfy requirements. This was necessitated by the introduction of massive PAVN forces into the South and "by the many changes created by the US military buildup and the expanded advisory effort." By this point, PAVN troops outnumbered NLF regular forces nearly two-to-one in South Vietnam. American PSYOP expansion lagged behind the rapid tactical force deployments by both sides.[54] By the end of March, MACV staff and field commanders "confirmed the need for more PSYOP personnel" at all levels and agreed to expedite augmentation of Field Force (Corps) PSYOP staff sections.[55]

Likewise, the NLF began to expand its propaganda system, especially internationally. During Operation Junction City in 1967, US Army units captured a COSVN Sub-Committee on Foreign Activities propaganda analysis and guidance memo prepared for Hanoi authorities in June 1966. While acknowledging that the bulk of COSVN's propaganda support came from North Vietnam, the author urged that "we must use all the available means of propaganda to kindle a widespread anti-war movement among the [American] people." Doing this required motivating youths to protest against the US crimes and request an "end to the war of aggression in South Vietnam." The author called for specifically targeting American soldiers' family members to "launch anti-war demonstrations and to request repatriation." This presaged a radio program begun the following spring by the Committee to Aid the NLF, in which Walter Teague hunted for deployed soldiers' families to tape messages for broadcast over Radio Hanoi and Liberation Radio.[56]

Another COSVN report to Hanoi noted the great help provided by the international press. It reported investigating "the possibility of employing the propaganda 'capabilities' of the world's organizations, newsmen, writers, and officials in foreign countries" and had arranged schedules for visiting foreign reporters. After the reporters returned home, the story noted that they "helped us a great deal in propaganda." The NLF also "furnished information, documents, films to a number of major international organizations and foreign personalities who sympathized with the revolution in South Vietnam."[57]

To support this key NLF and North Vietnamese psychological objective, the Hanoi Foreign Language Publishing House printed a series of Vietnamese studies pamphlets in English and distributed them to American university libraries. The target audience included educated Americans who were opposed to the war or who might be questioning US involvement. The series consisted of at least forty-seven pamphlets employing classic propaganda techniques, designed to inform opinion leaders so that they could properly indoctrinate and agitate the grassroots. The booklets contained detailed information on the North Vietnamese view on given subjects: *The Failure of the Special War, 1961–65*; *Initial Failure of the U.S. "Limited War"*; *The N.F.L.* [*sic*] *Symbol of Independence, Democracy and Peace in South Viet Nam*; and *They Have Been in North Viet Nam*, which detailed visits by foreigners opposed to the war. English language quality tended to be good and normally not overly bombastic, although each contained some grammatical errors and used British spellings. Pamphlets ranged from 88 to 200 pages in length, printed in 5-by-7-inch format. The paper quality was slightly better than newsprint, and the overall printing quality was good.

Like the *Vietnam Courier* newspaper, these pamphlets found their way onto the bookshelves of at least fifty-six university libraries. At least twenty-eight pamphlets remain today at the Willis Library at the University of North Texas. Due to changes in library checkout policies, it is impossible at this point to gauge how often library patrons read these pamphlets or how the libraries acquired them. However, the themes stressed in the booklets coincided with much of the messaging of the antiwar movement, spotlighting, for example, that Diem and the RVN were illegitimate puppets of the United States and that the ARVN was "a puppet army" composed of "inveterate criminals, capable like Hitler's SS troops of committing all kinds of crimes who burned down villages, disemboweled victims to take out their livers, ruthlessly massacred women and children."[58]

The CANLF bookstore in New York sold copies of most of these booklets. In May 1966, the committee received a telegram from Ho Chi Minh wishing the organization "good success" in the "struggle against the U.S. imperialist aggressors." In fact, the CANLF consistently followed the North Vietnamese propaganda line during its various shifts.[59] The organization also attempted to leverage its members' knowledge of American culture and political discourse themes to build an emotional connection between the movement and Americans. For instance, the CANLF symbolically linked the antiwar movement, the National Liberation Front, and the American Revolution. Teague ensured that a thirteen-star Revolutionary War flag always flew beside an NLF flag at protests "to prevent police from demanding we take down the NLF flag." It also had the possible effect of creating cognitive dissonance in viewers, questioning America's right to be involved in another nation's war for "independence."[60]

The North utilized the British philosopher Bertrand Russell to further isolate the United States from professed world public opinion. Radio Hanoi announced an upcoming "international tribunal" by Russell to "try Johnson and other war criminals." Thich Thien Hoa, an NLF member and president of the South Vietnam Buddhists Association, was one of the tribunal's members.[61] Russell appealed directly to American servicemen in South Vietnam "to accept personal responsibility for the criminal acts which are occurring every [day] against the Vietnamese." He accused the US Air Force of "bombing hospitals, tuberculosis sanatoriums, leper clinics, old-age homes, schools, and villages" and of using poison gas.[62] Russell claimed that the United States already controlled "60 percent of the resources of the world, but only has 6 percent of the world's population" and fought the war to gain more control. In his 1967 autobiography, Russell admits to advising the North Vietnamese on propaganda messaging.[63]

In a message meant to encourage war resistance among American soldiers, Radio Hanoi read a letter sent to the Chicago Women for Peace Club. It appears that the club then forwarded the letter to Hanoi for its use. The letter alleged a torture incident witnessed by the unnamed author. In response to seeing this supposed occurrence, the soldier wrote that he "struck it to the back of the guy from intelligence [*sic*]." "Ever since that day I have been sick in the stomach and haven't been out on patrol or anything. My sergeant tells me I am suffering from battle fatigue and might get sent home," the soldier wrote.[64] The nonstandard English indicated an invented letter, or one that had perhaps been translated back and forth.

Later, the station accused West Germany's defense minister, Kai-Uwe von Hassel, of providing the United States with arms "worth 300 million marks for the war in Vietnam" and of allowing West German pilots to join the US Air Force in Vietnam. Hanoi attacked the "dispatch to South Vietnam [of] the so-called hospital ship Helogoland [*sic*] provided with special military equipment and under command of a former Hitlerite naval officer who served aboard a Nazi minesweeper." This attack was part of a consistent theme in the construction of the chemical warfare narrative: tying the United States to Hitler's Germany. As noted above, the ship's real mission was to provide health care to poor people in Saigon.[65]

As Buddhist unrest arose again that summer across South Vietnam, the Hue Voice of Buddhist Salvation (which the people who took over the station used as their call sign) broadcast an appeal from Venerable Thich Tri Quang for people to move their family "altars to the streets," burn incense, and chant prayers. The goal was to provoke a violent response by the "Thieu-Ky clique," destroying altars that blocked the roads.[66] Radio Hanoi claimed that this happened. Supposedly, Vietnamese marines "penetrated into the Buddhist compatriots' houses, smashed the altars, intimidated entire households, broke open closets and trucks, and engaged in pillaging, thus creating a [terrible] and sorrowful scene of disorder. A number of First Corps soldiers who witnessed this heartbreaking scene shed tears and turned away."[67] If this event actually happened, it was because NLF agitators wanted it to happen. However, this incident is disputed by other witnesses. On 10 June, President Thieu sent POLWAR forces to seize the Hue station and establish a curfew to restore order. Liberation Radio claimed that "the Thieu-Ky clique had sent a psywar group to Hue to use the radio station to release false news and many fake communiqués."[68] As unrest continued, a new station came on line. It claimed that "a number of people still erroneously consider the Vietnam Unified Buddhist Church as representing all Buddhist faithful in the country." The Voice of Anticommunist Buddhist Force, a clandestine station, began broadcasting in mid-July. However, this appears to be a covert government station meant to suppress the ongoing turmoil. The station argued that religious leaders needed to abandon politics.[69]

DRV Reacts

At the end of June, the United States intensified its bombing of the North as part of continued attempts to signal American resolve. President Johnson authorized attacking petroleum and oil distribution facilities. Although the United States specifically kept North Vietnam's extensive dike systems off the target list, each time bombing occurred near one Radio Hanoi accused the raids of purposely targeting it. The North Vietnamese labeled the denials by the "Johnson-McNamara–Dean Rusk warmongering clique" as an attempt to "cover up its crimes by its deceitful offer for peaceful negotiation and its slanderous campaign about North Vietnam's aggression against South Vietnam."[70]

In the midst of these raids, Hanoi trotted out Captain Murphy Neal Jones, a US Air Force pilot shot down on 29 June, for a press conference on the intensive air attacks. "The air pirate still wore his khaki flying dress as he was brought into the conference hall," according to Radio Hanoi. "I am very much sorry for my criminal act against the Vietnamese people," Jones said. The station reported that Hanoi dripped with anger as Neal was later paraded through the city. "This towering 28-year-old American captain, who was supposed to represent the U.S. air superiority, stood meekly on an uncovered truck, his head bent." On 1 July, Hanoi rehashed the story in broadcasts to the South.[71]

Peking's New China News Agency discussed the "atmosphere of preparedness" in Hanoi resulting from the air strikes. "The antiaircraft units have begun an emulation drive to [defeat the enemy] and become 'invincible fighters' in defending the capital," the agency said.[72] Captured pilots were marched handcuffed in pairs "under armed escort through the streets . . . drooping their heads in the face of menacing fists and the formidable screams of a human sea which constituted a form of street tribunal to try the peace disturbers and child killers." Among those paraded through the city were Robinson Risner, James Stockdale, Kile Berg, and Gerald Coffee, most of whom had been captured more than a year prior but now were offered up for propaganda purposes.[73]

An article in the North Vietnamese magazine *Tuyen Huan* (Propaganda and Education) discussed the "struggle against U.S. imperialists' PSYWAR in the province of Quang Binh," North Vietnam. Ho Nhu Y, the province's deputy propaganda chief, acknowledged that "it should be said that the U.S. imperialists' war of destruction carried out by aircraft is essentially a PSYWAR act" aimed at demoralizing the citizens of North Vietnam. The article contrasted these increased attacks with the incessant calls for peace by President Johnson.[74] The United States had recently restarted the Fact Sheet campaign. Among the leaflets dropped over cities in the Red River Delta at that time, one asked, "What is the Future?" while posing alternate courses of "either more bombs and more dead sons and brothers, or an honorable negotiation." MACV intended the theme to sap the will for continued sacrifice in the face of RVN/US "willingness to negotiate."[75]

Ho Nhu Y criticized letters sent from the South "conveying their 'regards' and 'sympathy' to the compatriots and children in the North." These likely refer to the letters collected by the VIS chief discussed above. Ho especially attacked the "American pirates and their lackeys" for dropping "psywar goods" over his province. The goods included items such as "thread and needles, soap, toys, clothes, American butter and milk." Among other gifts were "radios pre-set to receive Voice of America." Ho reported that the people of Quang Binh Province destroyed the assistance and had remained unaffected by the "deceitful propaganda."[76]

While the article claimed that the PSYWAR was ineffective, Ho wrote that "a human body, no matter how healthy it may be, is vulnerable to microbe attack if it is not adequately protected and nurtured with new 'vitamins.'" The Vietnamese people required constant indoctrination to "engrave in their hearts and bones their hatred against the American pirates and their lackeys."[77] For counterpropaganda purposes, province officials had found it better to attack propaganda in general, rather than singling out any particular product, lest they draw attention to the enemy's "subtle and clever schemes." Only when a product was deemed "positively dangerous" would it be targeted for counterpropaganda.[78]

The North Vietnamese *Nhan Dan Daily* discussed the "PSYWAR Trick of the American Aggressors" on 18 July 1966.[79] The article argued that President Johnson's professions of peaceful intent only proved that the United States could not win the war as well as the bankruptcy of the "Special War." The author stated that the Americans continued to request "peaceful negotiations" as a cover for expanding the losing war and represented a "dangerous scheme of the U.S. psychological war in Vietnam."[80] However, a translated report from the chief justice of the North Vietnam Supreme People's Court indicated that fear of agent and PSYOP activities against the North was growing. In order to maintain the "Proletarian dictatorship," the court urged "repressive and preventative measures" to combat "espionage and commando activities, psywar, and 'moral action' activities, and organized anti-revolutionary activities." At this time, the North also increased its cooperation with the Stasi, East Germany's state security service, which it viewed as the best internal security organization.[81]

While the North complained about the American propaganda program, the US Army questioned its effectiveness. The Army Research Office at Duke University conducted an evaluation of American PSYOP training in July 1966. The investigators analyzed the program of instruction, facilities, and the US Army's PSYOP school at Fort Bragg. The study determined that a one-size-fits-all course could not work. The study's authors recommended breaking the course into a PSYOP officer operator course and a PSYOP staff officer course. Although some course overlap would result, field-grade officers requested more material on behavioral science and PSYOP planning and coordination. Entry-level officers desired more hands-on training.[82]

6th PSYOP Battalion Growing Pains

One glaring training problem surfaced in relation to the war in Vietnam. The six-week Military Assistance Training Advisor Psychological Operations Course at Fort Bragg was too short to be useful, especially for officers destined for immediate deployment to Vietnam. A study ordered by the Army Research

Office recommended abolishing this course. All officers should instead attend the ten-week course, according to the authors. They also noted that it was "imperative that all personnel assigned to psychological operations be instructed in the language, culture, beliefs and customs of the area" in which they were to work. However, a shortage of language-trained soldiers remained, which the US Army could not quickly resolve. The 6th PSYOP Battalion used local hires to bridge the gap. The selection of officers for PSYOP training was deemed a problem as well. PSYOP was viewed as a career detour, and the army often selected officers for positions without assessing their suitability.[83]

Leaflet production since February at the 6th Battalion's printing plant had exceeded 200 million leaflets by the fall. During the first part of the reporting period, four presses arrived from Hawaii to supplement the overworked presses. However, incessant heavy use caused the multilith presses to break down frequently. As a result, the staff began working on a modified table of organization and equipment to consider the greatly increased workload. By August, the 6th PSYOP Battalion sent forward to the Department of the Army a proposed table of organization and equipment for a PSYOP group and new PSYOP battalion formations.[84]

Generally, PSYOP companies divided into two detachments, one for heavy printing and the other for field operations. Each of the companies produced roughly 1.75 million leaflets and posters per week. Coupled with leaflets produced by the battalion and the 7th PSYOP Group, they dispersed more than 20 million leaflets monthly in each corps. These numbers may seem excessive. However, PSYOP doctrine called for density of up to thirty leaflets per 1,000 square meters. Thus, an area of one square mile would require about 78,000 leaflets. Using this estimated coverage, monthly leaflet dissemination for each company was about 256 square miles. In reality, density likely needed to be higher due to the impenetrable jungle canopy, the dispersed populations, and the effect of the humid environment, which quickly degraded leaflets. These were factors not addressed by PSYOP doctrine.[85]

The 244th PSYOP Company continued its support of tactical operations in I CTZ. Ongoing tactical operations fully engaged all loudspeaker and interrogation teams, and demand steadily increased. The 246th PSYOP Company found that support requests "more than doubled over the previous reporting period. Printing facilities of this unit were deluged," and equipment failures compounded printing shortfalls. The company attached two mobile teams, HB loudspeaker and HE audio-visual, to the Australian Task Force.[86] The 245th PSYOP Company experienced similar demands and "even though operating at maximum capability all requests for support could not be filled." Its HB and HE teams were likewise fully employed.[87]

During the quarter ending 31 July, I Field Force reported utilizing two field loudspeaker teams in direct support of Operations Beauregard and Hawthorn. Several problems were encountered during Operation Hawthorn. The target audience was highly susceptible to PSYOPs, since the PAVN soldiers had "just completed the arduous trek along the Ho Chi Minh Trail and were tired, afraid and still unsure of the terrain." Nevertheless, equipment and coordination issues between aircraft and PSYOP field teams degraded the overall effort.[88] Teams HB 1 and 2 of the 245th PSYOP Company provided general support to the US 1st Brigade, 101st Airborne in Kontum Province during Operation Beauregard, which netted six ralliers. An after-action report noted that PSYOP saturation alone was not effective. The program "must be introduced into an area where strong tactical pressure has been applied." Additionally, the payoff from such an operation could take four to six weeks. PSYOP was not an immediate answer, and a quick assessment of the results was impossible. Nonetheless, over the three-week operation, 13 million leaflets and forty-one broadcast hours produced sixty-three ralliers, exceeding the number of captured, killed, or wounded Vietcong.[89]

Operation John Paul Jones, a follow-on operation that summer to protect the rice harvest from appropriation by the Vietcong in Phu Yen Province, included one team from the 245th PSYOP. Nine million leaflets and fifty-eight hours of broadcasts produced just two ralliers. The report identified the need for PSYOP interpreters and translated copies of all leaflets to improve operations. The 1st Brigade also requested the expedited augmentation of a PSYOP staff officer and NCOs to facilitate future planning.[90]

PSYOPs had received a high priority from General Westmoreland, and the general's preference appears to have filtered down to lower levels, at least on paper. For instance, in July the US 1st Infantry Division published its PSYOP annex to the division standard operation procedures. This document detailed the division's PSYOP themes, procedures, and capabilities in order to integrate Civic Action and PSYOPs into all its operations. The division's psychological objectives included generating defection and "disaffection with the NLF cause," reducing enemy combat effectiveness, creating dissension between cadres and the rank and file, and increasing "suspicion and security precautions." The chief target audiences were VC provincial battalions and local force companies.[91]

One important PSYOP mission identified by division commander Major General William E. Dupuy was explaining the use of riot-control munitions such as tear gas. Another was simply explaining why the American troops were in an area. The division commander ordered that the PSYOP team leader be the brigade commander's adviser and attend all planning sessions. Mobile PSYOP teams' command element (the HA team) typically consisted of one officer and one enlisted man to assist in planning, dissemination, and operating loudspeaker equipment. They provided tactical staff expertise by analyzing local target groups and vulnerabilities. The division's standard operation procedure required that all units submit "changes in enemy vulnerability, significant results of friendly operations, significant changes in enemy propaganda themes or intensity of effort and feed-back" from previous operations for analysis.[92] The division held extra loudspeakers for "exploitation of developing PSYOP situations." Each brigade maintained 500,000 leaflets of various themes for a basic load, while the division maintained 5 million more.[93]

Unfortunately, commanders often understood General Westmoreland's command to use PSYOPs as an order to use it more frequently. Measurement of efforts rather than effectiveness spurred commanders to request larger drops than warranted by a particular target. There was a tendency by commanders to believe the "higher the number of leaflets dropped the more successful the psychological operation." More, however, did not necessarily correlate with more effective.[94]

The I Field Force, the US Army command in II CTZ, required that all units provide feedback on PSYOPs because information to evaluate effectiveness was "at best, weak." Normal intelligence summaries were late and focused on "order of battle, not psychological operations intelligence." To acquire this information, a PSYOP interrogation questionnaire was developed to obtain data needed to determine enemy susceptibilities and "the effectiveness of the PSYOP program." A standard PSYOP situation report was developed as well for this purpose. Despite these changes, problems still arose with PSYOP intelligence gathering. The ultimate solution was to integrate PSYOP intelligence personnel into the overall intelligence process, ensuring, for instance, that a PSYOP team had access to newly captured guerrillas for rapid debriefing.[95]

Poor theme selection (such as "stressing hunger when a VC unit had just taken all the rice from the nearby villages") or using the wrong language for a target audience also decreased PSYOP effectiveness. A related problem was dropping leaflets on ridgelines rather than valley floors normally used by the Vietcong. As an example, failure to understand correct leaflet dissemination techniques led to a request for a million leaflets that a I Field Force report noted would cover "an area of over 200 square miles of the South China Sea" with leaflets. As a result, I Field Force ordered that "drops be based on guidance from a trained PSYOP officer" to mitigate these problems. The I Field Force also limited quick-reaction leaflet requests to 50,000 "so as to speed printing time." Units were also requested to haul their own leaflets because the 245th PSYOP had no organic aircraft for distribution. The 6th PSYOP

Battalion alone produced 132,579,000 leaflets during the quarter; despite this, a production backlog remained.[96]

A new 26th PSYWAR Detachment arrived in Qui Nhon on 19 August 1966. The unit, consisting of two officers and twenty enlisted personnel, moved first to Nha Trang. The detachment moved to Saigon in November to provide additional printing support to the battalion, using a heavy mobile press capable of printing 8.5 million leaflets per week. There seemed to be no standard detachment number system, and often the same number was reused for future detachments.[97]

The largest PSYOPs campaign remained the Chieu Hoi program. By late summer, "20 new leaflets and a library of 35 loudspeaker tapes had been developed, with about 3,000 copies of the latter distributed" to the 5th Air Commando Squadron and PSYOP operators. Daily Radio Saigon and weekly Voice of Freedom broadcasts, JUSPAO cultural/drama team performances, and films helped publicize and explain Chieu Hoi. Increasingly, intelligence indicated that the NLF had ordered and conducted assassinations of Chieu Hoi returnees, and documents showed continued targeting of Chieu Hoi centers. Along with increased ground fire at PSYWAR aircraft, the level of concern that the NLF had for the program was apparent.[98]

Beyond the narrow focus of Chieu Hoi, the broader PSYOP campaign to influence fence-sitters continued. US Special Forces Team A-245 in Dak To, Kontum Province, reported success in its PSYOP program. A campaign of MEDCAPs and school construction reportedly shifted the targeted village from neutral to pro-CIDG. One measure of effectiveness was that forty personnel absent without leave returned to the force. Many other A-teams reported successful CA/PSYOP activities that assisted in building a positive relationship with local people. Teams also began to distribute radios to key communicators. Detachment B-41 in IV Corps received 100 hamlet radios, which it quickly gave away to help spread news about the upcoming 11 September Constituent Assembly elections. Locals enthusiastically received the radios. The teams also placed posters in all villages and camps in the area to explain voting procedures and to support the election.[99]

The 5th Special Forces Group teams presented movies and cultural/drama shows in late August for a combined 7,000 people in the Tien Phuoc area. Vietnamese PSYWAR soldiers ran this program entirely and, according to the report, "exemplified near-ideal counterpart cooperation." One team stated that "an additional 2,000 people attended the showing of PSYOP and western movies on 28 and 31 August." In Da Nang, an LLDB PSYWAR officer initiated a one-hour weekly radio program of military news, recruiting appeals, and music. All Special Forces camps reported biweekly distribution of PSYOP magazines and newspapers.[100]

Intensified Marine operations in I Corps in August conducted 10,655 patrols and ambushes, destroying enemy units with an estimated kill ratio of "five enemy killed for every Marine killed." Despite the increased combat, III MAF continued to concentrate its PSYOP efforts on County Fair operations. They placed special emphasis on exploiting VC "harassment and terrorism against the Vietnamese civilians." However, the Marines restricted the operations to the area directly impacted by the attack to avoid amplifying the effect of the VC terror message.[101]

A goal of many allied operations was to cut off the NLF from its support base that provided food, taxes, and recruits. This objective was being met. Liberation Radio made clear that it expected followers to overcome food shortages, "with a view to overcoming all thoughts of fearing hardships and of self-complacency with regard to the small quantities of items in stock." The broadcast said that cadres needed to "step up production in the liberated areas, in the areas of conflict . . . in order to have much food to feed the masses and intensify the resistance."[102]

While focused on their psychological war against the enemy, however, American analysts continued to discount propaganda against the US military. Themes used by the enemy included accusations that allied artillery and aircraft were killing innocent people, and equating the United States presence with French imperialism. Criticism of the use of noxious chemicals by US forces, inciting racial strife, accusations that the soldiers were puppets of Wall Street, and highlighting personal hardships in Vietnam also had high potential as themes for the North.[103] One announcer read a petition to the US government ascribed to two American pilots. Wendell Rivers and Raymond Merritt demanded that America get out of the war in order to "let the Vietnamese people settle their own affair." Rivers, an A-4E pilot, was shot down on 10 September near Vinh on his ninety-sixth mission. Merritt, an F-105 pilot flying out of Korat, Thailand, was shot down a week later. American analysts appeared slow to recognize the corrosive effects that these themes could have on the long-term morale of American troops.[104]

On 14 August, the Radio Stateside program reappeared, more skillfully produced than the original show. This time it alternated male and female announcers, who accused President Johnson of basing US involvement in Vietnam on lies. "Johnson lied in February 1965. That lie resulted in the sending of several hundred thousands of American troops to fight and die in a hopeless war in South Vietnam," they said, followed by popular music.[105] This accusation of lying leaders is a very potent theme in wartime because of the fear of risking one's life for nothing. The claim did not require much basis in fact, as the mere accusation had power. It is also a difficult charge to refute even under the best circumstances. However, the opaque way in which Johnson expanded deployments left him wide open to the charge. His attempts to send signals backfired as well, and the North capitalized on the policy shifts in its propaganda to the South. "Well, who is the one who is escalating, and who is the one who must deescalate?" Radio Hanoi asked. It listed all the troop deployments since 1964 and referred to this as "a shameless psychological warfare trick of the furious, lying, and stupid pirates." Messages such as these, repeated over various media, may have seeped into the American civilian, if not military, psyche over the course of the war.[106]

A series of captured VC entertainment troupe documents provide details on activities designed to rally rural inhabitants. A Tay Ninh group consisted of a cinema group, a photography cell, an artist cell, and a playwrights and composers cell. The latter was tasked with publishing a folk songbook, classical songbook, and a magazine. The group's orders called for reactivation of megaphone cells to make appeals to ARVN posts. They also called for secreting a man on busses to talk "about RVN conscription, the draft," and other hot topics as a way to spread propaganda. According to the report, the group published two newspapers, *Thoi Su Danh My* (Attack on the Americans) and *Co Giai Phong* (Liberation Flag). The Vietcong saw entertainment as a form of agitation. The guidance suggested agitation slogans for entertainment that included "kill the U.S., annihilate the puppets" and "we must break off the enemy 'Chieu Hoi' policy and destroy strategic hamlets and refugee camps." Not surprisingly, a key objective of the group was to "eliminate enemy entertainment." The Tay Ninh group admonished each village to "organize children's dances, popular entertainment groups and village entertainment groups if possible." However, the author recognized that the group needed to train better playwrights to accomplish this task.[107]

Another entertainment team cadre notebook captured by 1st Cavalry Division in September detailed the author's activities during the summer. The cadre's group had arrived in Binh Dinh Province after a difficult passage from the North. "Many actors and singers were sick, rice was short and the enemy operation made movement difficult," the cadre wrote. Their rations consisted of an insufficient one pound of rice per day, and "quite often the group was panic-stricken when they encountered the enemy."[108] He reported attracting between 7,000 and 8,000 spectators per show, although this number seems high.

On 24 November 1965, seventeen-year-old Huyen Thi Phen rallied to the government. She had been coerced the year prior to perform as part of a cultural and entertainment team by a twenty-year-old village VC PSYWAR chief in Vinh Long Province. Phen stated that she originally accepted the position as a dancer for the troupe because she was "fond of amusement." According to her interrogation report, Phen "could not endure the hardships" and lack of food any longer. Her husband, an assistant platoon leader in the district, had rallied shortly before her.[109]

One study of Chieu Hoi defectors determined that those who decided to defect became sensitive to Chieu Hoi appeals. This is not unlike car advertising, which mainly appeals to those considering car purchases in the near future. Outside factors influenced the decision, but the appeals transformed the decision into action. Increased military pressure was one such factor, and "the stepped-up Viet Cong conscription program" was another. With increased conscription, ideological conviction waned. Many resisted recruitment by the NLF. According to researcher Andrew R. Molner, "personal hardships and the contempt shown them by the veteran Viet Cong were among the reasons cited for defecting." He also noted that shortages of food and medical support were also mentioned often. However, per Molner, ideological conversion to government support was rated low as a motivator.[110]

Another study concurred with the low ideological component in the decision to defect, stating that "neither allied propaganda nor any latent desire to live under the GVN were determining factors in the choice to rally." Mostly, ralliers were tired and angry with the NLF. This study claimed that, "on occasion, Front troops carried Chieu Hoi leaflets with them so they could, if necessary, pretend to rally rather than be captured and imprisoned." Other reports suggested that the NLF trained draft dodgers to volunteer for service in the RVNAF to become espionage agents, and some American officers referred to the program as "VC R&R." However, the numbers belie the anecdotes, and so do subsequent studies.[111]

A thirty-two-year-old rallier in Rach Gia had served five years in the NLF. He rallied in September 1966 after collecting Chieu Hoi leaflets and listening to loudspeaker broadcasts about the program. After rallying, he joined a South Vietnamese Armed Propaganda Team. In an interview, he said that the VC taxes were nearly double those of the government and that discipline in the force was severe. He enthusiastically avowed that NLF propaganda was better than the South Vietnamese product. However, the NLF could not "put their words into action," while the GVN propaganda had facts on its side. When asked how the Vietcong were received in villages, he stated flatly: "People did not like us." The reason was that Front soldiers stole food from locals and damaged homes. In his opinion, the Vietcong had more bad soldiers compared to the ARVN, but everyone was capable of doing wrong.[112]

The NLF's Quang Tri/Thua Thien Military Region Party Committee held a conference in September to review the summer fighting and to plan for the future. Beginning with a litany of victory claims, the report indicated extreme shortages of guerrillas and cadres in the I CTZ region. Guerrillas in that region had a low "determination to fight" and a fear of the enemy, according to the committee. They unsuccessfully conducted political indoctrination and countersweep operations conducted to maintain control in the face of MACV search-and-destroy operations. VC troop health was a problem as well. Some units reported 30 percent sick, a further burden on the force. The committee found some success in proselytizing ARVN troops to defect, but this was spotty.[113]

South Vietnamese Elections

One of the components of success lacking in South Vietnam was an ideological attractor for the government. This began to change with the Constituent Assembly election held in September. Radio Saigon's domestic service transmitted Premier Ky's speech to the ARVN Psychological Warfare Conference, stressing the importance of "the first really free elections in Vietnam." Publicizing the Constituent Assembly elections in September was perhaps the outstanding PSYOP success to date. According to the 5th Special Forces Group, "The remarkable aspect of this PSYOP campaign was that it was conducted almost completely by the Vietnamese. Vietnam Information Service personnel worked tirelessly, utilizing all available media in a concerted effort to reach every potential voter," with the support of military and JUSPAO advisers. The RVNAF printed four million leaflets titled "Should I Vote on September 11?" on the eve of the election. Although the American press viewed the elections as a cynical ploy, the vote resonated with many Vietnamese who saw the country as being back on the right track.[114]

Both Radio Hanoi and Liberation Radio sought to depress turnout for the election, calling repeatedly for a boycott of the "Thieu-Ky clique's farce of deceitful elections." Using a "bandwagon" theme, they claimed that "representatives of Buddhism, the Catholic Church, the Evangelical Church, and the Cao Dai and Hoa Hao religious communities" were all against the elections. The broadcasters asserted that the election was "a mere instrument for the conduct of the U.S. aggressive war in South Vietnam." Despite these claims, turnout was heavy and the election seemed largely honest and free of fraud.[115] About 80 percent of eligible voters cast ballots on election day. Turnout was as high as 90 percent in Quang Tri Province.[116] This impressive turnout was in spite of the increased Front propaganda efforts to disrupt the elections, including harassment and terrorist actions that spiked on election day. MACV policy had precluded US PSYOP resources from influencing the elections, but afterward they did exploit the success.[117]

With the clear failure of its boycott, the North opportunistically shifted its messaging. Radio Hanoi attempted to turn the election success against the Republic of Vietnam by claiming that the massive security presence used to discourage VC interference was simply to "force the people in the cities and the area under their temporary control to go to the polls."[118] Liberation Radio referred to the election as a "shameful failure of fraud." Since the NLF "controlled" two-thirds of the population, the election was obviously a sham in their opinion. This notion was then amplified via such stations as Radio Havana, which claimed that "the regime had to mobilize this force to compel the people to vote."[119]

In the aftermath of its propaganda failure, the North also returned to the standby accusation of toxic gas usage. It continued the myth that the blood bank was engaged in biological warfare and expanded the theme of West German involvement tied to Hitler. According to Hanoi, "Along with these activities, they have wooed a number of their satellites into collaborating with them. Many chemical factories of the reactionary West German, Japanese, and Australian governments have agreed to produce toxic gas under American contracts." Unable to compete any longer in the ideological sphere, Hanoi again accused the "406th Bacteriological and Chemical Warfare Team" of nefarious practices.[120]

Conclusion

The election of September 1966 gave the Republic of Vietnam renewed popular legitimacy after three years of coups and military chaos. To be sure, it did not settle the issue, but it offered hope that the country was on the road toward a stable government. Contrary to the cynical views of many Americans, as well as North Vietnamese propaganda, most of the South viewed the elections as generally free and genuine. This was an achievement that the North had never accomplished. Saigon's task now was to create a functioning government in the midst of war.

All belligerents had some success in reaching their psychological objectives by the fall—for example, strong indicators that the NLF/North Vietnam international program increased access to the antiwar movement and raised questions of the effect this would have on the war. Their program against the US military also intensified, but few effects were observable as yet. Nonetheless, the need for increased indoctrination of both US and ARVN forces seemed apparent.

At the tactical and operational levels, the American PSYOP program made solid advances. However, the organization remained unable to meet demands, especially so for the 6th PSYOP Battalion. Nonetheless, the overwhelmed battalion made huge strides in tactical PSYOP support and innovation. More important, captured reports increasingly showed the strong impact of the program when coupled with combat pressure. What was needed was a way to quantify that change. At the strategic level, however, President Johnson's duplicity threatened the psychological war, and his attempts at political signaling continued to fall on deaf ears in Hanoi.

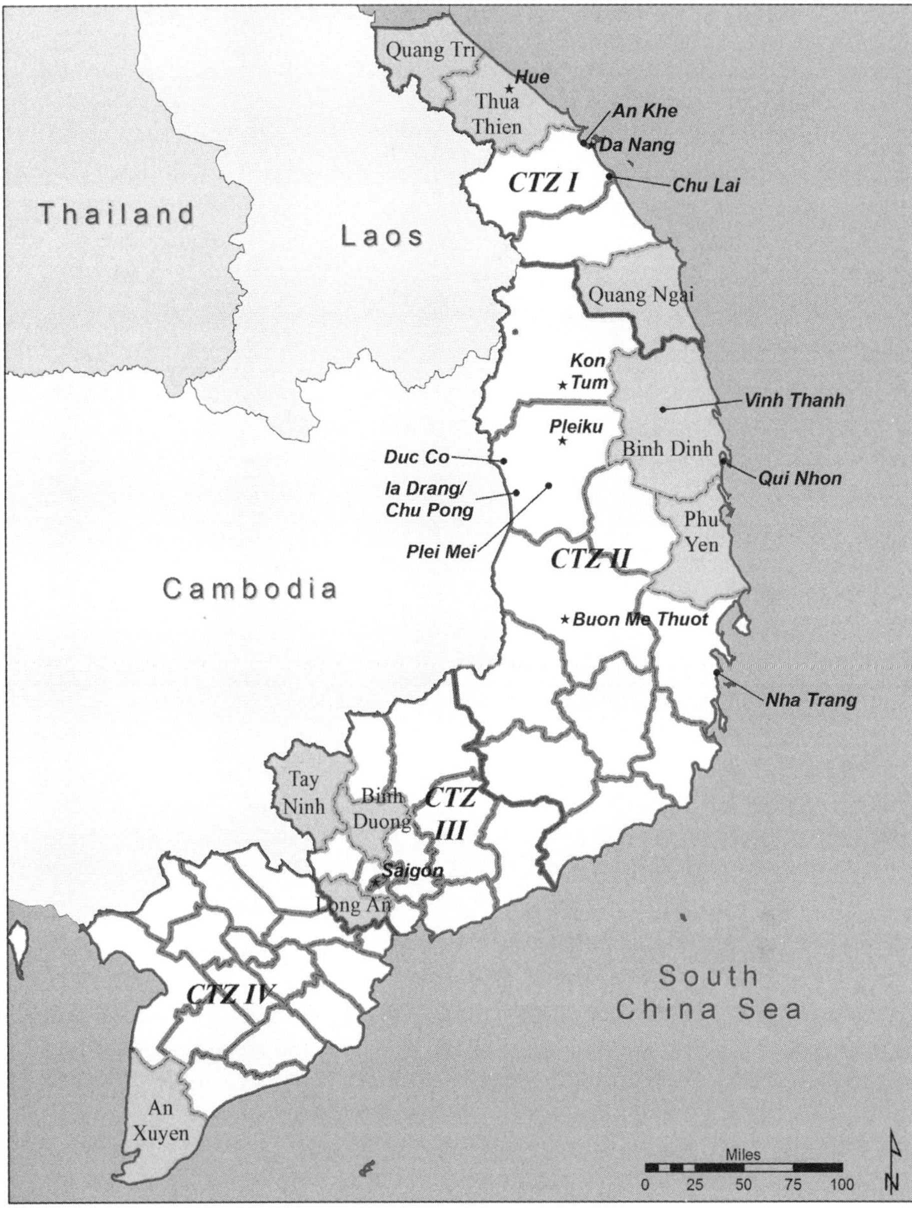

Map 10. South Vietnam, summer 1966 (chapter 9 locations)

10 Growing Pains, Fall 1966–May 1967

By the fall of 1966, the war was reaching a level of violence undreamed of a few years earlier. American military troop strength in Vietnam reached 385,000. Along with ARVN, South Korean, Australian, and other forces, nearly a million troops faced an army increasingly composed of Northerners mobilized for war in the South. Beneath the froth of combat, however, the subtler psychological war and pacification efforts grew as well. Each side could claim success and failures in this war. However, the flexible Northern program more and more failed to attract uncommitted Southerners. Increasingly, the NLF relied on threats and violence to maintain public support. The lies contained in Front propaganda were inherently unsustainable, however, "especially in the wake of the GVN triumph in the elections." One study showed that the VC structure was "vulnerable to imaginative and realistic psywar operations."[1]

The September constituent elections offered hope to the people of South Vietnam. These were followed in the spring by hamlet elections that returned local powers that had been taken away under former president Diem. In the quest to build national identification, the trends indicated positive change, though it was too soon to be certain. The South Vietnamese POLWAR and the VIS systems functioned, and here as well it was too soon to tell how effective they were. On the PSYOP front, the coming months represented a period of growing pains. Intelligence analysis was critical to effective psychological operations. Yet the system at all levels was not up to the task. Shortages in personnel, lack of focus, and limitations in language and cultural skills encumbered the intelligence process. Likewise, PSYOP equipment shortages and failures had to be corrected. Regardless, American PSYOP output was staggering. While 1966 had been a year of expansion and mobilization by all sides in the psychological war, 1967 proved to be a year of assessment of those programs. This was especially important for the Americans and the data-obsessed US secretary of defense, Robert McNamara.

Questions of Morale

The psychological operations program seemed to have some success at objectives such as weapons returns and the Chieu Hoi program that could be documented in numbers. Additionally, strong indications found in captured North Vietnamese, COSVN, and lower-level cadre documents show that the PSYOP program, in conjunction with increased combat operations, was breaking enemy morale. Signs of flagging morale in the North hinted at success in targeting the population there as well. However, that campaign's objective remained murky. The actions needed to be taken, and by whom, had not been articulated. Additionally, President Johnson's quixotic hope that Le Duan and Le Duc Tho would respond to his on-again, off-again bombing policy signals continued to be dashed.

In the face of combat and territorial losses, VC morale became a serious problem requiring constant monitoring and indoctrination by cadres. In the international arena, however, the North was having success in mobilizing front groups and utilizing access to worldwide friendly broadcasts and propaganda dissemination to target what Northern leaders perceived as a key target audience: the American population. Within the United States, access to key communicators in protest groups such as Tom Hayden, and linkages to groups such as the CANLF, gave the North and the NLF the ability to ensure that their themes and messages infused the growing antiwar movement.

American poll numbers in support of the war remained strong through 1966, and the protest marches had largely turned off the American populace. While the marches did serve to spread propaganda more widely within the movement and to keep the war in the public eye, the antiwar movement achieved no direct policy change and had come nowhere close to ending the war. In fact, one study found "approval of American bombing of Hanoi and Haiphong increased by 35 percentage points over the two-month period when this became American policy" during the summer of 1966. By the end of 1966, the American public still narrowly approved of the decision to send American troops to South Vietnam. A study conducted in March 1966 showed that, even among college students, the desire to escalate was greatest among women under twenty, and it found no support for a neutralist position. The subsequent drop in support for the decision to go to war in 1965 did not necessarily mean a desire to lose the war in 1967. Other studies after the war indicated that the demonstrations had a disproportionate effect on political elites, however. Meanwhile, in the movement's targeting of American troops, the jury was still out. Powerful themes had emerged, but it was unclear if they had any effect on soldiers.[2]

Intelligence Analysis Challenges

In order to properly assess effectiveness, both sides needed to collect data. To this end, MACV Directive 10-1, issued in September, required all units to submit a monthly PSYOP report. The goal was to monitor the number of leaflets dropped, the themes used, and the number of loudspeaker hours. However, these numbers reflected measures of effort. As for measurement of effectiveness, the report contained a narrative portion to describe effects related to areas of interest such as Chieu Hoi, North Vietnam, and tactical PSYOP support. While useful, this was not sufficient to gauge effectiveness. Measuring effects required a robust intelligence operation.

That next month, the US Army issued an updated field manual for PSYOPs. It greatly expanded doctrine for PSYOP support to counterinsurgency, covert, and unconventional operations and provided more detailed guidance for incoming PSYOP officers. The manual also argued that measures of effectiveness should not be based on quantity of effort. Instead, it stated that an analyst should study the target audience's response, reports from target audience members, observer commentaries, and indirect indicators to gauge the results of propaganda. Examples might include the enemy preventing access to the product or inoculating the audience against the message. Assessing the effects in this manner was termed "post-testing" to determine the "reasons a particular effect was achieved."[3]

In response to this need, the 6th PSYOP Battalion reorganized its intelligence section, which consisted of one "FC" team for Propaganda (Current Intelligence) and one "FD" team for Propaganda (Research and Analysis), into functional subsections. In a related development, South Vietnam's PSYWAR department directed the establishment of the PSYWAR Coordination Center, on which MACPD and the 6th PSYOP Battalion were represented. The RVNAF coordination center handled requests for printing, radio, and other PSYOP support beyond the capacity of the corps-level coordination centers. Each of these changes was meant to address observed problems and to increase the efficiency of PSYOPs in Vietnam.[4]

In spite of such improvements, PSYOPs still relied on an ineffective intelligence collection system. The overburdened American intelligence operation eyed troop attrition numbers, hoping to detect the elusive "crossover" point—neutralizing more combatants than the North could deliver to the battlefield. PSYOP Essential Elements of Information (the most critical information requirements PSYOP units had for the intelligence system to track) had "little influence on the intelligence information collection effort" that was focused on attrition.[5]

This intelligence system also suffered from poor dissemination of analytical materials, insufficient information storage and retrieval, and personnel assignment and utilization decisions, as well as a misdirected analytical effort. Consequently, a poor linkage of target analysis to the "development of leaflet, loudspeaker, and other media output based upon the identification of vulnerabilities in target groups" limited PSYOPs campaign planning. In Saigon, the 6th PSYOP Battalion's interrogators did not have official access to MACV interrogation facilities. To overcome this problem the battalion established a two-man liaison team to locate and develop sources of information in Saigon that could be exploited for PSYOP purposes. However, only about twenty-five personnel were directly involved in PSYOP intelligence.[6]

One analysis by the 7th PSYOP Group's 15th PSYOP Detachment characterized intelligence staffing as minimal compared to "the magnitude of the PSYOP effort in Vietnam which they are supporting is being considered."[7] The intelligence section's output consisted of target audience analysis and intelligence analysis products. However, the report stated that many of the current target audience studies were too broad to be of use at the tactical level. The intelligence section consisted of an intelligence community liaison section, an interrogation section, and a research and analysis section. Liaison with other units was determined to be critical to obtaining the information needed for the analysis section. Having people on the inside helped to ensure that information critical to PSYOPs did not go unnoticed. Soldiers worked unofficially with the Combined Military Interrogation Center and the Chieu Hoi Center. The study optimally recommended 220 soldiers working in PSYOP intelligence. This would enable more detailed propaganda studies and target analysis, as well as increased pretesting and post-testing.[8]

The report stated that all PSYOP products needed to be pretested on "Vietnamese closely akin to the target audience to obtain a measure of credibility and accuracy." Ralliers were one source of information, but the shortage of language-capable soldiers hampered this. The intelligence section had slots for two American translators. Nevertheless, US military replacements lacked sufficient Vietnamese language skills to be of immediate use, forcing the battalion to utilize an ARVN soldier. This took time and interpreters with the right attitude. It took about two weeks to train a soldier in the specific needs of PSYOPs. Interpreters for PSYOP interrogation teams needed a "basic understanding of psychological operations in order to be responsive to the interrogator's needs."[9]

To solve some of these problems, during the last quarter of 1966 the 6th PSYOP Battalion began a program fielding quick-reaction Psychological Operations Exploitation Teams. Consisting of a PSYOP specialist and interpreter, each team could rush to the "scene of action to exploit PSYOPs opportunities." The teams conducted fifty missions, producing sixty-eight different leaflets, seven posters, and twenty-seven tapes.[10]

A lack of personnel trained to repair PSYOP-specific equipment and a lack of "replacement parts for non-standard equipment" adversely affected the overall propaganda program as well. For instance, the 245th PSYOP Company reported that none of its five mobile audio-visual jeepsters was operational in January. These were three-quarter-ton Willys jeeps with an audio-visual shelter, of which the battalion was authorized twenty-three. Two of the companies that were assigned presses suffered frequent breakdowns, slowing production. The 246th PSYOP Company suffered equipment problems with plate-making machines and papercutters. Although these did not affect ongoing production, they limited the company's surge capability.[11]

In November, the 6th PSYOP Battalion instituted printing controls to focus on quality rather than quantity. A 245th PSYOP Company study of targeting showed that supported units were requesting large numbers of leaflets without considering the targets. It hoped that better planning and targeting would make the printing efforts more efficient. This change decreased product requests, helping the PSYOP companies' production keep pace.[12]

Surveys were one information source that JUSPAO used to help measure effectiveness trends. A subsequent Long An Province survey in 1966 indicated some shift toward government support over the previous year. This included opinion shifts on the presence of US and allied forces in Vietnam, which side bore responsibility for the war, and "airstrikes in North and South Vietnam." Large attendance at VIS-sponsored rallies and support for the Chieu Hoi program were positive indicators as well. Most people reported hearing about Chieu Hoi through "radio broadcasts and, secondarily, through leaflets and information cadres," indicating that the PSYOP messaging was breaking through.[13]

Residents of Long An tended to blame the NLF for the war and at this point expected the Republic of Vietnam to win. With more experience now dealing with Americans, respondents reported that the soldiers were "kind and jovial," stressing the "Americans' kindness to children." In one hamlet, most residents conceded that "the Americans are very kind to the people and very helpful to the poor."[14] They also tended to say that ARVN troops generally behaved correctly with inhabitants and did not steal. That positive image did not extend to the RVN's Regional and Popular Forces and VC troops, however. Additionally, despite acceptance of the bombing of North Vietnam as a cost of war, there was "unanimous criticism of civilian casualties" caused by air strikes in the South. Most understood intellectually the need for bombing;

however, many demanded that bombing missions be conducted only when based on clear intelligence and that the government "find ways of moving civilians to VC-free localities to protect them from such attacks."[15]

Combat Outpaces PSYOP Expansion

Combat operations far outstripped the planned capabilities of US PSYOP units. For instance, the 244th PSYOP Company in Da Nang disseminated 75 million leaflets by air during the fall. Of this total, it printed more than 21 million. The battalion's printing plant, out-of-country sources, and ARVN presses in the area filled the gap. Aerial loudspeaker time was nearly 270 hours for the period. On the ground, HB loudspeaker teams supported tactical operations while HE "audio-visual teams continued to reach a large number of people throughout the area," primarily supporting Civic Action.[16]

In II Corps, the 245th PSYOP Company continued to support tactical operations, including disseminating 149 million leaflets and more than 570 aerial loudspeaker hours flown by the 5th Air Commando Squadron. The company printed nearly 22 million of these leaflets. The 246th PSYOP Company disseminated more than 96 million leaflets throughout III Corps area, printing more than 25 million locally. Aerial loudspeaker time totaled 331.5 hours. The company also had seven HB loudspeaker teams as well as two HE audio-visual teams. The II Field Force credited the 741 ralliers counted during this period to the increased combat operations coupled with effective PSYOPs. However, the rapid increase of American tactical units in Vietnam threatened to place the demand "beyond the capability of the company unless additional PSYOP units are provided."[17]

The 245th PSYOP Company fielded seven HB loudspeaker teams, though the table of organization authorized only four. Personnel were taken from other sections to create the additional teams. Supported combat units utilized the teams to handle crowds and to issue Chieu Hoi appeals as well as for "revolutionary development" and "clear and search" operations. The 245th PSYOP Company also provided supplies to the ARVN 20th POLWAR Battalion in Pleiku "due to the ineffectiveness of their supply system." Meanwhile, the 246th PSYOP Company supported the US 1st, 4th, 9th, and 25th Infantry Divisions, 173rd Airborne Brigade, 196th and 199th Light Infantry Brigades, the Australian Task Force, and 11th Cavalry Regiment—far more than what the company was designed to support. However, until more PSYOP forces could be trained and deployed and a group-level formation created, the battalion struggled in the face of a surging tide of activity.[18]

The PSYOP teams in the field continued to entice ralliers. In November, the US 2nd Battalion, 502nd Infantry Regiment, 173rd Airborne, mauled troops of the PAVN 95th Regiment. A three-man team from 245th PSYOP Company conducted loudspeaker broadcasts to the surrounded regiment:

> Soldiers of the 95th, do you want to be buried in an unmarked grave? That is the only honor you will have left if you continue your senseless fight. Do you think that right? The soldiers of the US are everywhere. There is no escape. Approach the Americans with your hands above your head. Wave something white. Have your weapon muzzle down and you will not be harmed. This is your last chance and only hope. Life or death . . . the choice is yours.[19]

This script was a classic example of a surrender appeal: broadcasting while the enemy was psychologically weak and exploiting the target's fear of death. Any death in which the family would not be able to bury the body in the family grave was an important factor in Vietnamese culture. Lastly, the broadcast provided a way for the target to avoid this fate if he made the correct choice and followed the clear instructions. The incident netted thirty-six PAVN regulars, including a company executive officer. Most of them "indicated that the loudspeaker broadcasts had been the deciding factor in causing them to surrender."[20]

A directive on reducing civilian casualties stated that US commanders were now judged on both "efficient employment of men and resources and . . . the psychological implications of the manner in which the resources were used." In the opinion of the MACV commander, "An operation is not successful if the manner of its accomplishment creates an untenable psychological attitude among the population."[21] Failure to do this by some commanders, and a propensity to uproot society to win the war, attenuated support for the government despite gains on the battlefield. However, a dynamic set of competing values is difficult to balance. Failure to achieve security could lose the war as well.[22]

Targeting American Troops

Meanwhile, a November 1966 COSVN analysis of RVN/US propaganda, captured by the US 25th Infantry Division during Operation Junction City, provides details on the NLF's view of the campaign directed against them. The COSVN's political staff analysis found "a certain amount of permissiveness with regard to reading RVN publications and listening to radio broadcasts," which resulted in a "rather high desertion rate." The officer blamed this on "PSYWAR/Open Arms tricks" that deceived cadres and soldiers. The analysis cited Edward Lansdale by name as instrumental in this program. To counter the "tricks," cadres were instructed to "promote a deep hatred for the Americans and their puppets and to develop total confidence in the final" NLF victory and to frustrate the Chieu Hoi program.[23]

A 6th PSYOP Battalion analysis of VC propaganda targeting Americans at the time continued to assess its credibility as low. However, the messages did contain potent themes and symbols. For instance, one product targeting African American soldiers contained the message that "the negro soldier is fighting to protect the KKK." Ominously, the report noted increasing examples of VC leaflets "written either by [an] American or someone with an excellent knowledge of American military mannerisms and expressions." One example: "What's in it for you GI? McNamara says Americans will have to accept casualties. And that means you, brother, you won't find him sweating in the jungle or going home in a coffin. There aren't any bombs planted in the Pentagon, like there will be in your barracks, your base or the local bar."[24] The English language and military jargon had improved over earlier efforts. Other effective leaflet symbols included "photos of peace marches and draft incidents in the United States."[25]

Along this line, Bertrand Russell spoke over Radio Hanoi to "American Negro soldiers in Vietnam" on 18 November. Russell said, "It is clear that the same government and the same power structure which commits these acts against Negro citizens in the United States also directs the acts of cruelty against the Vietnamese." He called on soldiers to provide information for his tribunal while attempting to divide blacks from the US government and conflating their interests with the "oppressed" Vietnamese.[26] Radio Hanoi "acclaimed the war crimes tribunal set up in London at the initiative of Lord Bertrand Russell and pledged the South Vietnamese peoples' full assistance in its work." The station carried several similar stories about the ninety-four-year-old Russell.[27]

Targeting the South, Radio Hanoi asserted that a budding shortage of American pilots meant that the United States could not continue the air campaign. Due to heavy losses, squadrons were filled piecemeal, dissolving unit cohesiveness, it claimed. According to the broadcaster, one pilot said that "most of us have a wife and children. We often think of our family. After each flight, we sweat profusely out of fear. The atmosphere at the air force base is [very sad] because we think we may die at any moment."[28]

Despite these examples, American analysts found increasing indications that Front propaganda was "directed mainly toward countering American propaganda rather than creating new themes for their own use." This was a positive indicator that the US/SVN program had put the North on the defensive.[29] Additionally, not all Northern propaganda stunts worked as planned. Jeremiah Denton was listed as the 1,500th American pilot the North claimed to have shot down. At an October press conference, Denton defeated the meeting's propaganda value by blinking T-O-R-T-U-R-E in Morse code while mouthing the script: "This good treatment is derived from the kindness of heart of the Vietnamese government and people in spite of my vicious crimes against them. These crimes were carried out in obedience to orders from the aggressive American government."[30]

Vietcong Morale Slump

A cadre's notebook found during Operation Cedar Falls the following winter provided some useful insight into Front perceptions of how the war was going in 1966. According to this notebook, COSVN claimed five million liberated people, while the Republic of Vietnam had nine million. This reflected a drop of one million "people in the rural areas due to the presence of U.S. troops" and was five million less than its typical propaganda claims.[31] The results meant a steep drop in VC tax receipts and the potential recruitment pool. The report asserted that the Vietcong had problems winning over the people and filling recruitment quotas. It claimed that the guerrilla force had dropped to 180,000, far short of the 300,000 troops required. Additionally, there had been a decrease in the quality of troops.[32]

Along these lines, Liberation Radio notified cadres in the South of the winter–spring food production conference. The committee met to discuss the critical need to expand agricultural production. It praised those guerrillas who "cling to their rice fields." The NLF created an emulation campaign to increase production and "restore the economic foundation which has suffered from natural calamities and . . . to step up food production" for the war effort. The rice harvest protection operations cited above seem to have taken a toll on the NLF.[33]

The intensified combat during this period sapped VC morale. However, this did not automatically translate into support for the government. Rather, much of the populace merely returned to sitting on the fence to await developments. Government actions, such as the September elections, helped build support but were not sufficient and did not produce immediate results. The necessity remained to create a positive ideological reason to support the government beyond the material and security focus thus far. Nonetheless, after the elections, the basis for success seemed to be within grasp.

The Chieu Hoi program, from its inception through 1966, produced 48,031 returnees. The 20,242 that year represented an increase of 82 percent above 1965 (11,124), which analysts attributed to greater political stability, improvements to the program, "the increased PSYOP effort, and the military pressure which continued to deprive the VC/NVA of their food, ammunition, and safe haven areas."[34] Focusing propaganda appeals on "grievances, emotions, and aspirations," rather than ideology, proved most effective. Additionally, using former insurgents, rather than Americans, to create product improved the design.[35]

By late December, PSYWAR targeting of the North increased. Leaflets dropped on 25 December announced that Voice of Freedom radio broadcasts would begin supplying names of "North Vietnamese soldiers killed, wounded or captured in South Vietnam" as a way to increase listenership in the North. Many Northerners would never know their loved ones' fate without such notification. Broadcasting names in turn functioned as a hook for the actual messaging as more listeners possibly tuned in. Programming on the black stations often consisted of satirical commentaries. One program, "Talk to the North," used Communist propaganda examples to discredit the regime. Other programs digested anti-Communist books or sandwiched humor into propaganda, interspersing classical Vietnamese music with jazz. Reportedly, both the NLF and North Vietnam regularly attempted to jam the PSYOP programs but left the entertainment programs alone.[36]

To evaluate the campaign, in November Secretary of State Rusk requested information on the effectiveness of both the leaflet program aimed at North Vietnam and the overall psychological campaign directed against the civilian population and infiltrators from the North. A joint MACV/JUSPAO study the following month revealed that "the Hanoi regime fears the effectiveness of psychological warfare directed against it." The report quoted Ho Chi Minh as saying on 22 October 1966: "We must further heighten our determination to smash all spying and psychological warfare activities of the enemy, check and smash in time all maneuvers of the reactionaries and maintain public order and security in North Vietnam." The report noted that the PAVN's daily organ, *Quan Doi Nhan Dan*, "was even more explicit in expressing alarm over the effectiveness of PSYWAR directed against the North."[37]

The 6th PSYOP Battalion was printing about 64 million leaflets per month and had produced approximately 400 million leaflets since February. Over the course of the year, its companies had each increased printing output from about 6 million per month to more than 7.5 million, a 20 percent increase. The battalion's printing plant had increased output from about 33 million per month to about 44 million, a 25 percent increase. The battalion headquarters also assumed responsibility for placing the orders for all leaflets used in Vietnam that were printed outside the country. This amounted to a monthly order of 215 million leaflets, mostly produced by the 7th PSYOP Group's Okinawa printing facilities in Machiminato. The 7th PSYOP Group alone produced about 160 million leaflets per month. JUSPAO produced another 15 million per month, as well as nearly a half-million multicolored posters.

All told, roughly 350 million leaflets were produced for Vietnam monthly, likely more than 3 billion per year. To disseminate this, C-47 aircraft carried up to 5 million leaflets, while the smaller U-10 could deliver 100,000 leaflets. Although some of the leaflets were disseminated by hand and some were lost to spoilage, leaflet drops consumed many flight hours.[38]

The organization of the 6th PSYOP Battalion continued to evolve with the needs of the war. The unit paid the price of that active participation. As the 6th PSYOP Battalion celebrated its first year of existence, unit members had received thirty-seven Bronze Star Medals, six Combat Infantryman's Badges, and twenty-nine Air Medals. Staff Sergeant Roger Terwilliger, a PSYOP operative, and Specialist Darel Sills, a photographic expert, both assigned to the 246th PSYOP Company, had also posthumously received the Purple Heart for grenade wounds received while operating with the 173rd Airborne near Bien Hoa on 13 October.[39]

On 4 December, the Vietcong attacked the Saigon headquarters of the 6th PSYOP Battalion in the former Kinh Do Theater. The building had been the objective of several previous terrorist attacks prior to its occupation by the PSYOP unit. However, the 4:50 AM explosion blew the doors off buildings two blocks away and blasted the ceiling off the theater. The PLAF command hailed this attack and an earlier one against Tan Son Nhut Air Base. It claimed that while the Americans "were still panic-stricken" after the air base attack, "liberation forces used bombs to blow up a huge building downtown which was used by the U.S. officers as their psywar headquarters." Despite the claims, no one died in the attack, although eleven Americans were injured.[40] Radio Hanoi acclaimed the destruction of the PSYOP headquarters in Vietnam the same day. Wallace J. Moulis, commander of 6th PSYOP Battalion, responded that "it appears that we have been hurting them so much with our propaganda that they felt obliged to retaliate."[41]

Challenges Measuring Success

Throughout 1967, authorities struggled with ways to measure the effectiveness of the now wide-ranging PSYOP program. Beginning in January, MACV instituted the Hamlet Evaluation System in an attempt to measure who controlled the population. The system was designed to provide data to guide MACV in the pacification effort. Although fraught with subjective criteria, the HES did provide a wide array of useful data for trend analysis. In future years, the system was adjusted to reduce subjectivity. The HES did not specifically focus on PSYOPs, but data it collected was aimed at understanding the political and information environments at the hamlet level. Properly used, this could help guide the PSYOP effort. Despite its clear limitations, the HES provided better data than anyone else had, including the North Vietnamese.

A 6th PSYOP Battalion briefing conducted at this time reiterated that PSYOP effectiveness could not be measured by output alone. Interviews with ralliers and prisoners were seen as the key for assessing the overall program. "A great number" of ralliers had seen products, and many of them said that PSYOPs "influenced them, made them more concerned over their personal safety, made them aware that the truth had been kept from them, and in many cases provided the straw which made them rally to the GVN." PSYOPs could be effective, though, only when "integrated into operational plans and must be used to exploit the enemy's tactical defeats and strategic weaknesses."[42] However, JUSPAO reminded all PSYOP personnel that effective psychological operations for the Chieu Hoi program must be preplanned "and coordinated with artillery fire, tactical air strikes and ground operations." For best results against the Vietcong, combat and psychological operations needed to be coupled together.[43]

To illustrate ways to achieve this, JUSPAO released *PsyOps in Vietnam: Indications of Effectiveness*. This study illustrated the effects of PSYOPs through anecdotes. Because of the time lag between reception of a propaganda message and action based upon that message, measuring effectiveness often required looking at indirect indicators over time. This study assessed three. "What the other side says" translated captured documents; "How some results were achieved" contained anecdotes from US and ARVN psychological operations units; and "What the ralliers say" presented interviews with ralliers shortly after they returned to government control. The authors concluded that "psychological operations alone have produced few victories, but by playing on emotions generated by the stress of war . . . they can and have sparked a man or a unit into taking desired actions: to desert, surrender, mutiny, defect; to adopt passive resistance."[44]

Throughout numerous short extracts, five complete texts, and commentaries from the North Vietnamese press and radio, one key theme emerged from the documents. They contained consistent references to the fear that defectors would take intelligence with them that the government forces could act on. Several VC situation reports provide anecdotes of this occurring. In one instance, a security directive from the COSVN's Current Affairs Committee stated that "as a main part of the preservation of secrecy campaign, the counter-Chieu Hoi policy and mission should be enforced" and all cadre indoctrinated on this policy.[45]

The South Vietnam Liberation Army's political department produced one particularly detailed memo, "Counter Measures against Enemy PSYWAR and Chieu Hoi Activity." It stated: "Defense against [enemy] PSYWAR has recently been given added emphasis by all echelons; but the overall picture is one filled with relatively serious shortcomings. We tend to underrate enemy delusive propaganda themes. Many persons commonly listen to enemy radio broadcasts and read enemy publications."[46] The memo specified actions to be taken by cadres to counter the effectiveness of this campaign. Enemy PSYWAR sowed "bad seeds of pleasure-loving fear of hardships [and] fear of sacrifice" among Front supporters, debasing their "fighting spirit." The political department directed cadres to "point out the dangerous characteristic of the enemy plan which has the ability of ruining us politically and ideologically."[47]

Operations against the North

The US PSYOP campaign outside the RVN increased considerably as well. Forces dropped nearly 60 million leaflets throughout Indochina as part of the Ho Chi Minh Trail, Tally Ho, and Fact Sheet/Frantic Goat campaigns. With the limited bombing campaign against the North leaving most North Vietnamese untouched, the popular "rally" effect normally associated with bombing quickly wore off. As a result, people were left to confront shortages, a struggling economy, and a rigid police state without a scapegoat to blame. The Rolling Thunder campaign was designed to "apply steadily increasing pressure against NVN [North Vietnam] in order to cause Hanoi to cease its aggression in SVN [South Vietnam] and to make continued support of the Viet Cong insurgency as difficult and costly as possible."[48]

The Ho Chi Minh Trail leaflet campaign's goals were to "plant doubt in the minds" of infiltrators as to their survival prospects. The long, arduous trip southward by PAVN troops, under continuous threat of bombing, offered fertile ground to plant seeds of doubt. MACV analysts assessed that this program seemed to correlate with the increase of rallying PAVN soldiers. MACV's *Command History* credited this increase particularly to leaflets using the theme "Born in the North to Die in the South." The *Command History* warned, though, that "other factors, especially hardships, danger, and possible death, played an important if not a decisive role in the results of the campaign."[49]

US Air Force analysts assessed that the few targets struck in the Hanoi–Haiphong region had "considerable" physical and psychological impacts. However, the "gradual, drawn-out, and cautiously constrained air campaign" limited the effect. USAF leaders thought a stronger bombing campaign would increase the psychological effect. However, PSYOP effects cannot be measured like a simple linear equation. Additional dynamic factors beyond American control, such as the rally effect of bombing survivors, or (on the other side) reaction to an increasing police state, made it hard to predict the psychological outcomes of a policy change. The air force boosted pressure on the North throughout the year. However, it was left with a limited target set authorized by the White House. The authorized points always ensured that some high-value targets in the North remained undestroyed so that the administration could threaten them in the future as a "signal."[50]

Nonetheless, the US PSYOP campaign against the North began to produce observable results. Soviet diplomats in Hanoi reported "growing uneasiness among Vietnamese Communist leaders over the course of the struggle." The Soviet embassy's annual political report noted that the NLF could not continue the war without aid from the North and that the North "could not withstand American military pressure without the support of the Socialist countries whose aid comprised two-thirds of the DRV's budget."[51] Interviews with defectors and PAVN soldiers revealed that North Vietnam People's Security Forces (the internal security force of the North) prevented citizens from reading leaflets and that they "would go into an area after a leaflet drop and police up every leaflet they could find." Regardless, the interviewees stated that people still read the leaflets, although they "were unwilling to discuss them with their fellow countrymen or fellow soldiers," making effectiveness difficult to gauge.[52]

Reports about the theft of Soviet missiles shipped to North Vietnam showed the difficulty that the North had in balancing relationships with China and Russia. By 1967, relations between the two giants had deteriorated, and numerous reports indicated that it affected the transshipment of Soviet arms by rail through China. *Nhan Dan* "strongly condemned" the charge as an "odious slander by the imperialists." However, the event was likely true, given the level of vitriol between China and the Soviet Union. Regardless, it played into themes that SOG utilized to raise cognitive dissonance within North Vietnam about its two major supporters.[53]

SOG's black propaganda effort had grown considerably. The various programs employed about 32 US personnel and up to 800 South Vietnamese civilian and military personnel. The 1967 budget for these programs was approximately $3.7 million. Among the deception operations that SOG instituted was Project Urgency. This project took hardcore captured PAVN soldiers, enlisted them as "agents," and sent them back to the North—knowing they would report to the government on what had happened. Realizing that the regime would be paranoid about the trustworthiness of these men, SOG planted material on them tying them to the "resistance." SOG staff hoped "they'd all be killed or give false information that we planted."[54]

The similar Project Borden recruited North Vietnamese POWs as SOG agents and used them to "serve the objectives of the diversionary program." The idea was that, upon returning to the North, they would reveal all they had learned and unwittingly provide false information to the North. A related program, meant to add confirmation to the other deception programs, began later that year. Project Oodles broadcast false messages to fourteen phantom teams supposedly operating in the North.[55] PSYOP indoctrination in the SSPL program at Cu Lao Cham (Paradise Island) continued as well. During the previous year, 353 prisoners were captured and transported to the island. All but one later returned to the North. SOG also distributed 2,600 pretuned "Peanuts" PSYOP radios produced by the Counterinsurgency Support Office along with leaflet drops and gift kits.[56]

At the same time, the North boosted its international propaganda efforts and hosted increasing numbers of antiwar movement leaders in pursuit of its psychological objectives. The visitors typically received a tour meant to amplify and support DRV propaganda themes. These tours magnified the level of destruction created by the very restricted bombing campaign. Patricia Griffith, wife of a Cornell University professor, was impressed by the damage she saw during her Hanoi visit. She traveled along with professor Sid Peck, a Communist Party USA committeeman, among others. In a Radio Hanoi international broadcast, Griffith praised the "human dignity and beauty" she found in the North. "We were accepted as sisters and welcomed in every village as representatives of the American people, who they believe will surely aid them in their struggle for freedom and independence." Like most of the visits by foreigners, they were presented with "Potemkin Villages" throughout their tours.[57]

Radio Hanoi praised other Americans who condemned US policy in Vietnam. Martin Luther King Jr., according to Radio Hanoi, "affirmed that the actions of the United States in Vietnam, which were a new form of colonialism, have put the country in a state of moral and political isolation." Others quoted for the story were Senator Eugene McCarthy, Governor George Romney, and the historian and speechwriter Arthur Schlesinger.[58]

Against these efforts, the US Army had a modest counterpropaganda program that targeted American civilians and military members. *The Big Picture* debuted in 1953 as "a weekly, half-hour documentary format television program produced by the Army Pictorial Service of the Signal Corps." A mixture of public affairs, propaganda, and history, the program began as a vehicle for telling the army's side of the story at a critical period for a service in search of relevance in a nuclear world. Weekly shows, composed of "internal film bulletins and archival documentaries," played on up to 350 commercial channels nationwide over more than two decades. Distributed at no charge, it was first shown on ABC until 1959. In subsequent years, it played on various stations. *The Big Picture* also played on US military TV channels around the world.[59]

For example, *The Big Picture* aired on KERA public television in Dallas between about 1964 and 1968. The show played at 9 PM on Fridays, in a time slot right after *Hogan's Heroes* and *Gomer Pyle*, although it went head-to-head with *Star Trek* for one season. The target audience likely skewed toward an older demographic—possibly the parents of those protesting the war. The target age group would also include many World War II veterans, a group that might be more susceptible to the messages.[60]

It is difficult to track the number of shows covering the Vietnam War. However, an analysis of the titles suggests, unsurprisingly, that Vietnam coverage increased after the arrival of US combat forces in 1965. Thematically, the Vietnam-centered programs tended to focus on pacification operations, often narrated by Hollywood luminaries. Glenn Ford hosted the 1965 episode, on the 173rd Airborne Brigade, titled "The Unique War." This was adapted in 1966 as a troop indoctrination film. In 1966, John Wayne hosted a show from Vietnam, "A Nation Builds under Fire." This presaged his 1968 film *The Green Berets* on the same subject. The story selection in the *Big Picture* series in the United States indicated that the military understood the need to combat the corrosive effects of propaganda. These shows tended toward a fairly low-key counterpropaganda approach. They presented the army's view but did not directly attack any North Vietnamese or antiwar movement themes.[61]

Overextended

The ground war reached new levels of intensity in January. Operation Cedar Falls in the Iron Triangle was the largest combat operation to that point of the war, involving 30,000 troops. The objective was to destroy a Front base area and supply dump and to remove the civilian populace. The American forces could then "initiate a free fire zone and prevent enemy reuse of the area." The Vietcong had used the Iron Triangle as a base area since 1945, but since 1963 it had accelerated the construction of "a complex of cement fortifications, three-tier tunnel systems, ammunition depots, munitions factories, hospitals, troop rest and recreation areas, and communication centers."[62]

Two days prior to the beginning of the January 1967 operation, PSYOP troops targeted the region with 215,000 leaflets instructing inhabitants "to go to designated assembly points for evacuation before the shooting started," purposely forsaking the element of surprise. The goal was less about defeating the enemy than destroying a base area to prevent the enemy from using this critical region. Thus, surprise was not essential. To achieve the objective, civilians within a 100-square-kilometer region had to be moved.[63] During the operation, PSYOP soldiers conducted assessments of refugees. This provided information to refugee experts on the evacuees' needs that, if met, would produce the highest psychological payoff. Based on this information, schools were built for the refugees first. As a result of the operation, 455 guerrillas rallied to the government, with 306 more during the following six weeks.[64]

Even as Operation Cedar Falls continued, the 1967 Tet Chieu Hoi campaign began. However, it got off to a sluggish start due to slow GVN planning for the national Tet campaign. On 10 January, the Ministry of Chieu Hoi formed the Joint Tet/Chieu Hoi Campaign Task Force and the next day issued campaign plans to the regional and provincial Chieu Hoi service and VIS chiefs. The campaign's goal was to take advantage of family fealty during the holiday to increase desertion. Meanwhile, for more than a month all Northern venues broadcast statements announcing a Tet truce. However, the combat pressure on the NLF did not abate. The "U.S. forces and the country-selling lackeys" did not follow the NLF's truce "order and even violated it more brazenly and barbarously," per LBS.[65] The effectiveness of the surrender campaign quickly became apparent. At I Corps, 238 rallied in January, including eighty-four in Quang Nam Province alone. Because of the success, the commander of III MAF "directed all of his unit commanders to expand and intensify psychological operations during the forthcoming period of TET" in order to sustain the momentum into the post-Tet period.[66]

Despite the accomplishments of this campaign, many in the 6th PSYOP Battalion understood by the spring that PSYOP product quality needed to be improved. Pretesting of leaflet/poster texts and illustrations pointed to cultural design problems. For instance, a "photo of two elderly men, even though smiling, showed them squatting before a doorway to a house," but pretests indicated that viewers perceived them as prisoners. A smile, as comprehended in America, did not necessarily translate to all Vietnamese situations. Additionally, JUSPAO warned, other images and symbols meaningful to Americans had "absolutely no meaning whatsoever to a Vietnamese." JUSPAO produced a leaflet design guide in March, "Communicating with Vietnamese through Leaflets," in an effort to improve quality. One key suggestion was to design the leaflet in Vietnamese, rather than translate it from English, and to use Vietnamese artists attuned to local sensitivities. The study also warned that English-language colloquialisms often sounded "false or foolish when expressed in Vietnamese."[67] Furthermore, American PSYOP specialists were enjoined to avoid demonizing the enemy. Information gleaned from interviews was used to improve products. One such exploitable detail was the family hardship "caused by the absence of male members serving in VC units." In light of such lessons learned, the 6th PSYOP Battalion revised its leaflet design procedures.[68]

The 5th Air Commando Squadron's leaders also sought to improve that unit's effectiveness. They requested a study of the most favorable times to conduct aerial loudspeaker broadcasts. The 246th PSYOP Company provided an analysis showing that the hours 0600–0800 and 2200–dawn were the most favorable times because the lack of activity made the messages more easily heard. Additionally, as one rallier related, during those tranquil periods "they most often thought of their families, making them more susceptible to 'family' messages at those times."[69] The 6th PSYOP Battalion provided liaison officers to the 7th Air Force Tactical Air Control Center to coordinate PSYOP missions. The liaison attended daily briefings, targeted leaflet drops, and wrote the monthly "order for the leaflet-dropping aircraft."[70]

Tactical PSYOP support increased considerably that winter. The 245th PSYOP Company disseminated 145 million leaflets, of which the company printed nearly 16 million during the months prior to Tet. Company field teams conducted more than 550 hours of loudspeaker missions in support of ten US and Korean operations and provided revolutionary development support as well. One frequent tactic coupled Chieu Hoi appeals to "safety instructions to the populace" so as to avoid unintended casualties. Aerial loudspeaker sorties totaled 854 by the 5th Air Commando Squadron and 99 by army aircraft in II Corps alone. Due in part to the intensified Tet campaign, 3,347 rallied in II CTZ during the reporting period, an increase of 2,090 over the previous quarter. The increased rate of ralliers indicated success for the overall program there.[71]

The 244th PSYOP Company's Tet campaign appeared successful as well. The unit mobilized all PSYOP resources in I CTZ, coordinating among the PSYOP elements of III MAF, MACV Advisory Group, and Special Forces. The 244th's PSYOP effort seemed to result in a 20 percent increase in ralliers over thc previous quarter, showing the benefits of a "combined, concentrated effort aimed at neutral and enemy targets for a specific purpose."[72] In February, 279 rallied, and they consistently told interrogators that they had heard and seen Chieu Hoi appeals that stimulated their decision to rally. Ralliers also "showed a deeper understanding of the advantages offered by the Chieu Hoi program than those who had been interrogated during previous months."[73]

With the arrival of the US 9th Infantry Division in the Mekong Delta in December, MACV commissioned a survey in Dinh Tuong Province "to determine the population's attitude" about the new neighbor. The newly formed 19th PSYOP Company became operational shortly before the arrival of the division. The company quickly assumed the role of introducing this new American division to local inhabitants in hopes of mitigating potential problems. The survey found that residents most feared "inflation, possible civilian casualties caused by U.S. firepower and prostitution." This study provided a baseline to assess future shifts in public attitudes and to guide PSYOP activities. MACV also surveyed people living along the Bassac River regarding US Navy river operations. This study revealed favorable reception due to the increased security and MEDCAPs that the unit provided.[74]

Even as American deployments continued, additional allied PSYOP forces arrived. In August 1966, the Republic of Korea deployed a PSYOP and Civil Affairs company to Nha Trang, consisting of 134 soldiers and 5 civilians. They were equipped with six audio-visual jeepsters and fifteen loudspeakers as well as two multilith offset presses, although most of the equipment arrived much later than the unit. The company's three platoons were attached to the ROK Marine Blue Dragon Brigade and the ROK Army Tiger and White Horse Divisions. Throughout the war, ROK PSYOP units used tae kwan do as a way to build positive relations with Vietnamese, conducting training and competitions. As part of the radio expansion project, in early 1967 the Seoul Domestic Service announced Korean plans to also build radio stations in South Vietnam. Located in Saigon, Nha Trang, Chu Lai, Tuy Hoa, and Bien Hoa, they were built by the United States but staffed by South Koreans. These stations conducted anti-VC PSYWAR and provided entertainment to ROK troops. A Philippine Civic Action platoon arrived the month following the ROK unit. This platoon indirectly conducted PSYOPs in Tay Ninh Province, spreading information while conducting medical aid programs. In mid-1967, a Royal Thai PSYOP platoon arrived in Vietnam, rounding out allied PSYOP forces during the period of this study.[75]

By 1967, the 5th Air Commando Squadron, like the 6th PSYOP Battalion, was overextended. The Air Force split the squadron to create the 9th Air Commando Squadron. The 5th covered III and IV CTZs, while the newly formed unit covered the northernmost corps. The new squadron based its three flights at Da Nang, Phan Rang, and Bien Hoa, as well as elements at Nha Trang and the Pleiku airfield, to cover the region. Six loudspeaker-equipped C-47s arrived in Pleiku for the new unit that January. Eighteen improved loudspeaker-equipped, O-2B Skymasters were scheduled to be operational by the summer to replace the older U-10s. One drawback to the Skymaster, however, was its limited leaflet capacity. The weight of the speakers, coupled with the requirement to carry a person to kick out the leaflet boxes, limited the aircraft to about 50,000 leaflets per mission, much less than the 1.5 million that a C-47 could carry.[76]

The Johnson administration placed strict limitations on Cambodian overflights that limited distribution along the Ho Chi Minh Trail. Aircraft used the wind-drift technique, flying along the border to insert leaflets a few miles into Cambodia. However, the Joint Chiefs of Staff had stipulated that these drops be conducted at night, by cargo aircraft, and not below an altitude of 6,000 feet. On 13 March, the Joint Chiefs of Staff authorized Operation Camel Path flights 15–20 kilometers inside Cambodia in limited areas for a six-month test. Only Vietnamese-language leaflets targeting PAVN forces were authorized, however. The Americans wanted to avoid any hint of targeting Cambodians, which would threaten the delicate relations with Cambodia and Prince Norodum Sihanouk. The authorization specified that "under no circumstances will anyone having knowledge about these operations acknowledge that leaflet drops are being conducted over Cambodia." All other comments, even off the record, were prohibited; if pressed, they could say: "We have for some time been dropping leaflets in South Vietnamese border area. Given wind drift, we assume some of these have been falling inside Cambodia. Leaflets are directed at Viet Cong and North Vietnamese forces" (i.e., in support of the Chieu Hoi program). In the event a crash occurred inside Cambodia, navigational error was to be blamed.[77]

PSYOP Gimmicks

On occasion, soldiers used PSYOP stunts to achieve results. Although tricks can be useful, if done improperly they could discredit the entire PSYOP effort. Members of the US 4th Infantry Division built a contraption called Magic Eye, consisting "of two boxes which were equipped with impressive dials, gauges, colored lights, horns, and an antenna." People were told that it could detect VC members who passed by the device and that "those who confessed to being VC would not be mistreated." When known Vietcong entered the area between the two boxes, a hidden observer activated a switch that caused lights to flash and horns to sound. "In one community three VC confessed before reaching thc Magic Eye." It may have achieved some tactical results, but its utility on a broader scale was questionable.[78]

Another gimmick widely used by American soldiers involved the ace of spades playing card. Many American soldiers believed that the ace acted like a death card in tarot. Because fortune-tellers were widely believed in Vietnam, the idea of using the card was born. William Shelton, a Special Forces officer assigned to SOG's Command and Control North, recalled using the "ace of spades trick" as PSYOP strategy. Eventually its use spread throughout Vietnam. American soldiers left cards on dead bodies or at battle scenes as a threat. Later, playing card manufacturers got in the game, producing unit-specific cards.[79] (See photo 23.)

The 245th PSYOP Company requested an analysis of the usefulness of the ace of spades as a calling card. The study, conducted by the 6th PSYOP Battalion S2, concluded "that it would not be an effective/meaningful symbol." Culturally, most Vietnamese had little or no experience with Western playing cards and attached no value to the ace.[80] According to JUSPAO, "The notion was based on isolated instances of behavior among Montagnard tribes-men familiar from French days with the Western deck of cards." These would be the people with whom Shelton operated. Among the broader populace, however, the card did not "trigger substantial fear reactions" because Vietnamese cards used Chinese symbolism. A JUSPAO study warned: "Tampering with deeply-held beliefs . . . should only be attempted if one can get away with it and the game is indeed worth the candle." The cards were most often used by non-PSYOP soldiers who thought they were conducting psychological operations.[81]

Stunts aside, actual PSYOPs continued to be effective. A document captured during Operation Lam Son 67 cataloged ongoing problems PSYOPs caused the Vietcong. Marked for distribution only to higher-level cadres, the memo, titled "Countering of 'Open Arms' Activities to Be Increased" and dated 14 March 1967, was not for general dissemination. "The recruitment of soldiers and personnel was done without caution," it noted, and "defense against PSYWAR and Open Arms . . . was not rigorously undertaken. As a result, many soldiers and workers have lost spirit and become demoralized." Cases of dereliction of duty were widely reported, and nearly ninety ralliers left the unit that spring. Local elements reported having no appropriate plan to counter the PSYOP program. This document provided further evidence of the effectiveness of the Tet campaign. "In some areas, the number of soldiers, and cadre workers who deserted or defected to the enemy as a result of the appeals of their family," reached 80 percent, and the number of guerrillas and cadres reached 70 percent.[82] The memo urged cadres to "lay bare the enemy scheme, and strengthen the people's morale . . . motivate the people to turn a deaf ear to the enemy's propaganda, demagoguism and threats, not to read their leaflets, books, papers and posters (in weak areas). Every time the enemy drops leaflets, the cadre and servicemen should collect them and turn them over to the organization instead of keeping them."[83] These had been longstanding orders throughout the NLF that clearly were not being obeyed. The memo further ordered all echelons to study the enemy PSYOP programs and "improvise means to counter attack and help the Committee echelon to guide."[84]

The American embassy in Saigon later used this document in a press release for propaganda purposes. The release juxtaposed the memo with the incredibly high VC casualties. In March, a record 9,015 were reported killed, with 5,557 ralliers reported nationwide.[85] This compared to 2,917 ralliers for February.[86]

Contrary to the anecdotal evidence, a study conducted by Simulmatics Corporation for JUSPAO stated bluntly: "Chieu Hoi is not a VC R and R program." The study noted that very few returned to the NLF; as important, the NLF did "not usually trust anyone who has been in GVN hands." The author encouraged "efforts to compromise almost all ralliers" in the eyes of the NLF through effective PSYOPs.[87] Confessed ralliers were unlikely to return to the NLF.

Additionally, the NLF discriminated between rich and middle-class peasants, "though many VC leaders come from that background." Thematically, Simulmatics determined this discrimination, and a preference for officers in doling out "food, clothing, freedom, security and sexual outlets," opened exploitable cracks within the enemy, especially the PAVN. These fracture lines needed to be expanded. In a cautionary note, however, the author wrote that "PSYWAR should never diabolize or insult the VC," especially through drawings, which needed to be culturally respectful. Also, messages should criticize the NLF's "failure to carry out its promises" rather than the programs themselves, which most members perceived as "basically good."[88]

A VC platoon political officer who rallied in 1967 mentioned points from another high-level briefing on the desertion problem. He said that the Americans brought progress such as electricity, water pumps, and roads that the NLF could not compete with. The lure of material progress thwarted Front propaganda. The officer said that, despite telling people the socialist regime "had been reconstructing the North for many years . . . but the progress wasn't as good as that achieved by the South in two or three years of American aid." This cadre argued that delivering these "modern conveniences to the people was a very poisonous scheme of the Americans."[89]

In a related problem for the Vietcong, Edward Lansdale had returned to Vietnam as senior liaison officer at the US Mission. He reported later that year on the extensive nature of the NLF's tax system. Highway taxation was conducted between the first and fifteenth of the month. If this did not raise enough revenue, the road was mined for the remainder of the month. In another fundraising tactic, Front tax collectors assisted people in avoiding government red tape, then threatened to turn in the person as leverage to extract increased taxes as "protection" money. Most bars frequented by Americans also had to pay taxes in this manner, often reaching 25 percent of their income. Firewood purchased for American camps was priced so as to absorb such VC taxation. Oppressive actions such as this ran counter to the efforts to mobilize people to support the NLF.[90]

Likewise, the PLAF's political department issued a report ("Counter Measures against Enemy PSYWAR and Chieu Hoi Activity"), labeling the programs as a "manifestly dangerous plot." The report noted serious shortcomings in defending against these programs and that many people listened to "enemy radio broadcasts and read enemy publications." According to the report, desertion was rising at an alarming rate; worse still, "prior to leaving, some assassinated our cadre and men stole secret documents and later led the enemy into our area for destruction." Although some Americans feared that the Chieu Hoi program was simply a rest-and-recreation program for guerrillas, such was not the case. Internal security paranoia on the part of all elements—COSVN, PAVN, and PLAF—precluded them from trusting those who had any contact with the Chieu Hoi program. Small numbers may have infiltrated via the Chieu Hoi system, but it was not the norm. The author recommended heightening class consciousness and increasing indoctrination of troops and cadres to build a "high sense of resentment against the enemy."[91]

At its heart, this was the goal of agitation: to mold, exacerbate, and redirect any grievance in an opportunistic fashion. Doing so opened the target audience to manipulation. Focusing on hatred and emotionally fixing an enemy trumped logic and facts. The Vietcong saw hatred as the prime defense against enemy propaganda. To influence behaviors, agit-prop cadres employed "whatever arguments" were "likely to be effective."[92] Per one Front propaganda handbook: "We exploit every existing conflict. When we propagandize a person, we have to find out what his dissatisfactions are and then work on his dissatisfaction to incite him to struggle."[93]

Conclusion

As the war intensified during the summer of 1967, internal Front and North Vietnamese documents showed that morale was near a breaking point. These accounts cover the entire country and exceed the ability of a study such as this to fully discuss. Record numbers of ralliers, KIAs, and the effects of increased pacification portended a cascading collapse of the North Vietnamese position. By no means was the war over, however. Even in light of these trends, the war could drag on as long as the North had the will to send soldiers to invade the South. Nevertheless, without a game-changing move, such trends would make it impossible for the North to mobilize sufficient forces in the South to stave off eventual defeat. SOG's efforts to sap that will seemed to be having an effect in the North, as indicated by intelligence reports, ralliers, and diplomatic sources. Draconian internal security measures avoided complete collapse. However, how long could the "North First" faction be kept in check? Despite successes such as these, the US PSYOP program continued to search for solutions to a host of problems. Studies showed evident but imperfect improvement. US psychological operators likewise identified problems in intelligence, equipment, and organization and made corrections. However, reform was an unending process.

14. Private First Class Wesley Molandes, projectionist for the 245th PSYOP Company, explains to the villagers in broken Vietnamese what a movie was about. Phan Thiet, Vietnam, 16 October 1967. NARA 111-CCV-443-43942. Courtesy of Fold3.com by Ancestry.

15. Private First Class Gerald W. Welle, pressman, 246th PSYOP Company, prints thousands of propaganda leaflets and checks a finished one to ensure the 1250LW Offset Press is working properly in the printing shop at Bien Hoa, 25 kilometers northwest of Saigon, 4 October 1967. NARA 111-CCV-443-43741. Courtesy of Fold3.com by Ancestry.

16. Sergeant First Class John D. Holmes, PSYOPs sergeant, Headquarters, 25th Infantry Division, distributes leaflets to village children, expecting them to be taken to elders or parents during the operations in the village of Phuoc Vinh Minh, Cu Chi District, Hau Nghia Province, 14 June 1967. NARA 111-CCV-443-CC40497. Courtesy of Fold3.com by Ancestry.

17. Staff Sergeant Leroy Massie, cinematographer with Team C, US Army Special Photo Detachment, Pacific, films a PSYWAR operation in the village of Phuoc Vinh Minh, Cu Chi District, Hau Nghia Province, 14 June 1967. NARA 111-CCV-443-CC40496. Courtesy of Fold3.com by Ancestry.

18. Local populace watches a South Vietnamese government–provided TV in the Saigon marketplace, 23 January 1967. NARA 111-CCV-230-CC40609. Courtesy of Fold3.com by Ancestry.

19. ARVN Sergeant First Class Bui-Quoc Thanh, on a US Navy PBR, speaks over the loudspeaker during a Mekong River PSYOP mission, 13 December 1967. NARA 111-CCV-641-CC44516. Courtesy of Fold3.com by Ancestry.

20. A group of singers with the Civil Operations Revolutionary Development Support (CORDS) Band give a folk concert for the citizens of a small hamlet located approximately 10 kilometers northeast of Di An, 13 October 1967. Propaganda speeches were made between performances. NARA 111-CCV-443-CC43943. Courtesy of Fold3.com by Ancestry.

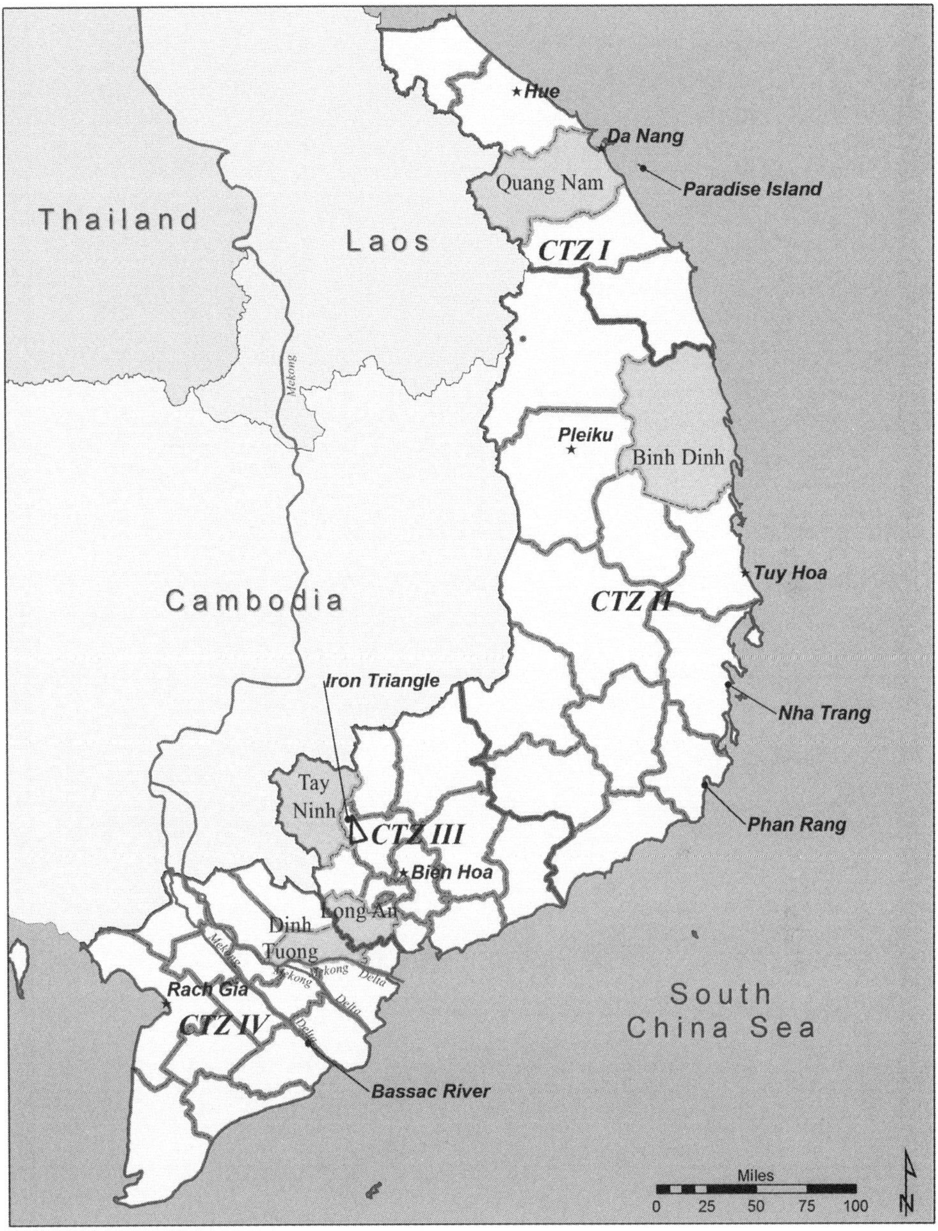

Map 11. South Vietnam, spring 1967 (chapter 10 locations)

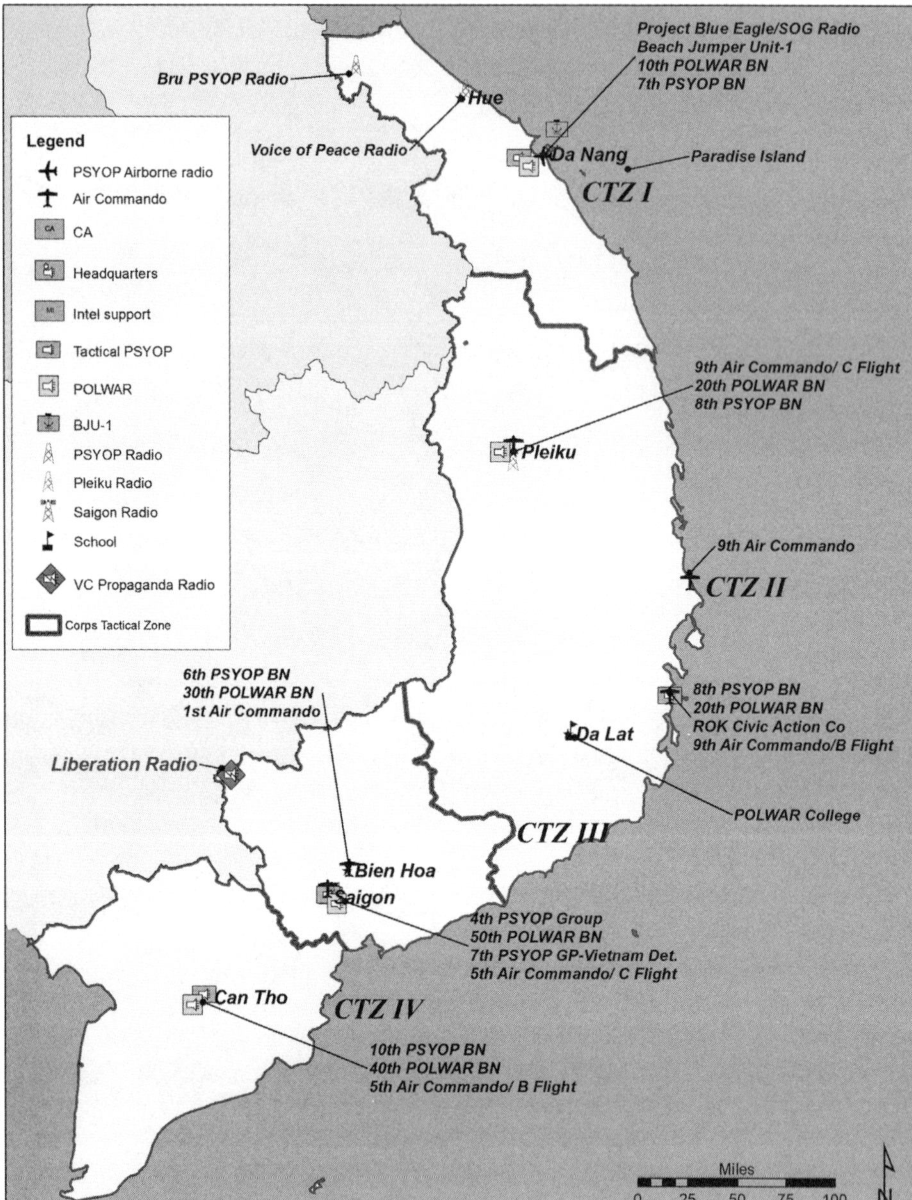

Map 12. PSYOP unit locations, early 1968

11 Measuring Effects, June–Fall 1967

By the summer of 1967, North Vietnam recognized the real dimensions of the morale problem among its forces. Despite bombastic propaganda to the contrary, often parroted in the world press, the Communists' war against the RVN was cracking. Internal documents, public opinion polls, anecdotal reports, and rallier numbers all attested to the scale of the problem. Limited reporting from the North tended to confirm developing morale problems there, as well. SOG continued to effectively agitate these concerns. In the coming months, these trends continued. The North needed to take a big gamble to change the game. By fall, the outlines of that gamble began to emerge.

For the allies, even though the war had clearly overtaxed the PSYOP effort, all levels strove for improvement. The decision to activate an American PSYOP group in Vietnam, consisting of four battalions, portended the solution to many of the identified problems. It also indicated the level of interest commanders had in PSYOPs. Perhaps the most important decision at this time was to create a single manager for pacification, in line with the March 1966 PROVN study. This change helped streamline the PSYOP organizational scheme.

"Hero" Nguyen Be

The North's propaganda program was not working smoothly by this point. Hero-emulation campaigns were a staple of Front propaganda, at least since the Battle of Ap Bac. Emulation was important for Hanoi as a method of encouraging people to continue to tolerate hardships and copy that hero. But such a tactic could backfire. Nguyen Van Be's case was a good example. The saga of Nguyen Van Be (not the same Be discussed earlier), began after Hanoi had praised the soldier as a hero worthy of emulation for blowing himself up while killing many Americans. After being captured by Americans, or so the story supposedly went, Nguyen Be "seized a mine weighing 10 kilos and shouted: 'Long live the National Liberation Front of South Vietnam, down with the American Imperialists.'" Rushing "like a thunderbolt" to sacrifice himself by destroying an armored vehicle, Be damaged "two M-113 tanks [and] killed and wounded 64 enemies, among them 12 Americans," according to the North.[1]

Be supposedly died in the 30 May 1966 attack. However, in March 1967, he was found alive in a South Vietnamese prison. From his jail in the Mekong Delta, he was whisked into a propaganda maelstrom. JUSPAO created the Nguyen Be task force to exploit the opportunity to discredit North Vietnamese propaganda. Starting that month, the United States and the RVN began a countercampaign using leaflets, posters, and radio and TV spots in an attempt to discredit the Communist hero. The revelation of a living Nguyen Be threatened an emulation campaign that the North had conducted for nearly a year. In response, Radio Hanoi charged that the "resurrected Be was a fraud" and challenged "Saigon authorities to produce Be before his own people, in his own hamlet." The tenacity with which the North held on to this "hero," despite clear evidence to the contrary, suggested that the North Vietnamese were short of heroes.[2] (See photo 21.)

In the ongoing propaganda battle that became known as the "Hero Be affair," Peter Arnett reported the story as a psychological defeat for the South. Journalists, unaware of the reason, were sent to Be's village before noon on 21 April to cover his arrival. "The homecoming of the 'dead' Vietcong hero was elaborately staged today except for the final detail," according to Arnett. "I am glad to be home amongst you," Be told puzzled villagers. Asked how many of the villagers he recognized, he responded, "None." As Arnett characterized it, "The covey of psychological warfare specialists, both American and Vietnamese, began looking concerned. Hero Be actually lived in another hamlet nearly 2.5 miles away. The psychological warfare officials suggested the mistake was made by hero Be himself. They said he had never before seen a map." Nonetheless, he had shown himself to be alive—hardly the defeat that Arnett described.[3]

To counter the US/RVN scheme in the Hero Be affair, the Vietcong turned to violence. The NLF kidnapped several men who admitted knowing that Be was alive and put others on wanted lists. Some were reportedly killed for saying he was alive. MACPD noted the risks involved with continuing this war of words. The Communists later attempted to transfer the emulation campaign to other heroes. This did not serve the purposes of the allies because it might help the Northern effort. However, the program continued to proceed while constantly weighing benefits and costs.[4] On balance, the Hero Be affair favored the South.

Each side used a preferred set of propaganda epithets for the other: *bandits* and *criminals* (describing the North) versus *lackeys* and *puppets* (the South). JUSPAO also described the NLF and Vietcong as demons. However, it avoided personalization because of the "hoped for political solution" that meant an eventual cessation of military conflict with the North Vietnamese government remaining intact. This seems to reflect the Johnson administration's ambivalence about prosecuting the war so harshly as to preclude a negotiated peace. It hoped that pulling its punches would avoid fracturing the society to such a degree that it could not be reconstituted after the war. The North had no such illusions. The Front fought the war specifically to radically change the society that existed in 1954.[5]

Polls continued to show that Southerners were not in favor of the fundamental change sought by the NLF. CBS News broadcasted the results of a South Vietnamese public opinion poll it completed in March 1967, contracted though the Center for Vietnamese Studies in Saigon. As with all polls conducted in a combat zone, the results must be read carefully. However, CBS felt that the findings had "reportable validity." The sample surveyed 1,545 persons between November 1966 and February 1967. Nearly two-thirds believed the Vietcong was losing the war and blamed it for the war's continuance. In fact, many saw the United States as the primary pursuer of peace. Three-fourths wanted to live in a postwar country that did not include the NLF in its government. Americans were most admired for their assistance in social welfare, especially among rural inhabitants. However, about 25 percent attached negative qualities to US troops, mostly related to drunkenness and the growth of prostitution. Many also expressed the desire that the United States avoid bombing villages, which was an ongoing source of friction.[6]

One effective PSYOP technique that US forces used to reduce this source of friction was to warn villagers of impending artillery bombardments and air strikes. This placed the psychological burden for deaths on the Vietcong. Studies indicated that such appeals were "highly effective in promoting disaffection between the population" and the NLF. Animosity toward the Vietcong grew, especially given the guerrillas' tactic of forcing the "people to stay in the area" to purposely cause civilian casualties.[7]

In addition, warning the Vietcong before and after B-52 bombing runs heightened the psychological value of those strikes. Survivors of a March B-52 strike in the Iron Triangle after Operation Cedar Falls stated that many of the unit's men "deserted after the airstrike and some of the deserters probably returned to the [RVN] government under the 'Open Arms' program." Civilians had already left the area in the months prior, due to the consistent PSYOP campaign of leaflets and loudspeaker broadcasts urging them to leave. The bombing largely destroyed the VC 271st Regiment, counting killed, wounded, and poststrike ralliers. The ralliers fled despite political officers' entreaties that they were needed to stay put and tend the fields.[8]

In reaction to continuing intercultural friction between US Marines and

Vietnamese civilians, the III MAF established the Personal Response Council and Personal Response Contact Team, which were "directed at improving the relationships between" Marines and civilians. As a result of this program, Marines noted an "attitudinal improvement in intercultural relations."[9] Along with this improvement, higher defection rates in I CTZ persisted into April as III MAF continued its support of pacification objectives. Marines integrated Chieu Hoi leaflets into daily patrols and used medical and dental teams to distribute PSYOP product into villages and hamlets. Seventy-three ralliers arrived that month, with many reporting "that their decision to defect was greatly influenced by the leaflets or loudspeaker appeals which described the advantages of the Chieu Hoi program."[10]

The ability of the 6th PSYOP Battalion to rapidly produce leaflets was illustrated by an event in March 1967. On 9 March, a leaflet request was received at 1600 hours. The 244th PSYOP Company sent the completed artwork by courier, along with specific requirements, to the battalion's printing facility. At 0700 the next morning, "one million leaflets were printed, cut, packed for shipment and delivered to Tan Son Nhut for loading on a special mission aircraft." Less than a day after the request was put in, the leaflets poured into the jungle from above. The rapidity of this turnaround far exceeded most current PSYOP capabilities.[11]

Renewed Effectiveness Studies

In April 1967, General Westmoreland again directed that a study of MACV's PSYOP effectiveness be conducted. The report concluded that the objectives of the programs reflected "the broader national objectives," and it gave sufficient latitude to allow for "creativeness and ingenuity." However, "the major weakness was the lack of research and analysis capability" due to personnel shortages in critical skills. MACV anticipated that additional "highly-trained personnel" would arrive soon, leading to a "quantum improvement in PSYOP."[12]

To better exploit the wealth of data that was accumulating from captured documents and interrogations, the MACV PSYOP directorate activated a research and analysis branch. In addition to standard target analysis functions, the branch evaluated "the psychological effects on target audiences of current and proposed MACV military operations and policies."[13] However, exploitation of intelligence remained "inadequate and limited because the best, most accurate and timely sources of information" were ralliers, who were not "readily available for timely exploitation."[14]

The 6th PSYOP Battalion used every resource it could muster with an eye toward understanding the effects of its operations. Intelligence and propaganda analysis benefited from Psychological Operations Exploitation Teams' (POET) access to valuable feedback information. These were small teams sent on quick-reaction missions by the battalion. For instance, in the Thuong Duc district, a team "revealed the effectiveness of VC propaganda concerning Allied defoliation operations." The propaganda convinced inhabitants that the spray caused "illness and spitting of blood if inhaled." To counter the effects, the NLF sold inhabitants nylon facemasks for 15–20 piasters each. Other feedback gathered by PSYOP exploitation teams fine-tuned optimal aerial broadcast time as between 2200 and 2300 hours, because "families are gathered together and have the opportunity to discuss what they hear."[15]

The 6th Battalion's intelligence section compiled a biweekly psychological intelligence summary for PSYOP personnel in each corps. This product helped fill a gap for operators. One finding that spring, based on rallier interviews, was the usefulness of appealing to guerrillas' relatives to approach "their sons and husbands, explaining their right to rally under the Chieu Hoi Program."[16]

That May, Premier Ky began the Doan Ket (National Reconciliation) plan. In essence, this broadened the Chieu Hoi program and sought to build an ideology of support for the government, rather than simply being against the Communists. One of the goals was to bring in higher-level Front members, thus breaking the Vietcong's ability to continue the war. One way to do this was to offer them positions within the government. Doan Ket became a major PSYOP campaign after May. To support this, JUSPAO produced five leaflets and a poster to introduce the program. Beyond that, the plan envisaged using all available communications means—radio and television, newspaper, film, and special events—to spread information on Doan Ket. The call came at a particularly vulnerable time for the NLF as the military tide ran against them.[17]

A report by the Party Committee of Saigon–Gia Dinh stated that the Chieu Hoi program and subsequent use of ralliers against the NLF were a serious problem. Ralliers had compromised secrets and acted as guides, resulting in forty-three attacks on VC bases in the region and causing a "considerable" loss in men and material. Further, the PSYOP program had helped cause dissension, resulting in "several incidents of firefights, bloody quarrels, theft and debauchery" among NLF members. The committee tasked all units within the NLF to boost activities against the PSYWAR program.[18]

The party committee was likely referring to the Kit Carson Scouts program. Under this program, former VC guerrillas led patrols and identified hundreds of other "Viet Cong through facial recognition procedures." The 1st Marine Division formed trial units in October 1966. All Chieu Hoi ralliers were "screened to determine their suitability for the important task." By the end of April 1967, the program employed thirty-two scouts with an additional thirty-one undergoing training. Due to initial success, the III MAF planned to employ 100 additional scouts, or 50 per Marine Division, and possibly as many as 450 by June 1968.[19]

The Combined Intelligence Center Vietnam produced an analysis of Front propaganda activities in May. Among the NLF's propaganda goals, the report noted, were discrediting the RVN and raising the morale of its side, meeting manpower requirements, causing dissension within and between ARVN and US troops, and lowering the morale of allied soldiers. The Front propaganda program was "marked by a flexibility of approach and a thorough understanding of potential target groups," according to the intelligence center. However, because of sustained US and ARVN attacks, the NLF had reverted to "protracted war" themes abandoned after the 1963 coup. The theme suggested that cadres had to hunker down for a long war. Along with this, propaganda against rural development, derisively referred to as the "Lansdale-Lodge Plan," increased, along with anti–Chieu Hoi propaganda. Almost all of these themes were defensive in nature.[20]

Clearly feeling the impact of military operations, Liberation Radio admitted that the attacks that year had weakened Communist fighting ability. The enemy continued "to carry out their pillage-all, burn-all, and kill-all policy during their military operations," according to the LBS. "They pitilessly pillage and destroy foodstuffs, especially rice and paddy, and kill domestic animals in our rear areas, from eastern Nam Bo to central Trung Bo," and seize rice in liberated areas. Liberation Radio urged listeners to smash this attempted economic sabotage. Hyperbole such as this was unlikely to have wide traction and unwittingly indicated the effect that allied operations had on food production. The station's legitimacy was also hurt by statistical inflation. The LBS claimed the killing of 8,000 enemy troops during Operation Junction City; the actual number was less than one-tenth of that.[21]

Intensified Front International Operations

At the same time, the North and the NLF intensified their international propaganda efforts. Both entities used the "truce" theme to exemplify President Johnson's intransigence, despite his own calls for de-escalation. This tactic helped confuse the issue in the minds of the target audience. According to the LBS, the pacification effort was a program "to massacre our people and destroy all their houses and rice fields and to herd our people into concentration camps where they are pitilessly repressed."[22] Losing on the battlefield, Hanoi attempted to focus international attention on alleged US/GVN atrocities and to "utilize world opinion to drive the US out of Vietnam." To perform this task, the NLF expanded the number of overseas offices.[23]

Much of this attention concentrated on the American antiwar movement. The CANLF announced selling more than 8,000 copies of *Vietnam Courier*, along with 4,000 books. It also helped to ensure that college libraries received these products. By 1967, production quality of the committee's *Bulletin* had greatly improved from the original mimeographed sheet. Among other prospects the organization used to spread its message, the CANLF *Bulletin* suggested that readers listen to Radio Havana every Friday for *Vietnam Today*, as well as Radio Hanoi's service to North America. It also announced that the North Vietnamese were requesting that all activist groups in the United States send "duplicate copies of their publication via the Front's office in Prague" so that they could track these groups' operations.[24]

CANLF chairman William Teague also gave speeches and helped organize protests in New York. In September 1966, the CANLF began showing Northern propaganda films such as *Nguyen Van Troi Will Live Forever*, which included a typed script to be read along with the Vietnamese language in the film. The guidelines included detailed instructions on when to turn down the volume and read a propaganda passage. These excerpts used a very stilted, propagandistic tone and were unlikely to have mass appeal. However, the target audiences included those already against the war or who were wavering, and for them the images and text likely had strong emotional and confirmatory impact. As soon as new films arrived, CANLF made copies for rent and sale, presenting them in at least twenty cities. The film *Toxic Chemical Warfare in South Vietnam* became available in 1967. That May, the CANLF announced a showing of three of its films on WGBH, the public television station in Boston.[25]

By this time, the CANLF became more stridently radical and aligned with Students for a Democratic Society. The committee began producing a radio program to be broadcast over Radio Hanoi, *Radio Free Allied Forces*, to "present the thoughts of the American peace movement to the young and ill-informed Americans in South Vietnam." It interviewed black soldiers who refused to serve in Vietnam for the program. Later, the CANLF sought out relatives of those serving in Vietnam. "Imagine the effect if they begin to hear 'Joe's mother,' 'Mike's whole family,' . . . all saying 'the war is wrong and the troops should come home now,'" according to the bulletin.[26]

The support for the antiwar movement was reciprocal. North Vietnamese prime minister Pham Van Dong thanked the Americans Staughton Lynd, David Deming, and Barbara Dellinger for a telegram he received "informing him of this massive drive of the American people."[27] *Nhan Dan* also hailed the spreading antiwar demonstrations across the United States. The paper "recalled that in this biggest ever demonstration against Johnson's war in Vietnam, 500,000 Americans of all walks of life held high portraits of President

Ho Chi Minh and flags of the NFLSV, and shouted slogans demanding: 'Stop the war in Vietnam now,' 'Bring U.S. troops home,' 'Don't burn children,' 'Vietnam to the Vietnamese,' and so on."[28]

Concurrently, the 245th PSYOP Company noted that while most enemy propaganda continued its standard anti-American themes, analysts found "increasing examples of VC leaflets which have been written either by American or someone with an excellent knowledge of American military mannerisms and expressions." Numerous reports indicate that leaflet quality continued to increase, both in composition and print quality.[29] The American antiwar movement seemed to have affected the quality of Northern propaganda. Messages targeting US troops contained a mix of hyperbole and effective questioning. But as always, without any solid numbers, it is difficult to determine effects.

Liberation Radio also directed propaganda broadcasts at target groups within the US forces, such as blacks or soldiers of a particular division. However, most were "directed at the US soldier in general."[30] Photos of antiwar demonstrations, "ostensibly somewhere in the U.S.," and pictures of an "American helmet on a cross marking a grave on the reverse side," continued to be common symbols. But now, as statements against the war from Senator Wayne Morse and others became more widespread, they assumed prominence as legitimizing messages. Other leaflets portrayed American military members who had taken action against the war. Exhuming the story of Lieutenant Steinke, who refused to fight, one portrayed him as a man of courage who stood tall and acted against the "unjust" policy in Vietnam. Slogans such as "Oppose the US aggressive war in South Vietnam" called for soldiers to emulate this action.[31]

However, North Vietnamese messages in English targeting the worldwide audience stressed how evil American troops were, undercutting messages targeting the troops that sought to divide them from American policy. This incongruity diminished the credibility of all such broadcasts. One international broadcast claimed that in Tay Ninh Province Americans had "disemboweled three puppet soldiers and plucked out their livers in retaliation against the punishment of three cruel G.I.s."[32] Another broadcast read a purported "letter" to an unnamed mom and dad: "Today we went on a mission and I'm not very proud of myself, my friends, or my country. . . . We burned every hut in sight. It was a small rural network of villages and the people were incredibly poor. My unit burned and plundered their meager possessions."[33] Meanwhile, a supposed message from a pilot said: "I think of the issues which the marchers and demonstrators have raised against the U.S. aggressive policy in Vietnam and Southeast Asia. I wonder how successful they will be. Will there be still wider support and momentum? How much opposition will they have to bear from official agencies and hecklers?"[34]

The NLF's propaganda and training section, the heart of its effort, was found at all echelons. Subordinate to the party committee, the section received "guidance from . . . the next higher echelon in order to ensure a unified effort" in support of COSVN and North Vietnamese policies.[35] The structure at each level was similar to that below. At the hamlet level, cells transcribed Radio Hanoi "special slow-reading broadcast" to rebroadcast the news via loudspeakers and leaflets, and they hoisted NLF flags to assert area control. One former VC propaganda operator stated that he recruited "attractive women" to subvert GVN soldiers.[36] In the absence of large-scale attacks, the propaganda and training section also used terrorist acts as advertisements "to build morale within the ranks." Often this tactic was used during periods of military reversals to prove that they were still unbeaten.[37]

Liberation Radio devoted most of its airplay to news. The station carried "exaggerated accounts of ARVN/US losses at the hands of the VC," as well as "exploits of the peasants in the 'fight for liberation.'" The station was under the direction of Lieutenant Colonel Le Chan. Radio Hanoi, along with Chinese Communist and Russian advisers, provided support. MACV intelligence sources indicated that it operated from a concrete trench near Tan Hung, Tay Ninh Province, in Base Area 353 (see map 12). An "underground communication trench reportedly extended into Cambodian territory to provide for emergency withdrawal." Due to the weak signal, it was often rebroadcast via Radio Hanoi facilities. COSVN routinely used its station "to put across the latest party line and send instructions to VC cadre. These instructions are slowly repeated to allow the cadre to write them down for use at that particular echelon." In the Saigon area, as of 28 September 1966, another clandestine radio station with the call name "Red Star" broadcast warnings to the Saigon populace of the consequences of cooperating with GVN security agencies. This appears to be the covert CIA station that notionally broadcast the dissident Vietcong views.[38] (See figure 11.1.)

On 22 May, National Security Advisor Rostow sent a message, marked "LITERALLY EYES ONLY," to the president regarding Project Compatriot. Disclosure of the project offered a potential propaganda boon for North Vietnam. First tested under the code name "Popeye" in Laos the previous October, this project entailed seeding the clouds in Laos and along the Ho Chi Minh Trail in hopes of increasing rainfall and impeding infiltration, washing out roads and trails, and causing the loss of up to 10 percent of the North Vietnamese rice harvest. Foot-deep standing water along highways in the target area impeding "infiltration and resupply movement" indicated the success of the

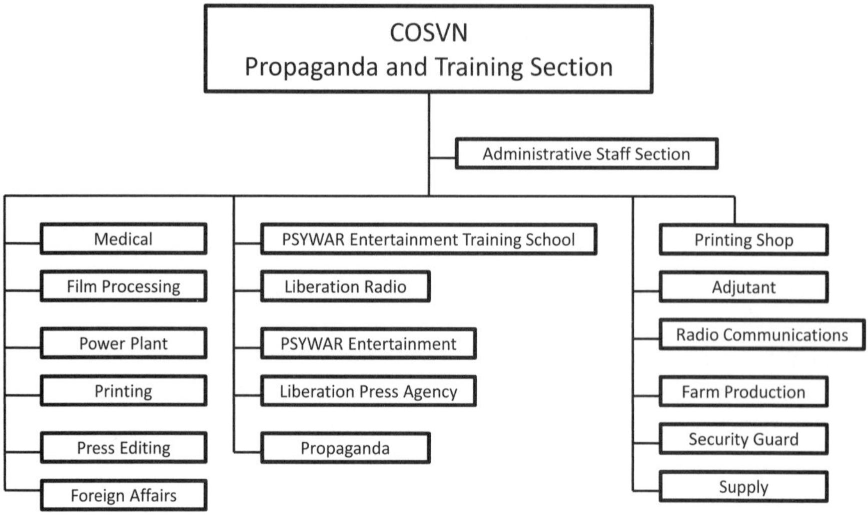

Figure 11.1. COSVN Propaganda and Training Section

test.[39] Rostow's purpose in writing the president was to have him consider the US response if the project became compromised. Given the already strong propaganda attacks on the use of tear gas, the accusation of "weather warfare" was a likely charge. According to Rostow, "We have already taken quite a beating on bombing, chemical agents, and even napalm. There has been far greater sensitivity to the use of these weapons in Vietnam because of the overall picture that we are a big and sophisticated nation making war on a small and backward one." Tests continued throughout 1967, though Rostow's fear never materialized.[40]

CORDS Improves Propaganda Coordination

A week later, MACV began critical adaptations to improve PSYOP integration in Vietnam. By that time, nearly 1,300 US Army, Navy, Air Force, and Marine personnel composed the American PSYOP force in Vietnam. The creation of the Civil Operations and Revolutionary Development Support (CORDS) pacification program alleviated organizational confusion to a degree. This organization merged the embassy's Office of Civil Operations and MACV's Revolutionary Development Support Directorate, allowing for greater coordination of military and civilian pacification and PSYOP activities. Presidential assistant Robert Komer became MACV deputy for CORDS that month. CORDS became the single manager overseeing the entire pacification program that the PROVN study had envisioned. Under Komer, four regional directors advised each Vietnamese corps. Province-level PSYOP personnel fell under the CORDS assistant province representatives for PSYOPs. The CORDS PSYOP program placed representatives throughout the country to support information programs, community TV programs, reading rooms, hamlet audio-visual programs, and the distribution of PSYOP products.[41]

At the tactical level, enormous growth in demand required novel responses. To help the 244th PSYOP fill demands for loudspeaker teams, fourteen US Marines were attached to the company, forming HB loudspeaker teams and adding one Marine intelligence NCO to the intelligence section. The company also instituted a three-day PSYOP orientation course for "PSYOP personnel to provide advice, prepare PSYOP plans, employ equipment, and be aware of PSYOP support available from the company." The 6th PSYOP Battalion reported, "The Marines have rapidly learned Psychological Operations—this, combined with their infantry training, makes them an effective valuable asset to this unit," suggesting a program with wider applicability.[42]

In another effort to integrate and improve PSYOPs, the 19th PSYOP Company began "co-locating field teams with ARVN PSYWAR companies in each ARVN Division Tactical Zone." It supplemented PSYOP programs and provided "a quick reaction capability to exploit incidents in each DTZ." The teams also provided liaison to the 19th PSYOP Company, gathered feedback on propaganda effects, and identified vulnerabilities for exploitation. Some of the teams were tasked out to other corps. The company suffered its first casualty when Staff Sergeant Pedro Cruz died in a 22 May ambush while providing loudspeaker support to the US 101st Airborne in II CTZ in Duc Pho, Quang Ngai Province.[43]

In II Corps, the 245th PSYOP Company continued to support American, South Korean, and ARVN operations and helped induce 3,298 Vietcong to rally that spring. More than 13,000 had rallied in the corps since January 1966. The company's seven loudspeaker teams and one audio-visual team were under the operational control of US units and the ROK 9th (White Horse) Division. Another HB loudspeaker team was attached to Task Force Oregon, a US Army force operating in I CTZ that later became the Americal Division. The company disseminated 370 million leaflets, conducted 1,430 hours of loudspeaker broadcasts, and oversaw 857 aerial missions using US Air Force assets and 116 sorties using US Army aircraft. Production on company presses reached 42,086,000 leaflets that quarter.[44]

As always, the enemy had a say in the success or failure of these operations. Due to hits by enemy automatic weapons, the 5th Air Commando Squadron directed loudspeaker aircraft to fly future missions no lower than 4,000 feet. "This has greatly decreased the effectiveness of tape missions," the 6th Battalion reported, "because in order for the message to be heard on the ground atmospheric conditions must be almost perfect." As a result, it received complaints from tactical units. One remedy identified was to "reduce the length of taped messages to a maximum of 25 seconds," which enabled the aircraft to fly straight "through at 3,000 feet rather than orbiting at 4,000 feet." However, this also limited the message's length.[45]

Other equipment problems plagued the units as well. The Kaiser-Willys jeepster chassis, used by the audio-visual teams, presented another challenge. The battalion headquarters requested that the Army convert to a standard three-quarter-ton truck chassis because the Kaiser chassis was too weak to bear conditions in Vietnam. It also requested field evaluation of the new MSQ-85 audio-visual truck-mounted shelter. MACV set the requirement for a total of twenty-three shelters for the battalion, the first six due to arrive in April.[46] (See photos 14 and 19.)

Without dedicated interpreter assistance, however, PSYOP teams were dependent on supported units to provide interpreters in order to communicate with civilians. This hindered team operations due to a dearth of linguists. The 6th PSYOP Battalion requested that interpreters be assigned, but due to shortages in qualified personnel this was unlikely to happen. PSYOPs were not the only element suffering from a lack of linguist support. The 4th Division Military Intelligence Detachment was also burdened by the scarcity of American personnel who were qualified as Vietnamese linguists. The division was due to receive 400 twelve-week Vietnamese linguists after October 1967, but this was hardly sufficient to provide enough language skills.[47]

Meanwhile, in III Corps, the 246th PSYOP Company conducted operations attached to US units and the 1st Royal Australian Regiment. A cadre notebook captured that June indicated that the Vietcong intensified indoctrination in the face of Australian pacification and PSYOP activities in Phuoc Tuy Province. Per the document, the 1st Royal Australian Regiment conducted "demagogic" activities by contacting adults; distributing candy, money, and gifts to children; providing medical care; and issuing medicines to the sick. The documents "admitted that the populace [evinced] their sympathy for [Australian] troops by saying that the . . . troops loved" Vietnamese children.[48]

The command committee of Ba Bien Province (RVN Phuoc Tuy Province) noted that the harsh war conditions along with PSYWAR and Chieu Hoi activities had resulted in eight guerrillas deserting with their weapons, sixty-two youths rallying, and twenty-two others being identified as government informants. Sixty-six others were listed as being "with the enemy." To counter this, the committee held 14 demonstrations and 217 propaganda sessions for 25,726 people in the Vung Tau area. This had resulted in fifty-one volunteers in the first quarter of 1967, a net loss of fighters.[49] The Australians also came under attack for the use of tear gas. Radio Hanoi accused the Royal Australian Air Force spokesman of excessive "cheek to allege that tear gas was a humane way of treating the Viet Cong [Vietnamese patriots] instead of bombing them."[50]

Vietcong Psychological Vulnerabilities

Minutes from a Long An Province district party committee meeting stressed the need to indoctrinate civilians to thwart ongoing US/RVN psychological operations. Guerrillas were ordered to attack pacification cadres, whom they termed the "velvet hand."[51] In Phu Yen Province, a Front district status report listed ninety-three defectors during the month of April, with thirty-seven more captured. The district listed 167 effectives remaining, meaning that the tally of killed, wounded, and defectors had cut the unit's strength by nearly half. The dead included the provincial propaganda chief. According to the report, "Allied PSYWAR and Chieu Hoi activities had a detrimental effect" on Front cadres and caused more damage to guerrillas than to military actions.[52]

As to VC psychological vulnerabilities, the US 4th Infantry Division's PSYOP personnel identified a lack of medicines and medical care, which led to high rates of malaria and pulmonary disease. Other sources found exploitable dissension between "NVA unit commanders and political officers, and between NVA cadre" and Southern guerrillas.[53] The 4th Infantry Division shifted to support long-range Civic Action projects to complement the Revolutionary Development program. Units expanded their Good Neighbor outreach, which visited forty-one hamlets five times per week. This program facilitated "cottage industries, health worker training, a veterinary medical program and advanced farming techniques." It aimed to provide pacified areas with the economic base to "achieve a viable society in this area." The division's audio-visual team accompanied the Civil Affairs teams on nighttime hamlet visits to show movies and perform other civic services.[54]

In June, the US 4th Infantry Division conducted operations to support the Republic of Vietnam's resettlement program near Pleiku. To facilitate that program, all newly assigned division soldiers received six hours of indoctrination on their role in CA/PSYOP efforts. The units dropped more than 50 million leaflets and conducted 100 hours of airborne loudspeaker and 85 hours of ground loudspeaker operations on themes emphasizing ARVN and American strength "designed to instill fear into the ranks of the NVA soldiers." It also relied on the standard Chieu Hoi appeals, requests for information concerning

PAVN/VC activities, and civil noninterference messages. One indicator of the effect of these operations was the "increased number of propaganda lectures" conducted by the Vietcong in that area.[55]

Regarding the effect of the PSYOP programs, CIA adviser Gary Williams noted when he arrived in Quang Nam the year prior that "he could drive from Da Nang or Hoi An into just three of twelve districts." By late summer 1967, Williams freely traveled everywhere in the province except for a couple of remote districts, and in his judgment the VC infrastructure had by then been severely damaged. The North Vietnamese Army still represented a serious threat, but it was "an invading Army," not an insurgent force. He remarked later: "We were winning."[56] However, a CIA history warned that looks could be deceiving. Even though villagers may have shifted preferences, the VC infrastructure remained capable. One CIA report found that, although gains had been made in 1966, "overall pacification gains were no better than 'marginal' after January 1967."[57] However, recent evaluation of Hamlet Evaluation System and other data also seems to validate the shift that Williams noted. Assassinations in the autumn did rise dramatically (a result of a shift in strategy by the NLF, described below). Nevertheless, in analyzing the distribution, most were in the contested and VC-controlled areas along the Song (River) Cua Dai. VC-initiated incidents mostly occurred within two miles of the river (see maps 16 and 17). Correlating multiple data sources overlaid in a way that analysts were unable to complete in the 1960s, it becomes clear where the war was being fought. Analysis of wartime data using geographic information systems (GIS) software shows that over the course of the year the trend was against the NLF's interests, especially in I CTZ.

A circular from the VC Thu Dau Mot Provincial Unit (Binh Duong Province) noted the linkage between PSYOPs and intelligence activities. The author wrote that the attempts to lower morale and induce desertion, along with "intelligence activities to collect information on the location and movements of troops," had been effective. In the previous year, "a number of persons, including Party members, either deserted or defected." The memo asked all addressees to increase indoctrination in an effort to stop this.[58] An earlier circular from the same office notified addressees that enemy PSYWAR and Chieu Hoi actions "were intended to lower the morale" of guerrillas and their families. Using "good treatment, threats and the influence of religious sects," families had been lured into forcing their relatives to defect.[59]

The United States conducted a survey of public opinion between July and September 1967 in each corps area. The authors cautioned once again that, due to the security situation, they could not achieve a truly representative sample. Thus, they urged the reader to pay more attention to trends than to details. Walt Rostow briefed President Johnson, highlighting key results of the survey. Rostow found the 71 percent positive opinion of the American military presence to be encouraging.[60] It was also promising that "the majority of the respondents reflect a highly derogatory image of the Viet Cong." Respondents described the Vietcong as "brutal," as "terrorists," as "traitors," and as "subservient to Communist China or Russia." They most frequently blamed the war's continuation on "colonialism and ideological and economic expansionism," pointing a finger again at the North. A few respondents were noncommittal, but most held a strong view. One negative indicator was that 40 percent in I CTZ and 33 percent in II CTZ answered that their lives had gotten worse during the preceding year.[61] Once again, most respondents blamed the Communists for the problems, whereas only 6 percent blamed the United States. Six in ten respondents favored the American presence in Vietnam, with the caveat that many wanted "greater respect for lives and properties, polite attitudes, no bullying," and 6 percent wanted an end to the spraying of defoliants. The numbers remained remarkably constant despite the increased combat.[62]

Regarding media access, the survey found that loudspeakers were a prime source of information on district events, while radio provided a national view. Printed material was less useful, according to the survey. In III CTZ, the survey found that three-fourths of the population listened to radio, almost exclusively to Radio Vietnam. The ARVN's own radio channel managed respectable ratings, with at least 59 percent reporting occasional listenership. By this point, almost the entire population lived within radio broadcast range of a Southern transmitter. Television increased the coverage as well (see map 8). Thirty-four percent of those surveyed read the very popular JUSPAO-produced *Huong Que* magazine. The survey found wide knowledge of the Chieu Hoi program, which indicated that the PSYOP program had spread to the hamlet level. In addition, the survey results showed that there was "good potential for people in secure hamlets appealing to relatives in Viet Cong ranks to rally." Respondents also expressed satisfaction with the spring 1967 hamlet elections, which boded well for legitimacy of the incoming government. However, Vietnam Information Service cadres generally continued to receive low marks from respondents.[63]

The MACV's *Command History* from that year noted increased numbers of Chieu Hoi program returnees, although most continued to be "farmers and low-level NLF members forced into service," while the hardcore members remained loyal. Low-level members had less commitment to NLF goals and "were more easily swayed by family sentiment and the local population."[64] Another problem was a lack of commitment on the part of some Southern officials to the program.

However, looking beyond the simple measure of Chieu Hoi ralliers, the study indicated that the broader themes of the PSYOP program were taking root. As operations drove VC cadres out of villages, those cadres lost access to the population. This cascaded into a lack of "opportunities to endear themselves to the populace" by helping villagers, keeping GVN personnel out, and spreading propaganda.[65] Quoting one Front cadre: "A number of people displayed a pacifistic attitude, were bribed, and served among the enemy as henchmen because we failed to stay close to and indoctrinate them in time." The Vietcong had lost its ability to influence the population.[66]

Radio War

In late 1966, MACV began considering a plan to produce single- or double-channel miniature radio receivers for PSYOP use. The plan called for dropping the radios over North Vietnam in October in conjunction with the expected commencement of PSYOP radio broadcasts from Pleiku. These radios were configured to receive only PSYOP broadcasts, aimed at reducing VC and PAVN morale. By mid-March 1967, plans for a six-month test program were set to start in July. The reception frequency was set close to Radio Hanoi's, using a technique called "spoofing" or "ghosting."[67]

Despite the clear shift on the battlefield, the North intensified its victory claims in its propaganda. Liberation Radio claimed that in the face of "successive defeats" during the winter–spring battles, "the U.S. imperialists [became] increasingly more cruel," leading them to resort to "fascist measures."[68] A tour of "journalists, writers, photographers, and cameramen from socialists and other countries" was led to a Red River dike section northwest of Hanoi, which, according to Radio Hanoi, "was seriously damaged by a U.S. air raid on 13 July." The visitors were shown four bomb craters near the dike as proof of systematic targeting of irrigation infrastructure for destruction.[69] The broadcaster claimed that the United States had "more than doubled their raids on the water conservation system in North Vietnam" to attack dikes, "dams, pumping stations, canals, and other irrigation works." Actual targeting would likely have produced more profound results than four nearby craters.[70]

Quoting USIS stories, Liberation Radio pointed to "an alarming decline in war morale in Washington, where the gloom is deeper than at any time since the massive U.S. intervention in early 1965." A *U.S. News & World Report* story read for broadcast said: "After two years of U.S. buildup, the reds are still gaining in troop strength and firepower and regaining the initiative."[71] It used stories such as these—often taken out of context or otherwise selectively edited—to support the allegation. Continued exaggerations (such as the claim of killing 90,000 enemy, including 5,000 "wicked pacification agents") hurt North Vietnam's credibility with the uncommitted.[72]

JUSPAO printed excerpts from the June 1967 issue of *Tuyen Huan*. This was a North Vietnamese indoctrination magazine for midlevel propaganda cadres, providing themes to use on their own troops. *Tuyen Huan* was the propaganda element of the agit-prop system. According to the magazine: "In addition to official radio stations operated by the U.S., the enemy also uses broadcasting systems which pose as the voice of such and such parties and which broadcast from the South to the North with the aim to sow confusion in the mind of the people." It also claimed that "the enemy uses word-of-mouth to whisper false information" and spins combat reports to "shake our people's mind."[73]

Tuyen Huan also discussed the use of poison-pen letters and gift kits, which MACV-SOG distributed. "The letters' contents generally deal with questions concerning the internal situation of our country, the policies of our Party and Government, and demands for clarification of these policies," although the real plan was to "cause the readers to have doubts about our Party and Government." As for the gift kits, the magazine reported that "the enemy believes our people are having difficulties with the shortage of supplies." The author stated that cadres should collect and promptly burn such things. The publication also provided counterpropaganda guidance to cadres as well, saying that they should "organize quick collection of enemy leaflets right after they are dropped." The cadres were instructed to read the leaflet and make a timely determination whether the argument needed to be countered or ignored. The radio broadcasts seem to have presented a greater threat. The article ordered that "it is necessary to educate our cadres, Party members and population so that they refrain absolutely from listening to enemy radio broadcasts." This was a valuable indicator of effectiveness. The publication also noted that the North contained "undercover reactionary elements and hoodlums left over in our society," especially ethnic minorities and unreformed "former puppets."[74]

DRV Spy Fever

The North Vietnamese minister of public security, Quan Hoc Hoan, wrote that the United States had intensified "psychological warfare aimed at sowing fear and confusion within the DRV." As a result, a state of emergency and spy fever gripped the North over fears of commando teams.[75] By mid-1967, moderate members of the politburo fled the North in response to the crackdown, further strengthening Le Duan and Le Duc Tho's grip on power. The Le Duan faction began the Revisionist anti-Party Affair at the 1963 party plenum to marginalize the opposition. This antirightist campaign climaxed with the arrest of the former vice minister of education and director of the Marxist Institute of Philosophy, Hoang Minh Chinh, on 25 July 1967. The purge, which included other pro-Soviet moderates, allowed for the passage of Resolution 14 in July, which gave approval to the general offensive and uprising strategy, culminating in the Tet Offensive. Fear of any deviation from the Le Duan/Le Duc Tho line had increased the regime's paranoia. SOG's program fed into this fear.[76]

In a reference to the Paradise Island operations, *Tuyen Huan* noted how "the enemy usually sent commando agents to kidnap our citizens" and used cajolery, "torture and terrorism to extract" information from them. The kidnapped personnel were released "with a quantity of goods designed to fulfill his wicked schemes." Given the level of knowledge about SOG activities, arguments in favor of surfacing the program seemed valid. However, even if Hanoi understood the outlines, the paranoid nature of the regime likely caused an overreaction to the threat.[77]

To further spread disinformation and black propaganda, SOG created Project Soap Chips. Soap Chips "placed forged letters on the bodies of PAVN soldiers in Laos and Cambodia" and inserted "slanted information about the home front into the letters" to lower troops' morale. Under Project Benson Silk, SOG counterfeited North Vietnamese currency (see photo 24). However, US law expressly forbade attempts to undermine the North Vietnamese economy. To narrowly stay within the letter of the law, SOG added a detachable text section of the leaflet, which allowed the recipient to cut it off in order to spend the counterfeit bills. Counterfeit leaflets such as these were not unique to this particular conflict and had been used in Vietnam since at least the 1940s. Millions were dropped over the North in 1966 and 1967. The use of the attached propaganda tag gave the United States plausible deniability against the charge of openly counterfeiting the currency of a foreign adversary. One possible problem with the counterfeit leaflet drops over North Vietnam was that they could provide the regime an easy scapegoat to blame for high inflation. SOG reconnaissance units also left bundles of cash on dead bodies to make it seem like they must have been working with the enemy to have so much money.[78]

Among the more interesting and successful of these black programs was Eldest Son, which began in August 1967. SOG reconnaissance teams in Cambodia frequently found caches of ammunition that they had no time to destroy or remove. The Counterinsurgency Support Office modified AK-47 and 12.7mm ammo and 82mm mortar rounds to cause misfires and explosions. Teams left the sabotaged ammunition mixed in with the caches after missions. The program's goal was to put this defective ammo into circulation in the PAVN. Afterward, planted news stories pointed to this being Chinese-manufactured ammunition. The program included the use of forged memos and radio broadcasts to bolster the deception. As guerrillas were injured or killed using the doctored ammo, general distrust of Chinese-supplied ammunition would grow. American and ARVN units received "reports of dead Communist troops either clutching exploded AKs or surrounding split mortar tubes." William Shelton remembered that the amount of sabotaged ammunition was small—but it did not take much. A single magazine with a single bad round could be enough to cause problems for any unit.[79]

Communications intercepts "revealed that the doctored ammunition was of concern to NVA soldiers." Subsequent stories appearing in *Stars and Stripes* and on AFVN of American soldiers injured by "trophy" weapons gave credibility to the deception. Sabotaged American equipment was also distributed, including radios meant to keep the enemy from touching American KIAs. Attempts to lace caches with inedible rice were less successful.[80]

In June, briefings on Civil Operations and Revolutionary Development Support discussed ongoing PSYOP program problems. Among them were VIS performance, limited surge print capacity, and manpower shortages. Shortages meant that twelve provinces had no full-time PSYOP representative, although CORDS had recently assigned two full-time advisers to the Vietnamese Information Service at the ministry level.[81]

Despite the problems noted above, American PSYOPs continued to make great strides. For example, the 6th PSYOP's intelligence section wrote five area studies for the 244th PSYOP Company that summer. Compiled from available intelligence data and reanalysis based on specific needs for PSYOPs, the area studies added "analysis of enemy psychological operations and conditions offering opportunities for possible exploitation."[82] On 29 July the 6th PSYOP Battalion received seven additional audio-visual teams, which supplemented the 244th, 245th, and 246th PSYOP Companies. In addition, eight Modular Audio-Visual Units arrived in October for evaluation. These consisted of an audio-visual truck-mounted shelter, radio, tape recorder, and film projector. These were designed to replace the MSQ-85 discussed above.[83]

The American PSYOP force arrived with very little knowledge of Vietnamese history and culture. Nearly two years of combat operations—not to mention years more of advisory activities—had greatly expanded the knowledge base. Additionally, JUSPAO issued a steady stream of guidance for field operators to use in formulating operations.[84] Typically, guidance consisted of a couple of pages of background material on a topic and direction on messaging themes.

In February 1967, for instance, JUSPAO issued guidance on defoliation operations. It stressed the use of similar herbicides "in the U.S., Great Britain, the USSR and many other countries employing scientific methods of agriculture" and that the defoliant agents left "no residual effect on the soil or future vegetation from the herbicide treatment." Per the guidance: "Defoliants used in RVN are non-poisonous; even food and water affected by the spray can be consumed without danger." It noted that the chemicals were used on more than 400,000 acres annually in the United States. At the time this guidance was issued, scientists thought that the agent was safe. However, by 1969 evidence emerged of potential problems, and the US military phased out use of Agent Orange by 1971. The unknown factor was PCB contamination of the agent, introduced during production. In the meantime, the NLF continued the widespread use of this increasingly credible theme in its propaganda.[85]

Conclusion

JUSPAO provided guidance that summer in 1967 to all PSYOP troops in Vietnam on exploitable themes to use against NLF cadres. This instruction was based on interviews with ralliers, which showed that Front cadres saw themselves as professionals in a bureaucratic system. Up to this point in the war, the NLF had offered cadres avenues for upward mobility to satisfy career goals. For many, these ambitions were more important than the cause. However, the belief that they were working for positive social objectives merged with personal goals to form a strong motivator to stick with the NLF. But as the movement developed, Communist class consciousness led to the bypassing of many original members of the NLF in favor of younger people from preferred social classes. Fear of a protracted war, loss of popular support, and diminished opportunities for advancement within the NLF, or the feeling that the NLF treated cadres unfairly, fractured the façade of unity. Low-level cadres were more easily targeted using these themes than the more committed and connected upper-level cadres. Propaganda in support of the Doan Ket national reconciliation program offered the possibility of enticing more cadres to rally because they would not lose their social status by rallying to the government. Stressing themes related to elections, land reform, and other programs undertaken by the government of South Vietnam that diminished the need for the NLF's social reform program also had potency. The major factor holding midlevel Front members from rallying that fall was the impending victory that North Vietnamese and NLF propaganda increasingly proclaimed. Defecting at that point in the war would jeopardize one's career.[86]

As summer turned to fall, the war was at a crossroads. In response to the combat and psychological defeats of the preceding year, the North was forced to gamble on a risky offensive. Purging the Communist moderates had given Le Duc Tho and Le Duan a free hand. At first desperately overburdened and facing a steep learning curve, the allies' PSYOP program now seemed to be having an effect. Morale in the NLF was in steady decline, and the idea of South Vietnam as an independent nation was taking hold. The coming elections offered an opportunity to consolidate this shift.

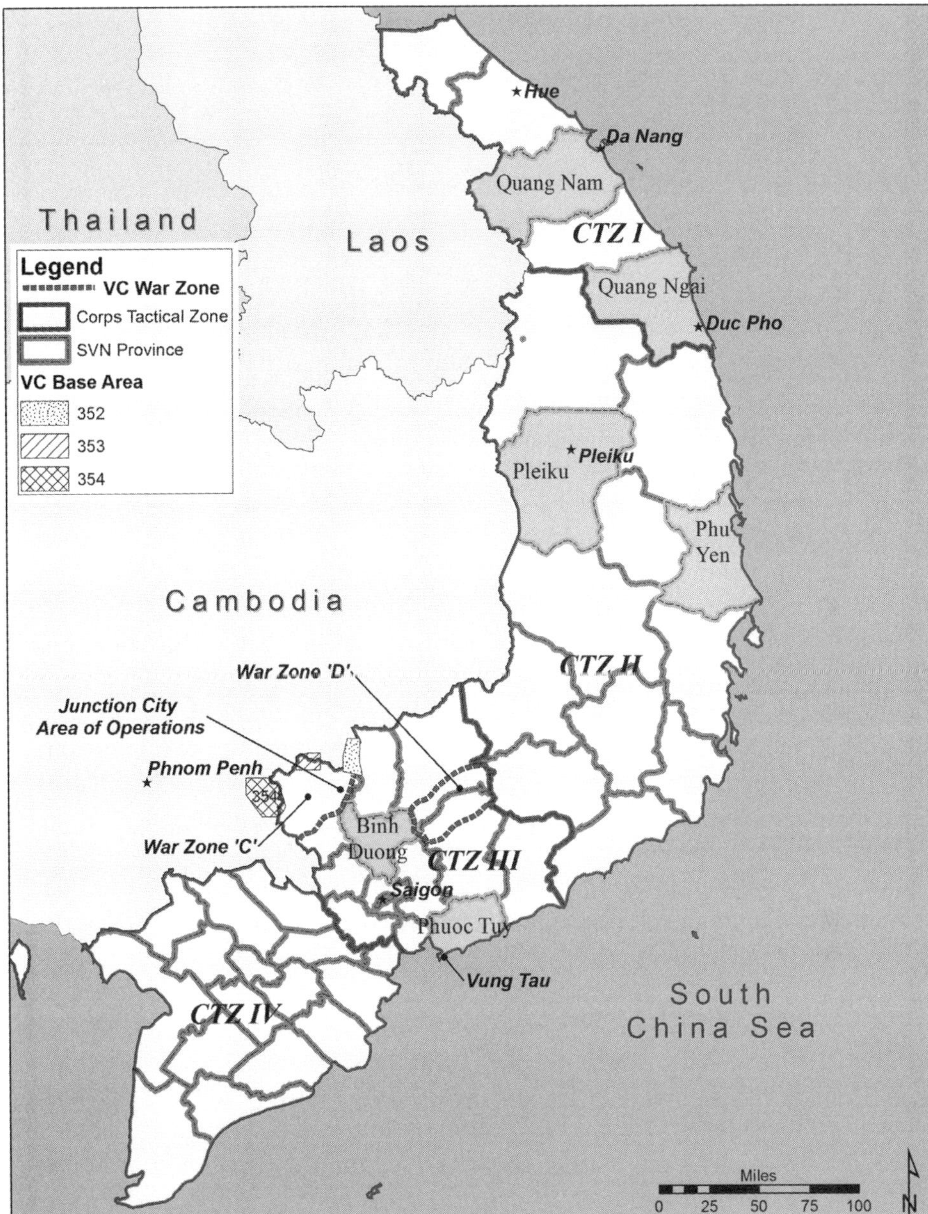

Map 13. South Vietnam, fall 1967 (chapter 11 locations)

12 The 4th PSYOP Group and the Tet Offensive, Fall 1967

As fall began, the war in Vietnam had come to a fork in the road. The combat and psychological defeats of the past year had forced the North to gamble on a risky offensive. The North Vietnamese made the decision during the summer of 1967 from a position of weakness, not strength. Morale in the National Liberation Front was solidly in decline, and the idea of the Republic of South Vietnam as an independent nation was taking root. The coming elections seemed likely to confirm this trend. However, Le Duan had been a high-stakes player before. Indeed, his decision in 1960 to go to war in reaction to President Diem grinding down the VC infrastructure, and the decision in 1964 to up the ante with the hope of winning before the Americans could become decisively engaged, were similar gambles. Incessant North Vietnamese propaganda claims beginning that fall of imminent victory held midlevel cadres in the movement for a couple more months and temporarily slashed the number of ralliers. To implement the plan, the DRV ordered increased assassinations and drafting large numbers of youths. Meanwhile, American intelligence analysis problems muddied the waters as to what these strategic changes meant. The period ended with the creation of the US 4th PSYOP Group in December, followed within weeks by the opening salvos of the 1968 Tet Offensive.

Manpower Problems Impact All Sides

In international broadcasts, Hanoi highlighted claims of fighting between American and RVN forces. One incident in Quang Tri supposedly claimed the lives of twenty-five GIs and twenty civil guards. In response to an attack, Hanoi stated, the Americans "gunned down or bayoneted any civil guards they could find." At the end of the orgy of violence, "the GIs then chopped them into segments and displayed them in the streets." An analysis of Front propaganda showed that exaggerated claims of victories and enemy infighting such as this maintained morale among guerrillas and civilian supporters in the short term. However, to maintain morale in the face of a succession of defeats, the NLF's claims became progressively fantastic. These boasts were unlikely to convert anyone with access to accurate facts.[1]

As American troop strength grew in 1967, Radio Hanoi increasingly used the theme of ARVN relegation to pacification duty. It claimed that this change resulted in rapid disintegration of the "puppet troops." Radio Hanoi alleged that the ARVN had "lost all their fighting ability and were no longer able to resist the increasingly more violent attacks of the Liberation Armed Forces." The broadcaster alleged that ARVN's decline had led to increased desertion, while the United States bore the brunt of combat. In reality, the ARVN was not taking a backseat in the fight. It had suffered nearly 60 percent of the combat deaths in 1967.[2] Though still suffering from defects as a military force, it certainly was not just sitting on its hands. The desertion rate remained high, however; comparing the first five months of 1967 with the same period in 1968, it dropped from 10.9 to 9.6 per 1,000. Importantly, when the 1968 Tet Offensive began, a record number of ARVN deserters rejoined their units—more than 12,000 men in February and March alone.[3]

The South Vietnamese were not the only force afflicted with desertion, though. The level of manpower problems that the NLF encountered that summer was exemplified by a battle in Quang Tin Province. Some of the 217 guerrillas killed in the attack were as young as twelve. A Front battalion chief of staff who rallied in July "disclosed that morale of the 'fighting troops' of his regiment had dropped since the beginning" of the year. He stated that this was due to effective ARVN operations, air strikes, "conflict between northern and southern cadre," lack of popular support for the movement, and supply and recruitment shortages. Long An Province officials also reported a considerable drop in recruitment in 1967 due to the pacification effort. A CIA Covert Action Station report from neighboring Kien Hoa Province in the Mekong Delta confirmed the recruitment problem. There, new recruits were fourteen and fifteen years old. Much of the province was nominally under VC control, but ARVN and US operations were steadily clearing the rivers and roads.[4]

Civilians increasingly reacted against harsh VC tactics. For example, the 27 August shelling of Can Tho killed dozens and achieved the opposite of the NLF's intent. Rather than making the residents angry at the government for failing to protect them, the population characteristically responded by rallying to the government against the attacking force. In some ways, this was a harbinger of the post–Tet Offensive 1968 effect. A CIA report from neighboring Phong Dinh Province found people enthusiastic about government Revolutionary Development Teams. Census-grievance reports revealed that "even VC cadres have complained about the difficulty and lack of popular support in areas where [Revolutionary Development] teams have been operating."[5]

In the North as well, the war took its toll. A Chinese source in Hanoi reported over the summer that stevedores at the Haiphong port were "extremely slow due to their apparent poor physical condition," which he blamed on a lack of food. Despite some contradictory reporting at the time, this appears to be the beginning of what became a real problem for the North by 1970. The economic and social dislocations caused by the single-minded focus of all resources toward the war effort had begun to register.[6]

Despite the negative trends for the North that summer in 1967, a Rand Corporation study found a high degree of cohesiveness among mid- and upper-level Front members. Among the factors enhancing this were the "sense of mission," the "ability to avoid . . . excessive frequency or duration of combat," reported "trust in their leaders" and the cellular control system's ability to maintain cohesion, an "adequate relationship with the villagers," and "if not expectation of victory, at least nonexpectation of defeat." This study was based on a few interviews from the previous Rand study and two other studies. As such, this report may be more useful in understanding the baseline popular attitude at the beginning of the year than in describing the status by the summer of 1967. Regardless, the hope of the imminent victory promised by Front propaganda held many in the NLF by the fall.[7]

The bright spot for the Northern propaganda campaign remained the international audience. The DRV and Front continued to expand linkages with supporting groups overseas. Radio Hanoi reported that students in an NLF-controlled area of South Vietnam sent a letter to the Student Nonviolent Coordinating Committee in the United States. The students said they had followed "with keen interest and deep sympathy the struggle of the Afro-Americans against racial discrimination fiercely raging now in the United States." The students professed unity between their struggle "against the U.S. aggressors," and the committee's protests against "the U.S. ruling clique's savage acts of repression against the just struggle of the Afro-American students and people."[8]

Stokely Carmichael, the recently replaced national chairman of the committee, visited Hanoi in August. Carmichael met with Northern leaders for several days and assisted with their propaganda program. He "stressed the warm militant solidarity between the U.S. Negroes and the Vietnamese people and expressed absolute confidence in the final and complete victory of the common anti-U.S. imperialism struggle waged by the U.S. Negroes and the Vietnamese people."[9] On Radio Hanoi, Carmichael stated: "We are here today in your country, not in the uniform of the imperialists but in the uniform of humanity, to tell you that we are building a strong movement in the United States—a movement where black people, young black people like myself, are refusing to serve in the imperialists' armed forces." After his return from North Vietnam, Carmichael became prime minister of the Black Panther Party.[10]

Northern Propaganda Shift

During the fall, South Vietnam planned a series of elections to constitute the new Republic of Vietnam. Presidential and senate elections were to take place in September, and parliamentary elections would follow in late October. In a Liberation Broadcast Service appeal to the people of South Vietnam, however, the station criticized the election as a farce: "It is the U.S. imperialists and their henchmen including the Thieu-Ky clique who have used bombs, shells, and bayonets to tear down the houses and destroy the tombs of compatriots, to carry out their burn-all, kill-all, and destroy-all policy, and to herd millions of our compatriots into prisons, strategic hamlets, concentration camps and into their new life hamlets of the present day."[11] Once again, however, Front propaganda failed to prevent a large turnout at the polls in September. Afterward, in a broadcast ("The Rigged Results of a Deceitful Election Farce"), the LBS accused the Republic of "fascist and terroristic measures," in which the "Thieu-Ky clique mobilized more than half a million police and security men to fulfill the so-called task of standing guard around the voting centers."[12]

Although 86 percent of eligible voters in Kien Hoa Province participated in the elections, the CIA station reported evidence in September that the Vietcong was conducting training on a larger scale and moving supplies and equipment into the province in preparation for "their expected spring-winter offensive." Additionally, CIA sources indicated that the Vietcong had escalated its internal security campaign in an attempt to remove "peace loving cadres." North Vietnam purged Southern cadres who were demoralized by the Chieu Hoi program and the protracted war, replacing them with Northerners specializing in antipacification operations.[13]

Intelligence analysts increasingly detected a shift in North Vietnamese plans and supporting propaganda lines during the summer and fall. The NLF and PAVN had completed indoctrination at all levels throughout the period to ensure preparedness to implement the change of strategy brought about by Resolution 14 in July. This indoctrination filtered into propaganda themes. In September, at the time of the RVN elections, the NLF held a congress to vote on issuing the Northern directives as the new Front policy. The North promptly hailed this vote, which included the key plank of creating "a broad national democratic coalition."[14]

Prior to this period, the North had continued to stress the need for a negotiated settlement and prepared cadres for a protracted war. Propaganda now began priming cadres for the final phase of the war. After the congress, propaganda described the upcoming winter–spring offensive as "the decisive phase of war" requiring an "all out effort, regardless of losses." The first references to the formation of a coalition government appeared at this time, suggesting that the "historical phase" of the war had arrived. The upcoming offensive was presented as a "turning point." The plan for the offensive was to "defeat a major portion of both U.S. and GVN military forces" and instigate political turmoil in the cities, culminating in a "general uprising" that would result in a coalition government and the withdrawal of US forces. In advance, all VC units were pressed to expand liberated areas, and Special Action Units were to "engage in widespread terror and sabotage" in the cities, unleashing a ruthless assassination campaign.[15]

The NLF made some gains that fall, per CIA analysts, because of the strategy change. One report stated that the Vietcong may have "regained control of portions of the population in at least Quang Ngai, Phu Yen, and Khanh Hoa provinces," regions that had been cleared that year. However, analysts mistakenly conflated the shift with intensifying operations against Revolutionary Development Teams. The NLF had increased these activities as part of the change in strategy, not simply as a reaction to expanding pacification efforts as the CIA surmised.[16]

That same month, the CANLF also began to shift its tone in conformity with the new NLF policy. The committee was fully involved in the escalating American protest movement that fall, encouraging the formation of a coalition government and the rapid pullout of American troops, parroting the NLF line. Teague was on the organizing committee for several major protests, including the October protest at the Pentagon. The CANLF sold copies of the newly published Front political program, which detailed the strategy change. At the fall protests, members sold an estimated 15,000 copies of the *Vietnam Courier* issue containing the platform. Additionally, Teague's North Vietnamese contacts requested that he acquire film footage of the demonstrations to be sent back for propaganda purposes.[17]

Radio Hanoi praised the series of protests across the United States in October, sending messages of support. Among the major events were the self-immolations of Florence Beaumont and Hiroko Hayasuki and the launching of the antidraft movement. All Hanoi-affiliated stations carried these stories.[18] David Dellinger, a recurrent visitor to North Vietnam, stated at a press conference before one protest march that "the antiwar movement had moved from simple dissent to active resistance," indicating the radicalization of the movement. According to Radio Hanoi, speakers at the march urged "U.S. troops in Asia to stop fighting." Later, protesters marched on the Pentagon. Dellinger and Norman Mailer were among those arrested during the demonstration.[19]

In tandem with the expanded Vietcong assassination campaign from October to December, the Front prepared for the 1968 winter–spring offensive in line with North Vietnamese directives. The NLF increased selective terror incidents in an attempt to reassert control, as the plan had called for. The effects of this terror campaign are clearly visible in map 14. The data reveal an apparent lack of any attempt to draw allied forces out of the cities prior to the Tet Offensive, as is often asserted. Reviewing the Vietcong Initiated Incidents (VCII) and Terrorist Incident Reporting System (TIRS) data map overlays, the violence continued in the contested populated regions. In Quang Nam, for instance, the violence took place almost exclusively within a ten-square-mile area, less than ten miles from Da Nang. This is in accordance with the North's professed strategy discussed above. The goal was to rapidly prepare conditions for a coalition government. Meanwhile, the allied forces were attacking into VC-controlled areas, forcing them to defend. (See map 15.) Fewer attacks or assassinations occurred in the green (government-controlled) or neighboring yellow (government-influenced) hamlets. Looking at the HES, TIRS, and VCII overlays from earlier in the year shows a clear progression, pushing the Vietcong back toward VC Base Area 116 south of the Song Cua Dai River. The Front was steadily being pushed away from critical resources, men and rice.

A document captured in Thu Dau Mot Province (RVN Binh Duong Province) noted falling morale among the Vietcong. The memo from Nguyen Tung, assistant secretary and assistant political officer of the eastern Nam Bo region, revealed that PSYOPs greatly affected cadres and men who did not "realize the necessity of countering Allied PSYWAR activities." Such people "contacted their relatives working for GVN, listened to the BBC, VOA and Saigon radio broadcasts, and read publications and leaflets published by the GVN." To counter these actions, Nguyen urged cadres to intensify indoctrination, surveil "cadre and troops whose relatives work for the RVN government," and improve soldiers' living conditions. Despite the fall in morale, rallier numbers began to drop off in October due to propaganda proclaiming the imminent victory of the NLF and increased internal disciplinary controls.[20]

Farther south, the US Navy Mobile Riverine Force conducted Operation Coronado to clear the Mekong Delta and rivers bordering Kien Hoa Province in November. The recently created 19th PSYOP Company had responsibility for this region. The company field teams, along with the ARVN 403rd POLWAR Company, supported the ARVN 7th Infantry Division in nearby My Tho. The company also supported the ARVN 9th Division at Sa Dec and the ARVN 21st Infantry Division at Bac Lieu, and it had detached teams operating with the other three US PSYOP companies. The primary mission of the 19th PSYOP Company, however, was to support the US 9th Division and the Mobile Riverine Force. The PSYOP teams used speaker-equipped PBRs (Patrol Boat, River) from the US Navy's Task Force 186, "broadcast information programs to civilian river traffic by day," and harassed guerrillas at night. Reports indicated that the harassment had the desired effect of angering the cadres.[21] The roads and river areas were being cleared; according to the HES/TIRS/VCII data, assassinations and Vietcong incidents were lower than in neighboring provinces. (See photo 19.)

Tactical PSYOP units continued to struggle to meet demand and overcome logistical issues, though. The 244th PSYOP Company reported that the arrival of US Air Force O-2B aircraft during the summer greatly enhanced the PSYOP program. Loudspeaker time increased by 25 percent, the report noted. However, a lack of spare parts meant that only four of the eight aircraft were operational at any time. The 244th PSYOP Company also controlled the operations of an armed propaganda company, which conducted twenty-four missions in support of US and ARVN operations.[22]

Although the ROK PSYOP Company received two light mobile printing presses, they were "still not operational due to lack of expendable supplies" such as ink and paper. It continued to rely on the 245th PSYOP Company for printing support. In one instance, the South Koreans requested assistance from the 245th PSYOP Company for their Nha Trang–based PSYOP company in "printing of a booklet on Tae Kwan Do." The 245th provided the necessary supplies, and the 6th PSYOP Battalion replenished their stocks.[23]

On 22 October, South Vietnam held elections for the lower house. The NLF largely ignored these elections, unlike previous elections, in which the Front attempted to thwart elections and called for boycotts, only to be humiliated by large turnouts. VC propaganda in the provinces continued to focus on establishing a coalition government, even backing dissident candidates, which might help build support for a coalition government. To ease guerrilla recruitment shortages, the NLF "instructed all village cell, squad and platoon units to recruit females for proselytizing activities and to boost morale." They were to be used in noncombat positions. This appears to be part of an internal deception plan for the Tet Offensive. The lower echelons were not told the real reason for the expansion, which was to enlarge the force for the coming offensive. Taxation also expanded. Businessmen were pressured to provide loans to the NLF, "on the promise that it will be returned in 1968."[24]

One campaign that MACV identified that fall—the so-called beautiful-girl scheme—targeted Americans soldiers. The Vietcong recruited attractive young girls to work in "bars frequented by Americans." After a period, they returned to areas under Front control to disrupt combat operations. "The plan was to solicit sexual advances from Allied troops" and then "clutch the man tightly . . . while calling for help." The hope was to anger the people at the

"American rapist." Another twist involved Vietcong searching houses during combat operations to "steal precious items." Once the Americans had left, the cadres gathered the villagers to ask if "anything was missing." When the losses were discovered, they would denounce US forces for stealing the "property of the people." The underlying motive of most Front propaganda was to raise and exploit anger. The United States and RVN exploited anger as well, but not as widely or as effectively.[25]

Dominant Front propaganda themes exploited self-proclaimed victories and the Vietcong's ability to attack anywhere. The assassination campaign that fall legitimized this theme. Using the protests as a backdrop, VC propaganda claimed that the "Americans were tiring of the war" and that the NLF "would win just as they did over the French." Information gleaned from press reports and organizations such as CANLF supported and validated these claims.[26]

One such victory, for example, took place when the US Navy pilot (and future US senator) John McCain was shot down over Hanoi on 30 October. The North immediately seized on the propaganda prize, who had parachuted, injured, into the city's Truc Bach Lake. Hanoi reported details of his family life, asking: "What glory had he brought by his job to his father, Adm. John S. McCain Jr., commander in chief of U.S. Naval Forces in Europe? His grandfather, Adm. John S. McCain, commander of all aircraft carriers in the Pacific in World War II, participated in a just war against the Japanese forces." Radio Hanoi accused the younger McCain of "participating in an unjust war." The North also claimed to have shot down thirty-five planes over a five-day period.[27]

Effectiveness Measurement Challenges

During the summer MACV Command and Staff Conference, General Westmoreland directed MACPD to "study methods of evaluating PSYOP effectiveness rather than quantities of leaflets and broadcast hours or similar measures of performance." The goal was to assess "PSYOP effectiveness against the intensity of the effort." The resulting report contended that the only useful measurement of effectiveness was Chieu Hoi rallier numbers. Westmoreland further requested that the Defense Advanced Research Projects Agency (DARPA) conduct a study to "formulate accurate parameters for measuring campaign effectiveness. To determine the vulnerabilities of target audiences. To provide detailed examples of how the vulnerabilities identified in the study might be exploited." Westmoreland requested additional DARPA studies on novel methods of PSYOP utilization, the use of folklore in PSYOPs, and methods of pretesting leaflets.[28]

In order to increase the effectiveness of the leaflet campaign, General Westmoreland authorized intensified targeting of the heavily populated areas of Hanoi and Haiphong. As such, he ordered "leaflets dropped in the Red River Delta be increased to 60 million to overcome the large percentage lost due to terrain, water, and weather."[29] The effort reinforced "the effects of air strikes," but it also targeted areas to "accomplish psychological objectives not necessarily related to air strikes." This operation, designated Frantic Goat, replaced the Fact Sheet campaign that month.[30]

In October, MACV reevaluated the PSYOP effort in Vietnam. Based on input from all staff levels, MACPD found critical problems, which included "coordination of effort, adequacy and utilization of personnel and equipment" and balancing between tactical and strategic programs. The complexity of the PSYOP program was a profound challenge. Two PSYOP staff sections, MACPD and CORDS, coordinated with JUSPAO, which disseminated guidance and supervised the field efforts. This did not include covert programs. Although CORDS had simplified efforts, the disjointed organization continued to hamper message coordination. The report recommended revising the PSYOP structure and developing a comprehensive program to ensure the "indoctrination of US troops" in Vietnamese religion, culture, and customs.[31]

The MACV Psychological Operations Directorate sought to go beyond simple measures of the PSYOP effort to determine the types of analysis and data that could be used to make sound assessments. Anecdotal reports were useful but failed "to satisfy the rigorous expectations of the scientific method." The problem was that PSYOP activities were not reproducible in the laboratory. The authors suggested developing systems that allowed trend analysis and graphical presentation, harkening back to endeavors undertaken during World War I.[32] Just as earlier studies had found, the Chieu Hoi program was one area where a correlation between efforts and effects could reasonably be surmised, because the program "was entirely dependent upon information advanced through propaganda methods." Another area was "weapons turn in rates and the buyback program," which PSYOPs directly influenced.[33]

Population security levels could provide some evidence of success, according to the report, but such changes might be due to a number of factors. Another area the authors found worth investigating was the terror index—that is to say, the number of terror incidents in a given area. Plotting such incidents on a map could help in planning psychological activities and in determining who controlled an area. However, the graph shape for this data could be curvilinear because incident quantity would drop if either the government or the NLF took control of an area. Looking at the Quang Nam security situation seems to bear out this theory. (See maps 16 and 17.) Most of the assassinations that occurred during that fall took place within two miles of the Cua

Dai River, a highly contested area. Areas held strongly by either side had fewer incidents. Relying on anecdotes, the authors stated that "the relative ineffectiveness of the anti-election campaign by the VC is in itself testimony in support of the PSYOP program."[34]

Analyzing all the data then available, the authors determined "that a distinct improvement has occurred from August 1966 to August 1967."[35] However, as the Vietcong shifted strategies in the fall, the gains receded somewhat. The most important point of the study was that measures of effectiveness had to be refined during the operation planning phase so that "data may be collected as needed, not derived in an *ex post facto* manner." Effectively designed psychological objectives and supporting programs were keys to determining what information to look for.[36]

While acknowledging some improvements in all these areas, the authors noted that "the lack of research and analysis capability for PSYOP within MACV" continued to be a significant problem.[37] Although the PSYOP units had gotten better at pretesting product, a gap remained in post-testing. Post-testing is the method of assessing the quality and effects of propaganda products. The 6th PSYOP Battalion did some official post-tests of product effectiveness, but it had no standard process to do this. As such, JUSPAO and all PSYOP units depended on information gleaned from intelligence reports on the effectiveness of psychological operations campaigns. The MACV study's suggestions described above were meant to address the problem of information needs. However, military intelligence units did not collect information with PSYOPs in mind. Most of the intelligence that they collected was to determine the enemy's order of battle, not the needs of psychological operations. This fact highlighted the need for the clear intelligence requirements that MACPD described.[38]

By November, the PSYOP battalion's intelligence section began fielding Psychological Operation Reaction Teams (PORTs). Each PORT provided rapid-reaction intelligence support for military operations by deploying a propaganda officer, an intelligence analyst, and an interpreter to the supported headquarters. Much like the PSYOP Exploitation Teams used in field operations, these elements provided combat units with on-the-spot PSYOP analysis and gathered "ground truth" to improve battalion PSYOP area assessments.[39] A 6th PSYOP Battalion intelligence officer contended in an after-action report that teams should deploy during the combat operation planning phase to provide their expertise to the headquarters staff. At this time the teams could also gather additional information and determine intelligence requirements. Once the operation was underway, PORTs should travel throughout the area of operations to evaluate the PSYOP campaign's effectiveness. This would allow the teams to conduct field interviews of persons who might otherwise be missed by the standard intelligence system. The PORT concept offered a possible solution to the problem that MACPD had outlined earlier.[40]

To solve some of the analysis issues, the 6th PSYOP Battalion also established the Propaganda Development Center (PDC). The functions of the PDC were to "analyze, in depth, target audiences in Vietnam and develop propaganda with greater impact." The center contained separate sections for "plans and programs, audience analysis, propaganda development, coordination, testing and evaluation."[41] The PDC established a target-group database for identification and analysis. It also studied "media credibility factors for specified target audience groups." To assist with information gathering from VC and PAVN detainees, the battalion's S2 received official authorization for "occasional" access to the Combined Military Interrogation Center.[42] This allowed direct access to prisoners to aid in post-testing. Among the (unpublished) intelligence studies completed at this time by the Research and Analysis Branch were "Quang Ngai Psychological Area Assessment (Revised)" and "Psychological Vulnerabilities and Opportunities: The VC/NVA Supply Procurement System."[43]

The 15th PSYOP Detachment Target Analysis Section, of the 7th PSYOP Group, published a study in November titled "Report on Psychological Operations Intelligence in Vietnam." This was an in-depth critique of the PSYOPs intelligence system. Among the issues the study noted, the 245th PSYOP Company's intelligence section utilized two part-time workers among twenty-one authorized personnel. According to the report, the section did not have basic manuals and books on Vietnamese culture and history.[44] The company was understrength, limiting its intelligence capability. For instance, the 245th PSYOP Company detachment in Pleiku had twenty-seven authorized soldiers but only fourteen soldiers assigned, none of them intelligence analysts.

One problem across Vietnam—one that still bedevils intelligence services today—was "stovepiping" of information. Units of all types conducted intelligence analysis, but they fed the results only to their own chain of command. A fusion cell to compile and assess this wealth of information was lacking. Through personal initiative, the Research and Analysis Team initially accessed Saigon-area interrogation facilities to conduct PSYOP interviews and leaflet pretesting. PSYOP elements depended on building unofficial relationships with supported units to accomplish their duties. Research and Analysis Team members only later gained official access. The eight-man team produced a monthly "Psychological Intelligence Summary" for distribution to companies as well as other units connected with propaganda. The 15th PSYOP Detachment's investigators praised the 6th Battalion's research team's "professionalism, intelligence and aggressiveness" in its duties.[45]

On a cautionary note: the report stated that PSYOP units were only as effective as the commanding officer. Many commanders in the 6th PSYOP Battalion were former infantry officers with little background in psychological operations. As such, companies took on the conventional characteristics of an infantry unit. On the contrary, PSYOPs required unconventional personnel.[46] As for officer training, thirty-four of the battalion's officers had attended the long course at Fort Bragg, eighteen the shorter Military Assistance Training Advisor PSYOP course. Together this represented 65 percent of assigned officers. This level constantly changed with each addition and transfer, but it never reached 100 percent. The authors found that the 5th Special Forces Group's units were better prepared to conduct PSYOPs. With its string of remote camps, the 5th Special Forces had a good pulse on local affairs.[47]

For instance, Captain Lloyd H. Hinote wrote an end-of-tour memo to apprise his replacement of the activities conducted in the Khe Sanh region that fall. His area of responsibility encompassed 8,000 Bru Montagnards living near Khe Sanh. Most tribe members were illiterate, forcing Hinote to adjust the PSYOP products used in the area. The tribe had been relocated from remote villages to prevent the Vietcong from being able to use the Bru for support. Hinote had set up a T-368/URT high-frequency radio as a local Bru-language radio station to reach them. He had distributed approximately 350 radios in the area, creating a ratio of one radio for each twenty-three Bru, which Hinote viewed as an acceptable rate. One hundred and eighty radios remained to be distributed. At the time of writing his memo (December), Hinote shut down the station while awaiting construction by the Seabees of a new bunker at Khe Sanh base. Prior to that, the station had broadcast one hour per day and received positive reactions from the inhabitants. In order to establish "a regular listening audience," Hinote urged his replacement to focus on entertainment programming. Hinote suggested that his replacement broadcast from 1830 to 1930 hours, with a rebroadcast one hour later, and "purport to be under the sponsorship of GVN." This notion had to be subtle due to the "deep-seated mistrust" of the Vietnamese among the Bru.[48]

Although radio broadcasts helped, the most important means of gaining influence with the Bru was through face-to-face communications. Bru chiefs welcomed Hinote "with open arms" and supported his operations. MEDCAPs were among the most positive "psychological operation programs conducted in Vietnam," Hinote wrote. However, he noted that "it is up to the [PSYOP] officer to display to his superiors and contemporaries the usefulness of [PSYOPS]."[49] Another problem he experienced at this remote base was receiving PSYOP products. Hinote suggested—ironically—that "the most immediate foreseeable solution is to utilize the U.S. mail system with [PSYOP] materials mailed directly to the [PSYOP] officer." Despite these issues, Hinote described life among the Bru as the "most enjoyable tour for a psy op officer because he is left to his own imagination to [pursue] his own ideas and see the actual results of his efforts."[50]

In November, SOG initiated Project Oodles, the first of a series of notional programs (sending messages and supplies to make the North think there was a team operating in-country) that were meant to reinforce DRV fears about infiltration agent operations in the North. This deception created more than a dozen nonexistent phantom-agent teams. To facilitate the deception, SOG dumped blocks of ice attached to parachutes out of aircraft over North Vietnam. In the morning, the empty parachutes would indicate the landing of agents. These phantom "teams" then received radio communications by the same method as the real ones. Later, empty resupply bundles "were parachuted into the North to reinforce the impression that actual teams were on the ground." Such ruses kept the Northern security services busy chasing ghosts.[51]

Hamlet Evaluation System Reports

By December, the monthly HES survey had been ongoing for nearly a year. This was a controversial measurement of population loyalty based on subjective monthly reports. Expecting precise measurements from such a system during combat is unrealistic. Nevertheless, the system, when used judiciously, proved useful for looking at larger trends. Hamlets were rated on a scale from A (fully government-controlled) to E (government ineffective) and VC (controlled by the Vietcong). The scores were derived by the point value assigned to eighteen questions answered by the local adviser. The December HES report, after one year of measurement, showed that VC control had decreased from 24 percent to 20 percent. Populations living in fully government-controlled areas grew from 3 percent to 5 percent and A/B-rated hamlets grew from 23 percent to 30 percent. Both sides were fighting for control of the remaining 70 percent of the population that neither side fully controlled. Although this does not reflect a spectacular shift, the trend seemed to MACV leaders in Saigon to be moving in the right direction.[52]

The HES showed a wide variance in simple numbers. Only 1.72 percent of hamlets were considered fully under the control of the government, whereas 30 percent were fully Front-controlled. However, the estimated population controlled by the NLF was 2.1 million persons compared to about 10 million under some government control. The remaining 4 million were contested. The Vietcong held more hamlets, but those were less populated. The trend in II CTZ and III CTZ had remained positive in 1967; however, I Corps and IV Corps had struggled toward the end of the year as violence increased markedly based on North Vietnam's changed strategy. Interestingly, although the HES captured data on propaganda activities, the monthly report did not discuss this. It is unclear to what extent the 6th PSYOP Battalion made use of this data, as battalion personnel had limited access to HES data terminals in Vietnam. MACV later created a data system to combine information from the HES with PSYOP effort reports in an attempt to gauge effectiveness. HES was only one of many data sets collected at this time, each of which could have proven more useful given modern computer systems and geographic systems software.[53]

Comparing the HES report to the Vietcong's perception of population control as indicated in captured documents, the general accuracy of the much-maligned HES system becomes clear. It also must be noted that the HES phrase *Vietcong control* is misleading. If judged by the same criteria as "fully secured government [A] hamlets," the number would likely be much lower. For a hamlet to be considered fully under the control of the government, there had to be no VC activity that month and no guerrillas within a six-hour march. Using these same criteria for VC-controlled hamlets would yield very few, because almost the entire nation could be hit by government forces in a matter of minutes. In general, if the person who filed the monthly HES report listed a hamlet as VC-controlled, one can be fairly certain it was not under government control. For hamlets listed as fully government-controlled, this also was as true as it can be amid an ongoing insurgency. If one uses the data judiciously, however, trends emerge that enable visualization of the war on a new level. Therefore, and although the system may not have been perfect for its original purpose, it can still provide a deeper understanding of the war for the historian.[54] (See maps 15–17.)

As the second anniversary of the 6th PSYOP Battalion's activation in Vietnam approached, the unit had hit its stride, despite never being able to fully meet demand. The tactical teams conducted more than 380 hours of ground and aerial loudspeaker broadcasts that week and took part in more than fifteen named operations. Although personnel shortages strained the unit's capability, the battalion and its companies produced nearly 69 million leaflets in the first week of November alone. This still did not meet requirements.[55]

US forces were not alone in suffering from personnel shortages. The ARVN 20th POLWAR Battalion was authorized 519 soldiers but had only 388. The battalion had companies in Pleiku, Qui Nhon, Ban Me Thuot, and Kontum. ARVN logistical issues left the battalion with a chronic shortage of printing supplies. The 245th PSYOP Company continued to provide one day's supply per week, enabling the battalion to print 50,000 leaflets per week. Personality clashes also hampered the ARVN's efforts. For example, the ARVN II Corps intelligence chief was not on speaking terms with the POLWAR units in II

CTZ. Thus, everything had to go through an American, slowing the intelligence and analysis process.[56]

ARVN POLWAR battalions that year had focused on "shoring up support among the military and less on conducting PSYWAR against the enemy." Indoctrination of the servicemen centered on the motto "Country, Honor, Responsibility." Soldiers were "told why he was fighting, for whom and what goal, and why he would win." They received an education on the nation's 2,000-year history, focusing on heroes such as "Le Loi, Tran Hung Dao, Nguyen Hue, and the Trung sisters." They were also trained on proper treatment of civilians and "humane treatment of prisoners of war and defectors." The battalion trained the soldiers through a combination of lectures, dramas, booklets, newspapers, radio, and television. An indicator of success in this program was a precipitous drop by half of AWOL soldiers during Tet 1967. This moral rearmament program continued throughout 1967. In addition to the traditional propaganda operations, the POLWAR unit aided military dependents, operating "fifty-eight dispensaries, twenty-six maternity clinics, and 126 elementary schools." Only after Tet 1968 did POLWAR shift to focus on offensive operations.[57]

The psychological objectives forming the basis for the JUSPAO program included increasing "the Vietnamese people's participation with their government in the war against communist subversion and aggression" and increasing their participation in "developing Vietnam's social and economic progress" while acquainting them with American society. The election turnout indicated that this had occurred. Although these goals appear to be reasonable, increasing another "nation's sympathy and assistance to the cause of the RVN" may not have been a realistic goal for American PSYOPs. Additionally, concrete measures of effectiveness remained elusive.[58]

The 4th PSYOP Group Activates

The US Department of the Army had organized the 4th Psychological Operations Group on 18 September 1967. On 1 December 1967, after nearly a year of planning, the group activated in Vietnam. This greatly expanded the American PSYOP commitment to the war. Each PSYOP Company, in turn, expanded to a battalion-authorized 175 troops. The 19th PSYOP, 244th PSYOP, 245th PSYOP, and 246th PSYOP Companies, respectively, became the 10th PSYOP Battalion in Can Tho, the 7th in Da Nang, the 8th in Nha Trang, and the 6th in Bien Hoa. Each continued to operate within its particular corps.[59] (See figure 12.1.)

The battalions consisted of a direct support company with loudspeaker and audio-visual teams and a general support company containing printing facilities. The group's total authorized strength included 880 US military personnel, 133 Vietnamese civilian personnel, and 78 ARVN interpreters. Both the group and its battalions were organized around the conventional S1, S2, S3, and S4 staff sections, plus a Psychological Operations Development Center. With expansion to group status, the unit was 417 personnel short, and the first significant arrival of soldiers was not expected for three months. Of its 104 currently assigned officers, only 58 percent were classroom-trained in PSYOPs.[60]

One problem, most acutely felt in II CTZ, was the need for translators in the large number of dialects spoken. The 8th PSYOP Battalion operated in an area with thirteen Highland dialects and required "indigenous people, familiar with the language, customs, taboos, vulnerabilities and susceptibilities" to help develop material. The 7th PSYOP Battalion in Da Nang attempted to hire a Vietnamese illustrator to ensure a more culturally appropriate product in its corps.[61] JUSPAO encouraged the use of local artists to create products that avoided cultural discrepancies. For example, a JUSPAO report warned: "A local artist would know better than to show a father publicly mourning the death of his young child"; and "depicting a female ex-VC Hoi Chanh with a cigarette in her hand would impair her credibility as a communicator."[62] The group's PDC also deployed tactical analysis and survey teams to augment information available to the center, then used this information to write area and province PSYOP assessments. Overall, the expansion to a PSYOP group and the creation of PDCs addressed the major problems identified during 1967.

The 4th PSYOP Group's PDC produced immediate results. The center translated more than 600 documents that fall. As a result, intelligence analysts had greater access to enemy propaganda, diaries, and Front directives, which often contained information showing evidence of PSYOP effectiveness. Studies such as "Slogans and Symbols in South Vietnam," and more detailed area studies, had the potential to greatly increase the effectiveness of PSYOP products. The group also conducted leaflet design courses in each of the CTZs to improve product quality. In part, designers needed to remove American cultural norms from leaflet designs. Ensuring that leaflets were designed originally in Vietnamese was one technique. Another was to "verify the message and its meaning by a minimum of at least two translation checks and laborious testing of the leaflet on a representative target audience." Another lesson drawn from experience was ensuring that leaflets focused on a single theme. Cluttering products with multiple themes lessened their impact. Pretesting of leaflets on ralliers was important. However, ralliers often affirmed that a leaflet was good simply to avoid insult. The group suggested that testers phrase questions carefully in order to ensure an objective answer.

The JUSPAO suggested several more improvements in product quality.

Among the most important was maintaining credibility: "Every item produced should be viewed as putting the entire PSYOP program on trial." Even one bad product could compromise the entire PSYOP effort.[63] Likewise, every action had the potential to weaken the endeavor as well. Psychological operations personnel also needed to refrain from making promises that could not be kept. Just as important was preventing fellow soldiers from actions that, inadvertent or not, could anger the inhabitants and feed enemy propaganda. One soldier acting inappropriately could undo months of PSYOP effort.[64]

Still, effective PSYOP teams could break down barriers. The 6th PSYOP Battalion noted that "field team members have found that the most effective way of gaining the empathy of the population is to make initial contact with the children." PSYOP teams gave the children novelty items, such as "Chieu Hoi balls, candy, [and] kites." This helped break the ice, gained the "interest and confidence" of adults, and served as a "stepping stone" to their cooperation.[65] Although noted above, it bears repeating: children tended to act as hamlet "tripwires," and inhabitants often formed judgments about visitors based on their treatment of children.

For HE audio-visual and HB loudspeaker teams in the field, another critical lesson was that their first target audience was the supported unit commander. Teams needed to sell the value of psychological operations to the supported unit "before the 'operational' mission(s) of actual PSYOP support can be implemented." This was especially true at the tactical level, where young, enthusiastic American officers had yet to see the positive effects of psychological operations. Often these officers disdained the less-direct action of PSYOP teams. Selling propaganda required soldiers of a special caliber, people who were able to make the case for how they could aid the unit commander's mission when the primary concern was the lives of troops under command. In light of this difficulty, the requirement seemed to require well-trained PSYOP "artists" rather than "scientists." Another problem, the need for interpreters at the team level, remained unmet. Normally, teams received interpreter support from the attached unit. Teams could not develop the intimate relationship necessary for an interpreter to function in PSYOPs under such transitory circumstances.[66]

Tet Offensive Mobilization

In December, Front internal propaganda documents continued to proclaim the imminent defeat of the Americans and their possible withdrawal. A notebook found in January contained entries of a member of a Quang Tri/Thua Thien propaganda section regarding actions to take upon the upcoming withdrawal

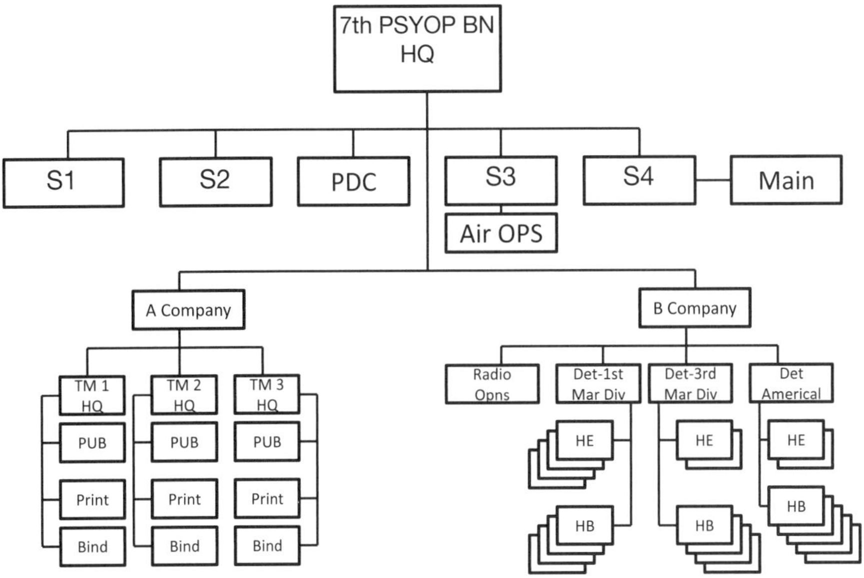

Figure 12.1. 7th PSYOP Battalion Organization

of US forces. Among the causes for optimism were the "deteriorating morale, political crisis, [and] conflict" between Americans and "puppet [RVN] authorities," as well as "the unpopularity of . . . President Johnson, the increasing national defense budget, the isolation of the United States in international affairs, the friction between the 'doves' and 'hawks' and the growing aspirations in the United States for negotiation and peace in Vietnam" and their serious defeats on the battlefield. Per the author, the "time is ripe for implementation of a general uprising to take over all powers in SVN." This was all part of the propaganda campaign meant to keep up Front morale in the face of ongoing setbacks.[67]

Cadres in Long An Province notified troops and sympathizers "that peace and a coalition government will follow the Lunar New Year Holiday." However, CIA analysts noted that failure to match these expectations could devastate VC morale and lead to large-scale defections. Since beginning this propaganda line in October, Chieu Hoi rates had dropped by nearly half in Long An. Around the country, "wavering higher-level VC cadre," one CIA cable noted, "do not want to lose the chance to reap their rewards for long service with the VC."[68]

Reports from Phong Dinh Province followed a similar line. Rallier rates had withered beginning in November. Along with that, reports of increased in-kind taxation and recruitment of younger people and females indicated that something was afoot. Front propaganda themes stressed a "general resistance campaign" that every person must take part in. Each hamlet also was directed to form an additional guerrilla platoon. The CIA continued to evaluate this as indicative of recruitment difficulties. Although that interpretation may be partly true, it was in the context of the larger numbers of fighters needed for the upcoming General Offensive/General Uprising dictated by Le Duc Tho and Le Duan.[69] Terrorism, assassinations, and surveillance of American and ARVN installations increased as well. Most of these activities took place in contested areas of the province, according to an analysis of HES and other data. Significantly, Vietcong-initiated incidents dropped off in late December to less than a third of November's activities in Can Tho. This seems to be part of the final preparation for the offensive.

North Vietnam provided the antiwar movement with special access, conferring legitimacy and power to protesters. As an example, the North allowed activists to carry letters from families to American prisoners of war. This was designed to manipulate POW families into supporting the antiwar movement. In at least one instance, Hanoi turned POWs over to activists for release, rather than proceeding through official channels. Tom Hayden, president of Students for a Democratic Society, flew to Phnom Penh, Cambodia, in November 1967 to receive the transfer of several American POWs, according to Radio Hanoi. Sergeant First Class Edward R. Johnson, a "black American . . . adviser to the puppet First Battalion, 31st Regiment," had been captured sixteen months earlier along with James A. Jackson. The NLF released them to Hayden, along with Daniel L. Pitzer, who was captured on 29 October 1963.[70]

At nearly the same time as this transfer, General Westmoreland was in Washington, addressing Congress and proclaiming that the war's end was in sight. Contrary to characterizations in the mass media, Westmoreland did not paint a rosy picture of the war. Instead, he argued that America had made progress and had turned back the enemy's attempt, begun in 1964, to conquer South Vietnam. According to the general, the light at the end of the tunnel was indeed visible—but not imminent—and hard days of fighting lay ahead. A detailed briefing at the Pentagon the following day reiterated this idea.[71]

The Vietcong continued to target propaganda at American troops, though it is questionable to what degree it was successful. Describing the effort, MACV command history noted: "How effective such VC/NVA propaganda is, directed against relatively sophisticated U.S. troops, is difficult to assess, but they keep trying." Certainly, in the number of defectors, the NLF's program was unsuccessful. However, the additional twin goals of dividing American and ARVN sympathies and decreasing American domestic support for the war may have been more successful. These themes clearly took hold during the post-1968 US drawdown. It may be impossible to determine the extent to which this propaganda still influences American perceptions of ARVN professionalism and bravery.[72]

The year ended with the defection to the Vietcong of US Private McKinley Nolan. He had left his unit in early November and was declared a deserter on 30 December. Along with Garwood, Nolan was one of only two declared American deserters to the enemy during the war. Nolan soon began broadcasting appeals on Liberation Radio. Leaflets with his photo and statements that "he is being well treated and appreciates the kind treatment he is receiving" from the NLF also appeared. The Texas native reportedly died during the Khmer Rouge period in Cambodia in the 1970s. The other deserter, Robert Garwood, returned to American control in 1979.[73]

Though they were the only two listed as defecting to the enemy, it cannot be said that the DRV targeting of American troops was a complete failure. In part, the US military did not do an effective job of accounting for and punishing deserters, according to a congressional report. A Senate report estimated the US military desertion rate at 4.9 per 1,000. However, the US Army was "unable to tell how many who deserted later returned to military control." AWOL numbers for all services in 1967 amounted to 134,000. The total number dropped from the rolls in 1967 was 40,227. Due to the multiplicity of reasons a soldier might leave, it is hard to ascribe the effect of the propaganda campaign in whole to this change. However, the numbers increased by 25 percent the following year. By 1969, 174 deserted soldiers lived in Sweden alone. At least one of these, Private Ray Jones of Detroit, had recorded an interview broadcast over Radio Hanoi in July 1967. He encouraged his fellow soldiers already in Vietnam to desert and to fight at home for the "freedom of your own people."[74]

1968 PSYWAR Plans

JUSPAO published the year-end guidance for the coming year on 28 December 1967. The goal was to prioritize the overall effort while allowing for regional variations. Helping the newly elected government communicate with the people was the most important priority. Improving the government's image with the people supported this objective. Increasing and maintaining "public confidence in the Vietnamese armed forces" was next in line. JUSPAO requested that American PSYOP units exploit "ARVN victories, courage, and civic responsibility" to support this goal. Related to this mission, PSYOPs needed to help build "a positive police image of public service" as well as explain the need for police controls to the people. The Chieu Hoi and Doan Ket programs were seen as key elements during the next phase of operations.

A final PSYOP goal for 1968 was to enhance the image of the United States by providing positive information and countering anti-American propaganda in South Vietnam. Per JUSPAO guidance, negative effects of the American presence "such as competition for goods and services that drive up prices . . . and occasional problems with troop conduct or war caused civilian casualties remain at tolerable level but need continuing PSYOP attention." Mitigating and preventing such problems was an important task for field personnel. The corrosive effects of enemy propaganda had finally been acknowledged.[75]

As 1967 closed, SOG once again requested approval to unleash a full psychological war against the North. Staff compiled a detailed plan of increased activities, including the surfacing of the SSPL as an active front organization. The plan was still under consideration as the Tet Offensive began. In light of the new political situation, Washington officials once again quashed the plan. The Johnson administration ceased nearly all PSYWAR activities aimed at the North by April 1968.[76]

Fifty-five percent of leaflets disseminated throughout the region during this period supported the Chieu Hoi program. The remainder supported other campaigns: the Ho Chi Minh Trail, B-52 bombings, Frantic Goat, and PAVN morale. The billions of leaflets dropped may at first appear to be excessive. However, there was logic to this avalanche of paper. Interviews with ralliers indicated a correlation between message repetition and soldier defection. Frequently, contact with family members pushed the fighter over the edge to believe the messages. In this sense, the PSYOP program prepared the individual to receive the action message from a more trusted source. Messages needed to be constantly provided in order to catch potential defectors at their weakest point. In a sense, Chieu Hoi was like advertising: nobody pays attention to a commercial, for example, except those who might need the product—and that audience is constantly changing.[77]

To support the long-term propaganda program, the 7th PSYOP Group's Radio Detachment (Provisional) deployed from Okinawa to Pleiku under Operation Yellowbird in December. The unit installed and operated a 50-kilowatt AM radio transmitter as a combined US/ARVN propaganda radio station. The ARVN's POLWAR arm provided the programming, "while the U.S. provided the equipment and supervised the technical operation of the station." Ten days prior to the Tet Offensive, the station began daily broadcasts of "six hours of news, music, and PSYOP messages to VC/NVA troops, their dependents, and sympathizers throughout the Central Highlands." However, Front forces attacked the station during the fighting that spring and destroyed it on

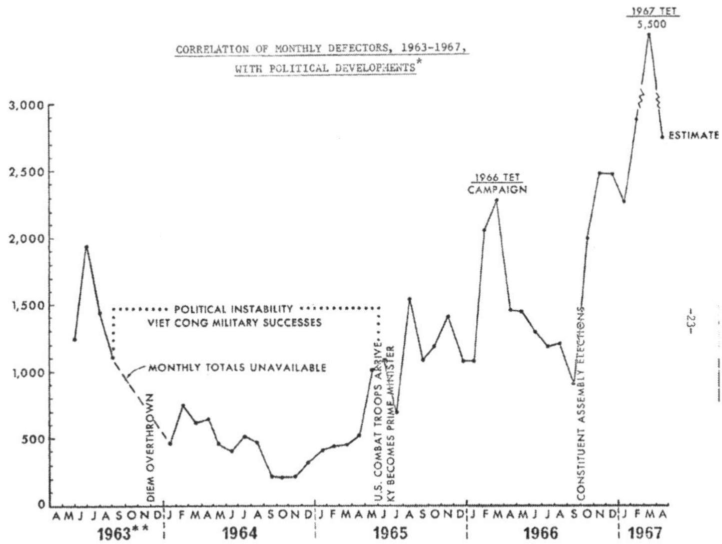

Figure 12.2. Chieu Hoi Rallier Totals

24 March 1968. Subsequently, the 4th PSYOP Group resumed broadcasts using a 400-watt AN/GRC-26 tactical radio to keep the station up and running (albeit with a reduced footprint of about twenty miles) until replacement equipment arrived.[78]

Conclusion

As 1968 dawned, PSYOP forces in Vietnam were engaged in the largest single campaign to date: the Tet Chieu Hoi campaign. At the same time, the 246th PSYOP Company cased its guidon and became the 6th PSYOP Battalion on 1 January 1968, the last of the original companies to expand to a battalion. Although rallier numbers had cratered after October 1967, no clear trend had yet emerged. It was unclear to analysts at this point that the cause of the downward trend was the North Vietnamese promise of imminent victory. The failure to keep that pledge would cost the NLF dearly in spring 1968, but that development still lay in the future. For now, the creation of a democratically elected government in South Vietnam, the upcoming Tet truce, and continuing statements of American support for the new republic offered signs of hope. Prospects for the future of South Vietnam and the PSYOP program looked bright as the Tet holiday approached.[79] (See figure 12.2.)

21. "The 'Late Hero' Nguyen Van Be" leaflet. Courtesy of psywarrior.com.

22. Republic of Korea friendship leaflet. AEHC, Peale Collection.

23. An ace of spades "death" card found by First Lieutenant Charles W. Sharman III, 1st Reconnaissance Battalion, 1st Marine Division. He sent the card home in May 1969, less than four months before he was killed in action. Image provided by Richard B. McCaslin.

24. One-Dong leaflet printed by 7th PSYOP Group for North Vietnam in August 1966. AEHC, Peale Collection, Terror. *Front*: Money is worth less and less. As the war goes on there will be less and less to buy. Prices will go higher and higher, your savings will become worthless papers. *Back*: Beware of another money reform such as that of 1959. You may lose all your wealth, fruit of your sweat and tears.

25. ARVN POLWAR Team operating in Cholon, Saigon, June 1968. NARA 111-CCV-443-CC50050. Courtesy of Fold3.com by Ancestry.

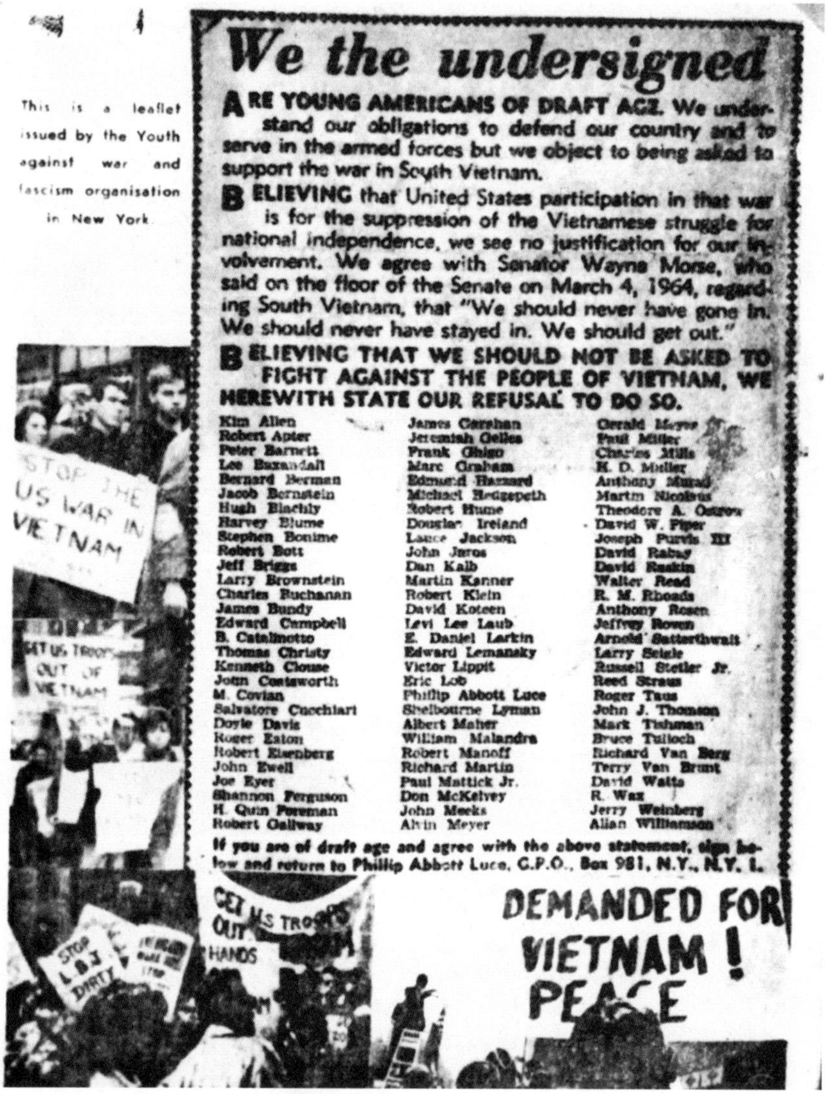

This is a leaflet issued by the Youth against war and fascism organisation in New York.

We the undersigned

ARE YOUNG AMERICANS OF DRAFT AGE. We understand our obligations to defend our country and to serve in the armed forces but we object to being asked to support the war in South Vietnam.

BELIEVING that United States participation in that war is for the suppression of the Vietnamese struggle for national independence, we see no justification for our involvement. We agree with Senator Wayne Morse, who said on the floor of the Senate on March 4, 1964, regarding South Vietnam, that "We should never have gone in. We should never have stayed in. We should get out."

BELIEVING THAT WE SHOULD NOT BE ASKED TO FIGHT AGAINST THE PEOPLE OF VIETNAM, WE HEREWITH STATE OUR REFUSAL TO DO SO.

Kim Allen
Robert Apter
Peter Barnett
Lee Baxandall
Bernard Berman
Jacob Bernstein
Hugh Blachly
Harvey Blume
Stephen Bonime
Robert Bott
Jeff Briggs
Larry Brownstein
Charles Buchanan
James Bundy
Edward Campbell
B. Catalinotto
Thomas Christy
Kenneth Clouse
John Coatsworth
M. Covian
Salvatore Cucchiari
Doyle Davis
Roger Eaton
Robert Eisenberg
John Ewell
Joe Eyer
Shannon Ferguson
H. Quin Foreman
Robert Gallway
James Garahan
Jeremiah Gelles
Frank Ghigo
Marc Graham
Edmund Hazard
Michael Hedgepeth
Robert Hume
Douglas Ireland
Lance Jackson
John Jaros
Dan Kalb
Martin Kanner
Robert Klein
David Koteen
Levi Lee Laub
E. Daniel Larkin
Edward Lemansky
Victor Lippit
Eric Lob
Phillip Abbott Luce
Shelbourne Lyman
Albert Maher
William Malandra
Robert Manoff
Richard Martin
Paul Mattick Jr.
Don McKelvey
John Meeks
Alvin Meyer
Gerald Meyer
Paul Miller
Charles Mills
H. D. Moller
Anthony Murad
Martin Nicolaus
Theodore A. Ostrow
David W. Piper
Joseph Purvis III
David Rabay
David Raskin
Walter Read
R. M. Rhoads
Anthony Rosen
Jeffrey Roven
Arnold Satterthwait
Larry Seigle
Russell Stetler Jr.
Reed Straus
Roger Taus
John J. Thomson
Mark Tishman
Bruce Tulloch
Richard Van Berg
Terry Van Brunt
David Watts
R. Wax
Jerry Weinberg
Allan Williamson

If you are of draft age and agree with the above statement, sign below and return to Phillip Abbott Luce, G.P.O., Box 981, N.Y., N.Y. 1.

26. Front side of a Vietcong propaganda leaflet found by US Marines in Quang Ni [Ngai] Province, September 1967. NARA 127-GVC-106-A189345. Courtesy of Fold3.com by Ancestry.

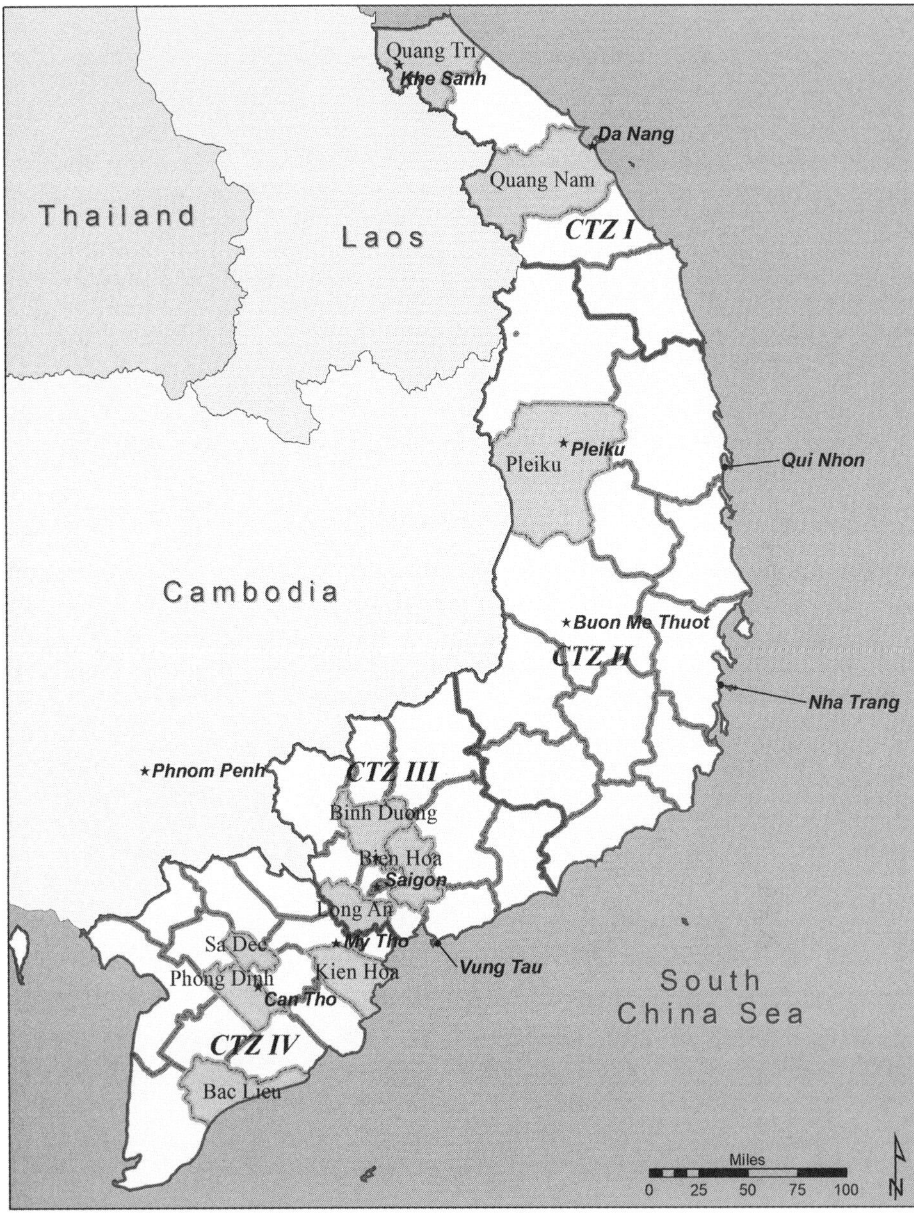

Map 14. South Vietnam, late 1967 (chapter 12 locations)

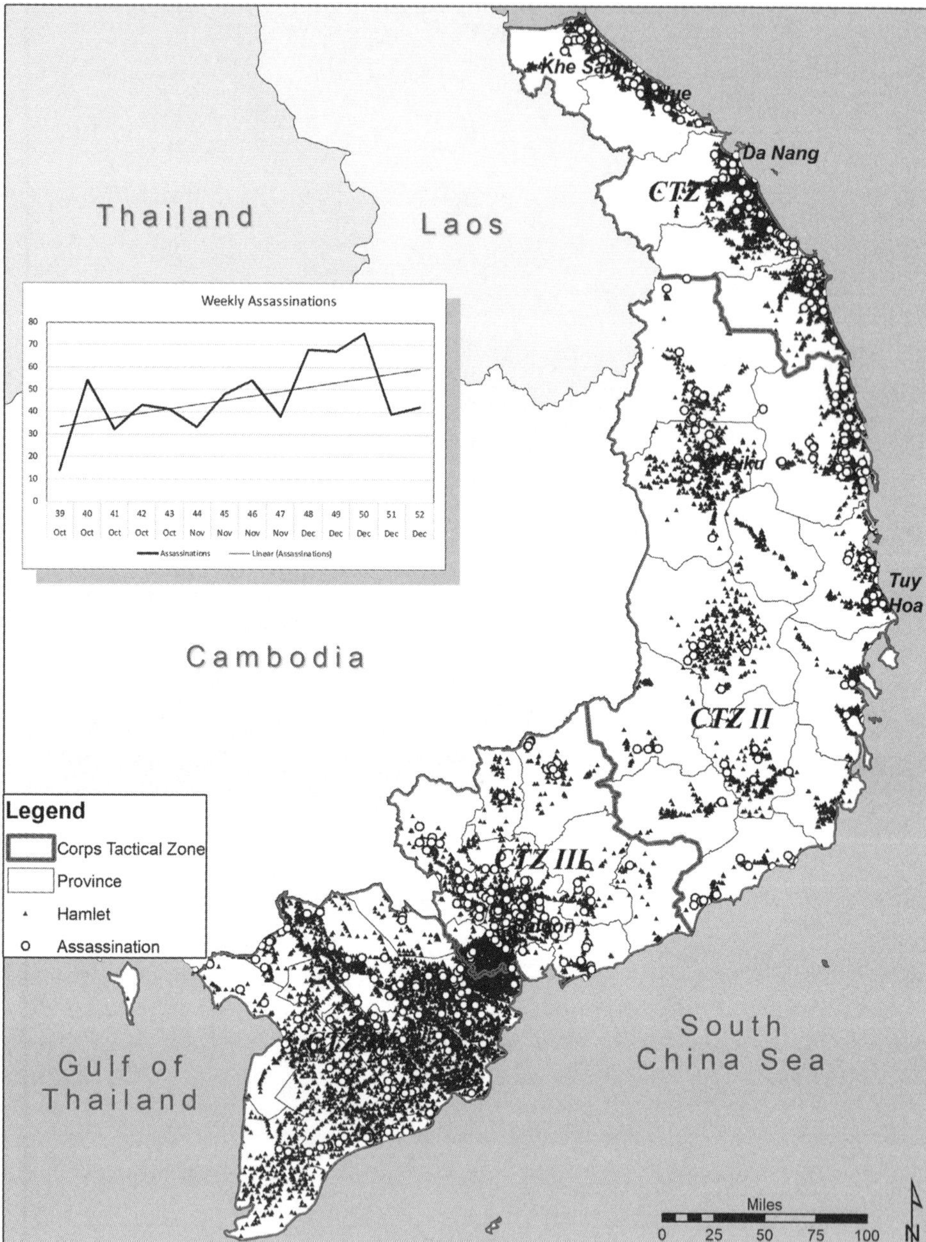

Map 15. Assassination campaign, fall 1967. This map shows the effect of the NLF's call for assassinations in preparation for the Tet Offensive. As indicated, most of the violence remained concentrated in the populated coastal areas and the capital.

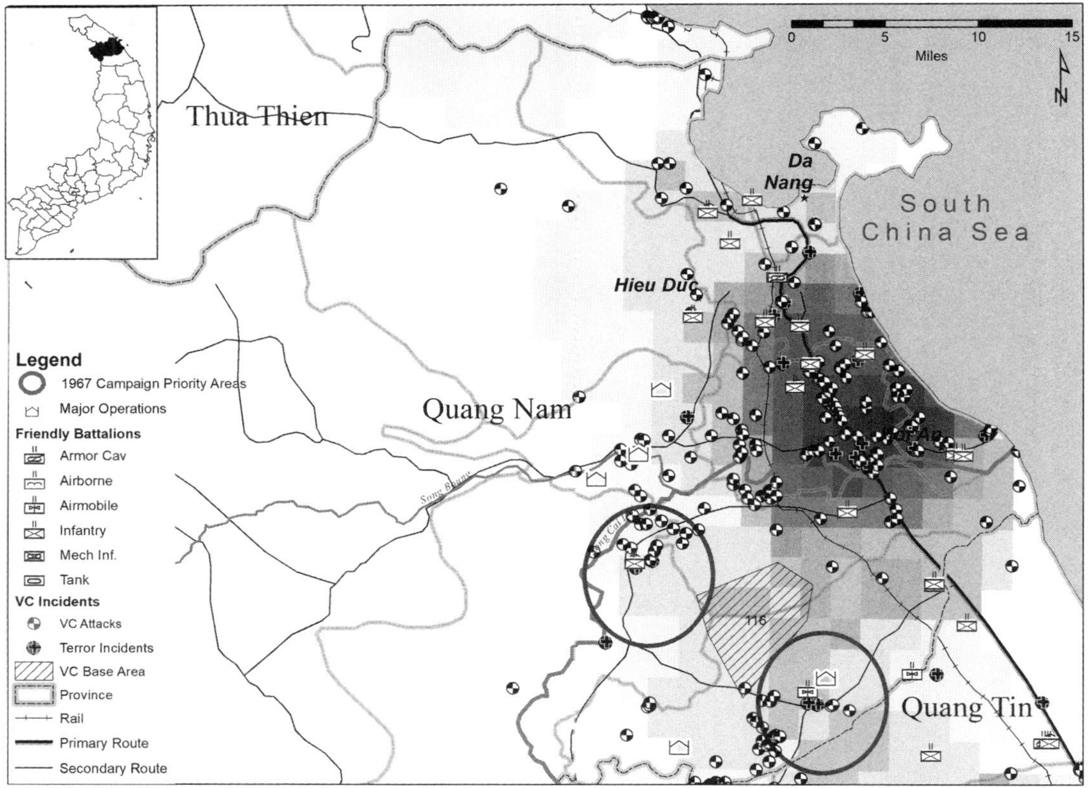

Map 16. Quang Nam Province map showing assassinations for the fall based on TIRS data. As depicted in the map, most Vietcong attacks occurred in the contested area along the Song Cua Dai River. Map includes friendly battalion locations from SEAFF during the period when US Marines fought to clear remote valleys.

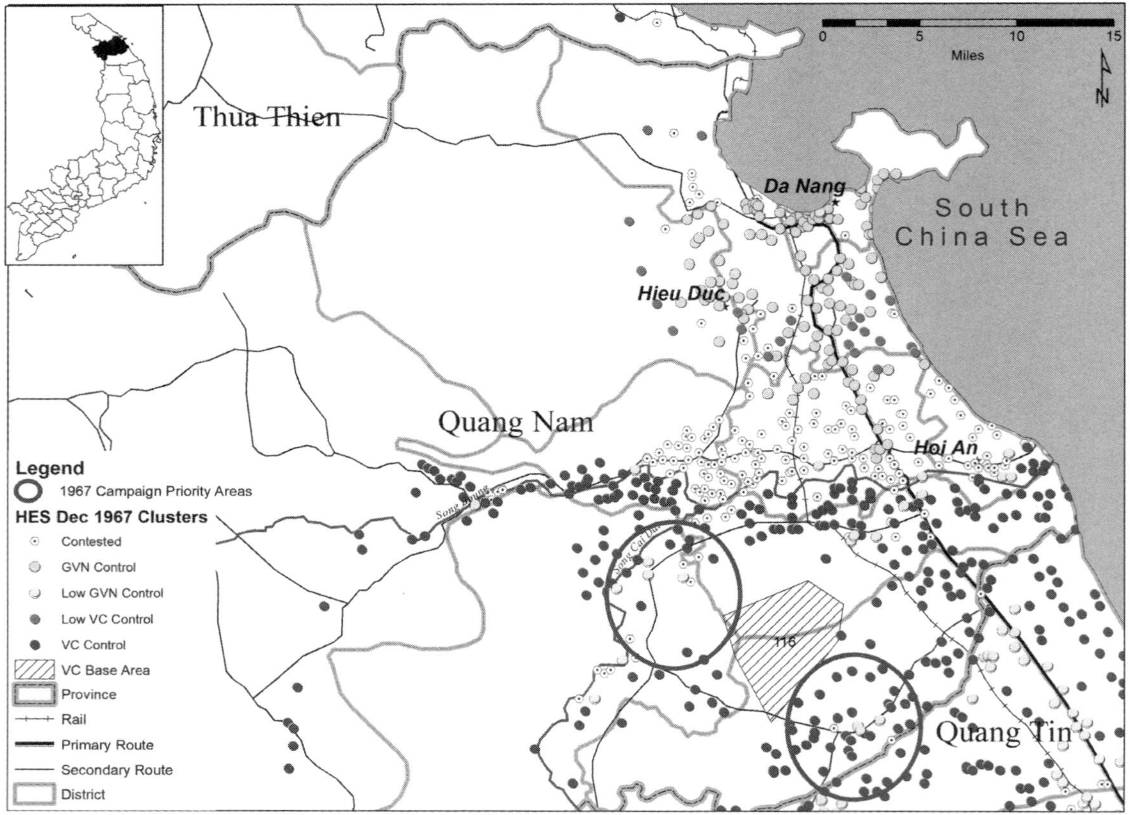

Map 17. Quang Nam Province map showing HES Security Control Status for December 1967. Map shows the clearing done over the previous year. In January, most hamlets around Da Nang were VC controlled. By December, the Vietcong were pushed south of the Song Cua Dai River toward Base Area 116.

Conclusion

Firecrackers boomed across South Vietnam in January 1968, heralding the end of the first year of the second Republic of Vietnam. The shooting had temporarily stopped. A Tet truce, proposed by the Vietcong, offered hope to the people of the South. Coupled with the recent elections, it seemed as if the war had turned a corner in their favor.

However, in the closing hours of January 1968, the crack of hostile bullets replaced the fireworks' joyous bursts and shattered those pleasant illusions. All over South Vietnam, civilians fled the shrapnel- and smoke-filled streets in search of safety. Areas of the capital, Saigon, became infernos. The National Liberation Front had launched a general offensive. One key target was Radio Saigon. Sappers took over the station, but it contained only the studios. Without control of the remote transmitters, the attackers were unable to broadcast. The studio soon became a smoldering ruin, filled with dead Vietcong.[1]

The Tet Offensive attacks hit no fewer than 5 autonomous cities, 35 provincial capitals, and 64 district capitals, targeting a total of 166 locations. The number sounds impressive, but roughly 12,000 hamlets and 244 district centers existed in South Vietnam. Brutal urban warfare took place in Saigon as ARVN and American forces sought to wrest control from the guerrillas in the capital. While Americans glued their eyes on Khe Sanh and Hue, Saigon was the strategic prize that the DRV sought.[2]

Alongside Southern combat troops, soldiers of the ARVN 50th POLWAR Battalion deployed to conduct tactical PSYOPs. Using loudspeakers, it aided desperate civilians fleeing to safer areas. POLWAR teams at the tip of the spear successfully encouraged many trapped guerrillas to surrender rather than face certain death. Many of the recently drafted guerrillas had minimal commitment to the NLF. One rallier said that they had been forced to fight, but "when their leaders were killed" they willingly surrendered to government troops. By the fourth day of the offensive, fighting centered on Saigon's Cholon District, where the Vietcong attempted to use women and children as human shields. A team of five VC insurgent attackers, persuaded by a POLWAR team to give up, provided valuable intelligence to aid in clearing the district. The POLWAR teams claimed to have avoided a possible massacre by getting the civilians out of the area. The POLWAR Civic Action troops operated refugee centers, and loudspeaker teams kept displaced persons informed of what was happening during a chaotic period.[3] (See photo 25.)

The fighting left tens of thousands homeless. Refugee camps ballooned in size as frantic civilians and orphans sought to reestablish their burned-out lives. The new government faced a severe test. Civic Action teams were instrumental in preventing disease and coordinating nongovernmental aid flowing in from around the world. Cultural-drama teams helped entertain the displaced. ARVN political warfare had come of age. In support of these activities, the new US 4th PSYOP Group also quickly shifted from the planned Tet Chieu Hoi campaign to one based on exploiting the failure of the general uprising.[4]

The North had spent six months preparing the offensive, moving supplies, drafting forces, and promising ultimate victory early in the New Year. According to many historians, the PAVN conducted attacks on outlying areas in order to cover the movement of troops. One argument claims that the "genius" of the plan was an ambiguity-reducing deception that fed "an alternative set of expectations" to American commanders. According to this line of reasoning, Hanoi initiated attacks "near the South Vietnamese border" to lead the enemy to believe that "a more conventional operation" was in the offing and to draw US and ARVN forces away from the cities.[5]

To be sure, PAVN and NLF attacks did occur against outposts such as Khe Sanh, Dak To, and Loc Ninh. However, as depicted in the maps in this study, the overall violence remained centered on the contested areas along the coast. Had the actual plan been to draw forces out of the cities in order to more easily overthrow the government, one would expect the opposite. In response to increased assassinations in the contested coastal areas, most friendly forces remained in place. Beyond a few high-profile troop movements to Khe Sanh, which captured the attention of the Americans, the only significant redeployment was toward Saigon. These shifts began as the meaning of the intelligence collected since the summer of 1967 became more clearly understood at MACV. Rather than a well-developed and successful deception, the Tet Offensive represented a tactical failure.[6]

Additionally, intelligence pointed the Americans to Khe Sanh precisely because it was a target of the General Offensive, not a deception to draw forces away. The goals of the offensive were to create conditions for a coalition government and the withdrawal of American forces. Northern leaders sought to replicate the successful Dien Bien Phu model, but they failed. Recent scholarship looking at DRV documents agree that Khe Sanh was not a deception, although it may have been something less than a full Dien Bien Phu–style attack. At a minimum, the hope was to overwhelm and annihilate an American outpost as part of the General Offensive, in order to convince American leaders to accept a coalition government. Importantly, those massed PAVN troops were not available to help take Da Nang, another key target of the offensive. Regardless, the urban uprising never occurred, and the forces massed at Khe Sanh suffered immense losses when confronted with Marine reinforcements and massed air power.[7]

The Tet Offensive, far from being a war-winning operation, was essentially a Northern gamble to stave off defeat. The decision reflected Le Duan's penchant for taking high risks when confronted with defeat, just as he had done in 1960 and 1964. In planning for the Tet Offensive, the North may have believed its own propaganda about Southern enthusiasm for the NLF, a challenge that all people involved in psychological operations face.[8] Bui Tin, who helped accept the South's surrender in 1975, later wrote that after 1968 the North did not "learn from the military failures of the Tet offensive. Instead, although we had lost the element of surprise, we went on to mount further major attacks in May and September 1968 and suffered ever heavier losses." General Westmoreland's decision to rush Marines to Khe Sanh in the weeks prior to the Tet Offensive thwarted a major Northern objective.[9]

Sifting through the abundance of data on the Tet Offensive is like searching a haystack for needles. Looking back, it is always easier to identify the important facts. The signs for the Tet Offensive were clear in hindsight going back to September 1967. Indeed, analysts fully expected a winter–spring offensive, just not in the form it took. CIA analysts misinterpreted key points at the time. They thought that the reliance on younger NLF draftees simply indicated a shortage of personnel and a morale drop. The CIA likewise misinterpreted the demand for Front tax payment in food in lieu of money. Analysts were correct about the problems the NLF faced, but they missed the larger fact that these changes indicated preparations for the General Offensive/General Uprising. The NLF sought to expand its forces in spite of shortages in manpower and supplies.[10]

Though a military debacle for the NLF and a crushing blow for the PAVN, Tet was an enormous political defeat for the United States. The wellspring of trust upon which President Johnson depended (and which he had continually abused) now ran dry. The images of street fighting in major cities, particularly Hue and Saigon, and the encirclement of the 26th Marine Regiment at Khe Sanh, juxtaposed with the administration's previous dissembling and its professions about the "light at the end of the tunnel," exhausted American popular support for his leadership of the war. Mischaracterizations of General Westmoreland's professions of visible light the previous fall only heightened the anger.[11]

The allied propaganda effort had grown exponentially to meet the challenges of war. By the end of 1967, MACV's PSYOP directorate had 74 soldiers assigned. The USIA had another 103 Americans and 331 local hires conducting psychological operations. The newly formed 4th PSYOP Group, though short-staffed, was authorized nearly 900 more. ARVN's POLWAR arm, also short, had an authorized strength of more than 2,600 troops. These numbers do not include people assigned to the SOG, CIA, and allied programs. All told, nearly 4,000 people were authorized to conduct the propaganda campaign in support of South Vietnam. However, a "continuous and uniform staff channel" was still lacking. As noted above, four channels of communication for PSYOP messaging existed: the military, SOG, JUSPAO, and the South Vietnamese. Predictably, this overly complex organization caused problems with "overlapping missions and responsibilities between PSYOP units and various other agencies involved in PSYOP."[12]

Personnel problems existed as well. The majority of 4th PSYOP Group officers were graduates of the unit or staff officer PSYOP course. However, most of these trained officers served in headquarters, not in tactical positions. Training for American PSYOP enlisted personnel was woefully inadequate. Most soldiers received no formal PSYOP training. The addition of a psychological operations enlisted course could have helped, but a separate PSYOP Military Occupational Specialty was the ultimate solution.[13]

Psychological operations differ from marketing. Selling defection is not identical to brand loyalty. PSYOPs are as much an art as they are a science. Message repetition is important, and one must seek measures of effectiveness. But just as important is the development of psychological operators who understand PSYOPs and can function within other cultures. As suggested above, most psychological operators arrived as amateur propagandists—with predictably mixed results. Effective PSYOPs require outside-the-box, nonconventional thinkers. Personnel need to understand and interpret complex situations. Another issue was the lack of cultural expertise. They also need the skills to convince their own people of the value that PSYOPs bring to the fight and the fearlessness to prevent fellow American soldiers from committing cultural indiscretions. Finally, throughout this period, the US military failed to meet the demand for Vietnamese linguists. Acquiring this complex skill set requires the right personality type. Such individuals could not be selected at random and the military cannot develop soldiers with these skills overnight. Edward Lansdale, Rufus Phillips, and Frank Scotton are the models here.

Although high-ranking American officers tended to understand the value of psychological operations, the tactical-level leaders tended to focus on issues of life and death and often disregarded or underestimated its value. Only skilled PSYOP personnel, coupled with visible effects, could overcome this.

One apocryphal story concerns an American artillery battery in which the commander was asked how his unit supported the pacification program. He responded by pointing at two howitzers, one with "Chieu" painted on it, the other with "Hoi."

From a woefully lacking base, tremendous improvement in intelligence occurred during a few short years. However, many of the successes noted above came solely from the initiative of soldiers throughout the PSYOP structure. JUSPAO's research program included surveys, pre- and post-testing of PSYOP products, monitored broadcasts, interrogations, and captured document exploitation. Additional intelligence came from MACV PSYOPs and J2 staff, the RVNAF's POLWAR department, the US 4th and 7th PSYOP Groups, Rand Corporation studies, and the CIA.[14]

However, American intelligence analysis was not without problems as well. The JUSPAO Research and Analysis section by 1968 had not created an adequate system to analyze captured documents as a means of assessing psychological operations effectiveness. The primary measure remained the numbers of leaflets dropped and hours broadcast. MACPD Directive 10-1 in December 1967 instituted a monthly PSYOP report from each corps. This format required that each corps provide the number of leaflets that it had received and disseminated, loudspeaker hours, newspapers and posters distributed, movies and cultural team performances, and County Fairs conducted. In addition to the numbers, the format required a narrative indicating an evaluation of effectiveness. However, a similar weekly report required only distribution numbers. The inability to capture usable measurements of effectiveness limited the program.[15]

Based on opinion polls, the most effective communication modes were face-to-face talks, cultural drama shows, and eventually television programs. Despite this, "printed material had the heaviest emphasis" due to the tendency of tactical "commanders to measure PSYOP effectiveness by the quantity of leaflets disseminated." This taxed the entire PSYOP system: printing, distribution, and dissemination for a lower payoff.[16]

One lesson—which the American side seemed to learn over time—was the importance of creating cognitive dissonance. Taking deeply held beliefs head-on raises defensive walls on the part of the target. Indirect questioning over a long period had a greater effect. This required consistent, generally truthful messaging. Ben Franklin, one of this country's first propagandists, understood the concept of cognitive dissonance, even if he had never heard the term. In his autobiography, Franklin discussed the value of the indirect versus the dogmatic approach in persuasive communications. He credited his ability to sway people to his views during the intense debates surrounding the American Revolution and the founding of the republic to avoiding dogmatic expressions. He wrote,

"I made it a rule to forbear all direct contradiction to the sentiments of others and all positive assertions of my own." The North Vietnamese often took a dogmatic approach.[17]

Despite allied problems, Hanoi began exhibiting a profuse fear of the US–RVN PSYWAR program as early as 1965. The program became the unwavering focus of Northern leaders, despite the limitations of US political constraints and SOG's mixed results. The North Vietnamese regime mounted an "often clumsy and heavy-handed" counterespionage campaign based on a fear of subversion.[18] For instance, the Lao Dong Party issued a decree in October 1967 that "listed twenty-one death penalty offenses against the state," nineteen of which could be tied directly to SOG's operations. If Hanoi's leaders knew the self-imposed limits to "encouraging violence and insurrection" in the North, the regime might have been less paranoid.[19]

Although the Republic of Vietnam's propaganda program was imperfect, so was the North's. The NLF's ruthless reputation, hostility to religion, and its "increasingly draconian demands on the villagers" eventually isolated it. The image of thousands of executed civilians in Hue after the Tet Offensive cemented this hostility in the minds of many in the South. Southerners often blamed the Vietcong for "allied artillery and air strikes that killed or dispossessed some and forced many others into GVN-controlled territory as refugees." The refugee phenomenon further decreased the population under Front control. While propaganda provided only limited opinion support shift *toward* the government, these episodes solidified the *fear* of a Communist victory.[20]

Exaggeration plagued the Northern propaganda program, sapping credibility. For instance, Radio Hanoi claimed that more than 1,000 enemy planes were shot down over North Vietnam during the year. The actual number was less than 300. Such overstatements undercut Hanoi's legitimacy with target audiences who possessed direct information about the war. However, for audiences such as the antiwar movement, the lies told by President Johnson mitigated this effect. Without direct information on the war on which to base opinions, antiwar activists were susceptible to Hanoi's promotion of cognitive dissonance.[21]

The year 1968 was pivotal in the psychological war. The effects of propaganda often take years to become apparent. While the Northern program had proved to be adaptable and opportunistic, its duplicity had become increasingly clear. Hanoi could fool some of the people some of the time, but not enough of the people enough of the time. The NLF increasingly depended on violence to maintain control. As a result, its support among the population cascaded downward. Earlier in the war, Radio Hanoi had warned listeners to ignore enemy PSYWAR just as they ignored the village dogs barking in the night. By 1968, the NLF had become the yapping dog that the villagers had learned, from experience, to disregard.[22]

American college-age youths were both highly susceptible and accessible to Northern propaganda. They were therefore a potential target audience for the North for diminishing American support for the war. However, the techniques used by the antiwar movement seem to have had little affirmative effect on the American public during this period. The impact on politicians is an important aspect of this story that needs to be explored further. As described by Robert McNamara in the film *Fog of War*, many in the political class had college-age children inculcated with the propaganda themes while attending prominent universities, leading to very awkward dinner conversations about the war. Whether these conversations affected policy may never be known. At the very least, the movement aided North Vietnam in legitimizing its propaganda. Typically, Hanoi wrote news stories, published or spread them in the United States using antiwar groups, and then published stories on the American people parroting the original story. The antiwar groups also created and improved on propaganda targeting American troops.[23]

Nevertheless, Hanoi remained determined to take control of the South. North Vietnamese leaders were not susceptible to change at the level of pain that President Johnson was willing to inflict. His mismanagement of the war squandered domestic support until it was too late. As it was, President Johnson ordered a cessation of agent infiltration and PSYWAR operations against the DRV in 1968 to meet a Northern requirement to begin peace talks. This requirement indicated the value that the North placed on the PSYWAR program. Captured documents also attested to the effect of the program directed against the Vietcong. The cessation occurred just as the program began to produce results and expansion plans were under discussion.[24]

Any PSYOP program needs to assess the results against the goals. Unfortunately, JUSPAO and other agencies never issued a single coherent PSYOP objective list. PSYOP doctrine at the time also did not have a concept for supporting psychological objectives. Well-written supporting psychological objectives help in the planning process by providing measurable and observable behaviors to indicate whether the program is working or not. With these criteria in place, it is possible to plan the intelligence collection effort and to track trends. Failure to do this made assessing effectiveness difficult. However, the program objectives can be deduced and some assessment surmised by reviewing documents such as the JUSPAO guidance.[25]

The goal to gain a secure environment was a very difficult task for psychological operations to achieve alone. Still, PSYOPs can help people comprehend and embrace change. Tiplines, weapons turn-in programs, and Chieu Hoi all supported this objective. The Vietcong had an estimated desertion rate of 5 percent in 1965, 7.1 percent in 1966, and a high of 20 percent in 1969—which indicates some US/GVN success in this area.[26]

There is also evidence of success in the goal of increasing popular support for the GVN. Polls and repeated high election turnouts suggest that people embraced the *idea* of a Republic of Vietnam if not a *specific* government official. However, the Vietnam Information Service was a weak link in the PSYOP system reaching civilians in the South. Long-term success depended on improving this program.

Gaining international support for the GVN as the legitimate representative of the South Vietnamese people was an inapt goal for American PSYOPs. Building the government's image needed to be done through South Vietnamese organizations in order to avoid lending credibility to the "puppet" label. In the face of the strong North Vietnamese and NLF overseas programs, the South Vietnamese program floundered. There is little evidence of success with this goal in the documentation reviewed. The Northern propaganda narrative flowed against it.

By 1968, evidence of the effectiveness of targeting North Vietnamese citizens to decrease support for the Democratic Republic of Vietnam began to accumulate. The morale breakdown in the North after 1968 was exacerbated by the agitation of very real problems there. However, due to the authoritarian nature of the regime, the agitation had little effect on the war. If the SOG program had continued, it might have seen more results and emboldened the North First faction. Additionally, the desired behavior change was never clearly articulated. Ironically, the bombing pause in 1968 helped fuel this anger. With no visible enemy to rally people against, it became more difficult to justify the failed economy and police state.

An important lesson drawn from the Vietnam propaganda war was the importance of dissecting the enemy's program in order to identify its psychological objectives and targets. Doing so enabled more effective countering of the enemy's narrative. Little evidence by this point in the war suggests success in the Northern goal of dividing American and South Vietnamese military personnel. Later friction and negative American views toward the ARVN may be related to this objective. But the United States was slow to grasp the potentially corrosive effects of Northern propaganda on American military morale. On a related point, one report found that "over 60 percent of the PSYOP team leaders replied that the actions of some U.S. troops were frequently detrimental to U.S./Vietnamese relationships." The types of incidents included "vehicle accidents, general misconduct of troops, and patronizing or disrespectful actions toward the Vietnamese people." Misconduct such as this played into Northern propaganda. The North was able to generate propaganda capital from offenses like prostitution, drunkenness, and military misdeeds.

An ACTIV report called for better troop indoctrination, even if such events were rare, to "emphasize how irresponsible conduct adversely affects U.S./Vietnamese relationships, the U.S. image and the PSYOP effort." There may have been some short-term impact of the Northern propaganda, but polling numbers attest to the long-term failure to turn the South Vietnamese people against the Americans.[27]

The North, likewise, had tactical successes in reducing the effectiveness of the government of South Vietnam. Communist ideology, with its focus on subjective and malleable truth, made it easy for the North to shift messaging rapidly to exploit opportunities. However, people live in the real world, where, over time, access to information reveals lies for what they are. Thus, the Northern program was often tactically successful but an operational failure. Hanoi's information war in the South was a long-term failure, although the associated assassination campaign did reduce government effectiveness. By 1968, the failures were becoming obvious. Polls, intelligence, and people's actions attest to this. The response of citizens in the South to the Tet Offensive and its aftermath further indicate the failure of Hanoi's messaging.

In the related goals of increasing world support for the Democratic Republic of Vietnam as the legitimate representative of the Vietnamese people, and gaining international support for the NLF as an independent movement representing the true Southern aspirations, broad acceptance of the messaging increased after the Tet Offensive. This made it much more difficult for the South Vietnamese international program to be effective.

It is difficult to assess the degree to which the Northern goal of decreasing American support for South Vietnam influenced the drop in Americans' opinions about the war. Approval numbers based on public opinion polling drifted downward throughout the war. In mid-1965, only 24 percent of Americans thought that sending troops was a mistake. And a plurality maintained a positive view of the decision until July 1967. Opinion on the war crossed over into negative territory in October 1967, when only 44 percent still agreed with President Johnson's decision. However, it is not clear to what extent the poll numbers simply reflected anger at the way President Johnson conducted the war rather than a desire to simply pull out. Perhaps some of the displeasure was with President Johnson's lies and seeming lack of commitment to the war. Some negative respondents may have wanted the war to be fought with more intensity.[28]

Part of the difficulty for both sides in understanding the war in Vietnam was the way in which it uprooted society. Americans who thought they understood Vietnam frequently sympathized with a country that had ceased to exist by 1967. Possessing an emotional connection to a nonexistent country often turned them against the war. This affected many who served in Vietnam during the early period, such as Frank Scotton and Daniel Ellsberg. They tended to analyze the war based on this emotional connection.[29]

In a related point, ARVN General Lam Thi Quang warned Americans early in the war about listening to the French on how to fight. He coined the phrase *loser's complex* to explain this, which he described as the "tendency of the losers to glorify those who defeated them on the battlefield."[30] Lam coupled this with the paternalistic "oppressed people's complex" in which the noble savage is always right. An oppressed people's backwardness proves that they are not fallen from grace or the Garden of Eden. Often Americans fall into both these traps in seeking answers to why the war ended in Northern victory. The Democratic Republic of Vietnam, and its surrogate National Liberation Front, were not particularly noble, and they were not right. Under the leadership of Le Duc Tho and Le Duan, they relentlessly used violence to unify the nation under a system that most in the South did not want. To do this they inflicted untold suffering on their own people by creating a ruthless police state and forced a war that left more than one million dead in its wake.

By contrast, the South and its allies now seemed for many the best hope of development and peace after the Tet Offensive. This vision of the nation contrasted sharply with that of the North Vietnamese regime. Indeed, the idea of a separate Southern identity and nation had a long pedigree. However, the newly elected government of South Vietnam had a narrow window of opportunity after the 1967 elections to develop a broad-based, democratic-leaning nation. Saigon made strides in this regard, but it could not act fast enough to stay ahead of the reaper. Rapid American drawdown and subsequent congressional withholding of war funding tipped the balance.[31]

Overall, the US/RVN PSYOP program was beginning to bear fruit by 1968. In the wake of the Tet Offensive, rallier numbers skyrocketed. In 1969, the PSYOP program reached its peak of 47,023 ralliers, including—importantly—a higher percentage of political ralliers. By 1970, a total of 194,000 had rallied. The Vietcong as a military force was defeated. The broken promises made by the North and the NLF prior to the holiday offensive, coupled with the massive draft of low-motivation personnel and the strict control measures implemented to ensure a large force for the offensive, led to a disintegration of the Vietcong. The post-Tet collapse in VC morale broke the back of the insurgency. Likewise, the political strength of the Vietcong infrastructure was diminishing. Terror brought war to urban areas that had largely known the conflict only at a distance until the Tet Offensive. The populace rallied to the newly formed republic, if not to the government of President Thieu. Record numbers of deserters returned to the ranks to defend the nation then—and again during the 1972 invasion. Significantly, the Tet Offensive proved the validity of the Chieu Hoi program. By and large, ralliers remained loyal to the government, and Armed Propaganda Teams and Kit Carson Scouts proved worthy of trust during the fighting.[32]

Progress was not linear, however. The mood of the population could shift rapidly based on its assessment of who was winning as much as ideological commitment. The decision to support one side or the other might mean life or death. Nonetheless, throughout the period 1965–1968, the trend was solidly toward support for the government. Unfortunately, any insurgency is like a fifteen-round boxing match. During the nine years of insurgency analyzed in this book (1960–1968), it did not matter who landed a good punch in 1961—the second round, so to speak. But it did matter who showed up in the later rounds. In the context of US participation in the Vietnam War, success in psychological objectives by 1968 was not enough. By 1975, one of the fighters had departed for the locker room.

Notes

Some Acronyms Appearing in the Notes

ACTIV	Army Concept Team in Vietnam
AEHC	US Army Education and Historical Center
CANLF	Committee to Aid the NLF
CARL	Combined Arms Research Library
CDEC	Combined Document Exploitation Center
CIA	Central Intelligence Agency
DPC	Douglas Pike Collection, TTVA
DTIC	Defense Technical Information Center
FBIS	US Foreign Broadcast Information Service
JCS	Joint Chiefs of Staff
JFKL	John F. Kennedy Library
JUSPAO	Joint United States Public Affairs Office
LBJL	Lyndon B. Johnson Library
LBS	Liberation Broadcast Service
MAAGV	Records of the Military Assistance Advisory Group, Vietnam
MACV	Military Assistance Command, Vietnam
NSAM	National Security Action Memorandum
NSF	National Security File
PROVN	*A Program for the Pacification and Long-Term Development of South Vietnam*
RH	Radio Hanoi
RVN	Republic of [South] Vietnam
TTVA	Texas Tech Vietnam Archive
USASOC	US Army Special Operations Command Archives
USIA	US Information Agency
USIS	US Information Service
USOM	US Operations Mission (the American embassy in Saigon)
VAC	Vietnam Archive Collection, TTVA

Introduction

1. *Staff Film Report 67–2: Chanh Luu Vietnam*, 27 min. (New York: Army Pictorial Center, 1967). This film provides combat camera footage and narration of the operation.

2. See Jeffrey Race, *War Comes to Long An: Revolutionary Conflict in a Vietnamese Province* (Berkeley: University of California Press, 2010 [1972]), and Frank Scotton, *Uphill Battle: Reflections on Viet Nam Counterinsurgency* (Lubbock: Texas Tech University Press, 2014).

3. David W. P. Elliott, *The Vietnamese War: Revolution and Social Change in the Mekong*

Delta, 1930–1975 (Armonk, NY: M. E. Sharpe, 2003), 182. Also based upon discussions with native speakers of Vietnamese.

4. Nhu Tang Truong, *A Vietcong Memoir: An Inside Account of the Vietnam War and Its Aftermath* (New York: Random House, 1986), v.

5. Department of the Army, *Field Manual 33-5, Psychological Operations* (Washington, DC: Department of the Army, 16 January 1962), 9; Garth S. Jowett and Victoria O'Donnell, *Propaganda and Persuasion*, 5th ed. (Thousand Oaks, CA: Sage Publications, 2012), 1–9.

6. Department of the Army, *Field Manual 33-5, Psychological Operations* (1962), 5.

7. Jowett and O'Donnell, *Propaganda and Persuasion*, 178–179; Benjamin Franklin, *Autobiography of Benjamin Franklin with Introduction and Notes* (1901; reprint, London: MacMillan Company, 1921), 87.

8. Jowett and O'Donnell, *Propaganda and Persuasion*, 24.

9. See Herbert Marcuse, "Repressive Tolerance," www.marcuse.org/herbert/pubs/60spubs/65repressivetolerance.htm.

10. Hillary D. Rodham [Clinton], "'There Is Only the Fight . . . ' An Analysis of the Alinsky Model" (BA thesis, Wellesley College, 1969), 8.

11. Ibid., 7.

12. Andrew MacKay, Steve Tatham, and Lee Rowland, "The Effectiveness of U.S. Military Information Operations in Afghanistan 2001–2010: Why RAND Missed the Point," Central Asia series (Shrivenham, UK: Defence Academy of the United Kingdom, December 2012), 3–4; Jowett and O'Donnell, *Propaganda and Persuasion*, 178.

13. Most of the information from this section comes from the author's experience conducting psychological operations. Additional details can be found in Department of the Army, *Field Manual 33-1, Psychological Operations* (Washington, DC: Department of the Army, 21 June 1968), and Jowett and O'Donnell, *Propaganda and Persuasion*.

14. *Field Manual 33-1, Psychological Operations* (1968), 1–3.

15. Ibid.

16. Department of the Army, *Field Manual 33-5, Psychological Operations, Techniques and Procedures* (Washington, DC: Department of the Army, 1966), 18–21.

17. United States Foreign Broadcast Information Service (FBIS), *Daily Report*, 23 July 1962, JJJ3, Radio Hanoi, "Let the Dog Bark, the Caravan Goes" (Arlington, VA: Foreign Broadcast Information Service, Joint Publications Research Service, 1960–1972).

1. Exciting a Spirit of Disaffection

1. Philip Grant Davidson, *Propaganda and the American Revolution, 1763–1783* (Chapel Hill: University of North Carolina Press, 1941), 226; Garth S. Jowett and Victoria O'Donnell, *Propaganda and Persuasion*, 5th ed. (Thousand Oaks, CA: Sage Publications, 2012), 80–85.

2. Stanley Sandler, *"Cease Resistance: It's Good for You!": A History of U.S. Army Combat Psychological Operations*, 2nd ed. (Washington, DC: United States Army Special Operations Command, Directorate of History and Museums, 1999), 11.

3. George Washington to Continental Congress, 11 May 1776, *George Washington Papers at the Library of Congress, 1741–1799* (Washington, DC: Library of Congress), memory.loc.gov/cgi-bin/query/P?mgw:1:./temp/~ammem_KXcS.

4. *Journals of the Continental Congress, 1774–1789*, ed. Worthington C. Ford, et al. (Washington, DC: US Government Printing Office, 1904–1937), 5 (1776): 653–656.

5. Lyman H. Butterfield, "Psychological Warfare in 1776: The Jefferson-Franklin Plan to Cause Hessian Desertions," *Proceedings of the American Philosophical Society* 94 (June 1950): 240.

6. Edward J. Lowell, *The Hessians and the Other German Auxiliaries of Great Britain in the Revolutionary War* (New York: Harper & Brothers, 1884), 222; Heinrich Carl Philipp von Feilitzsch, Christian Friedrich Bartholomai, and Bruce E. Burgoyne, *Diaries of Two Ansbach Jaegers* (Bowie, MD: Heritage Books, 1997), 24.

7. Timothy D. Johnson, *Winfield Scott: The Quest for Military Glory* (Lawrence: University Press of Kansas, 1998), 22–25; Walter E. Kretchik, *U.S. Army Doctrine: From the American Revolution to the War on Terror* (Lawrence: University Press of Kansas, 2011), 16–63; Richard W. Stewart, ed., *American Military History*, vol. 1 (Washington, DC: US Army Center for Military History, 2005), 137 (quote).

8. Johnson, *Winfield Scott*, 125, 168.

9. Ibid., 1, 179, 180 (quote), 188, 206.

10. Lawrence A. Yates, *The US Military's Experience in Stability Operations, 1789–2005* (Fort Leavenworth, KS: Combat Studies Institute Press, 2006), 3–7.

11. Henry Hotze and Lonnie A. Burnett, *Henry Hotze, Confederate Propagandist: Selected Writings on Revolution, Recognition, and Race* (Tuscaloosa: University of Alabama Press, 2008), 1, 14 (quote).

12. Ibid., 16.

13. Ibid., 14.

14. Ibid., 154.

15. John M. Gates, *Schoolbooks and Krags: The United States Army in the Philippines, 1898–1902* (Westport, CT: Greenwood Press, 1973), 103.

16. Ibid., 38, 56 (quote).

17. Ibid., 263.

18. Gates, *Schoolbooks and Krags,* 83, 149; Brian McAllister Linn, *The Philippine War, 1899–1902* (Lawrence: University Press of Kansas, 2000), 222.

19. Erving Winslow, "Report of the Executive Committee of the Anti-Imperialist League, February 10, 1899" (Boston: Anti-Imperialist League), archive.org/details/ReportOfTheExecutiveCommitteeOfTheAnti-imperialistLeagueFeb.101899; Linn, *Philippine War*, 222, 322; "Democratic Party Platform of 1900," The American Presidency Project, 4 July 1900, www.presidency.ucsb.edu/ws/index.php?pid=29587 (quote).

20. See Linn, *Philippine War*.

21. "A Lie Can Travel Halfway around the World while the Truth Is Putting on Its Shoes," from Quote Investigator (site maintained by Garson O'Toole), quoteinvestigator.com/2014/07/13/truth.

22. Linn, *Philippine War*, 327.

23. Paul M. A. Linebarger, *Psychological Warfare* (1948; reprint, New York: Arno Press, 1972), 63.

24. "WW1 Propaganda Posters," www.worldwar1gallery.com/propaganda/index2.html.

25. Linebarger, *Psychological Warfare*, 68.

26. George G. Bruntz, *Allied Propaganda and the Collapse of the German Empire in 1918* (Redwood City, CA: Stanford University Press, 1938), 31.

27. Committee on Public Information, "Four Minute Men News" (Washington, DC: Division of Four Minute Men, n.d.), 1.

28. Bruntz, *Allied Propaganda*, 33.

29. Ibid.

30. Clayton D. Laurie, "The Chanting of Crusaders: Captain Heber Blankenhorn and AEF Combat Propaganda in World War I," *Journal of Military History* 59 (July 1995): 460.

31. Ibid., 462.

32. Ibid., 463.
33. Ibid., 457.
34. Bruntz, *Allied Propaganda*, 63.
35. Laurie, "Chanting of Crusaders," 468; Jowett and O'Donnell, *Propaganda and Persuasion*, 113.
36. Linebarger, *Psychological Warfare*, 70.
37. Ibid., 68; Laurie, "Chanting of Crusaders," 479.
38. For details see Sophie Quinn-Judge, *Ho Chi Minh: The Missing Years, 1919–1941* (Berkeley: University of California Press, 2002), and William J. Duiker, *Ho Chi Minh: A Life* (New York: Hyperion, 2000).
39. Quinn-Judge, *Ho Chi Minh*, 22, 24–26.
40. Ibid., 4 (quote), 252.
41. Ibid., 42; Frank Logavall, *Embers of War: The Fall of an Empire and the Making of America's Vietnam* (New York: Random House, 2012), 10–16.
42. Quinn-Judge, *Ho Chi Minh*, 64–66.
43. Duiker, *Ho Chi Minh*, 101.
44. Quinn-Judge, *Ho Chi Minh*, 86 (quote); Duiker, *Ho Chi Minh*, 131.
45. Quinn-Judge, *Ho Chi Minh*, 96 (quote), 101.
46. Ibid., 103.
47. Ibid., 116.
48. Duiker, *Ho Chi Minh*, 155.
49. Ibid., 187.
50. Quinn-Judge, *Ho Chi Minh*, 192, Duiker, *Ho Chi Minh*, 221.
51. Quinn-Judge, *Ho Chi Minh*, 172 (quote); Duiker, *Ho Chi Minh*, 228.
52. Quinn-Judge, *Ho Chi Minh*, 218, 224 (quote).
53. Ibid., 230.
54. Ibid., 232.
55. Duiker, *Ho Chi Minh*, 255.
56. Ibid., 258.
57. Quinn-Judge, *Ho Chi Minh*, 247; David W. P. Elliott, *The Vietnamese War: Revolution and Social Change in the Mekong Delta, 1930–1975* (Armonk, NY: M. E. Sharpe, 2003), 46; Duiker, *Ho Chi Minh*, 254, 411 (quote).
58. Quinn-Judge, *Ho Chi Minh*, 246, 247 (quote).
59. Laurie, "Chanting of Crusaders," 480 (quote); Alfred H. Paddock, *U.S. Army Special Warfare: Its Origins*, rev. ed., Modern War Studies series (Lawrence: University Press of Kansas, 2002), 8. Paddock provides the best resource for understanding the constant reorganization of psychological warfare bureaucracy during the war.
60. Linebarger, *Psychological Warfare,* 91, 93 (quote).
61. Paddock, *U.S. Army Special Warfare,* 11; "Report of the 6808th Allied Information Service Field Consolidation Team (PROV)," the Psychological Warfare Division Supreme Headquarters Allied Expeditionary Force (n.p., n.d.), 14, www.psywar.org/psywar/reproductions/6808AIS.pdf.
62. Linebarger, *Psychological Warfare*, 98; Paddock, *U.S. Army Special Warfare*, 7–14 (quote), 15.
63. Duiker, *Ho Chi Minh*, 283 (quote).
64. Dixee R. Bartholomew-Feis, *The OSS and Ho Chi Minh: Unexpected Allies in the War against Japan* (Lawrence: University Press of Kansas, 2006), 153.
65. "Some Results of Recent Leaflet Operations," *Leaflet Newsletter* 1, no. 8 (Washington,

DC: Office of War Information, Intelligence and Leaflet Unit, 29 June 1945): 3, available through the Combined Arms Research Library (CARL), United States Army Command and General Staff College, Fort Leavenworth, KS.

66. "OWI Propaganda against Japan: A Summary," *Leaflet Newsletter* 1, no. 11 (Washington, DC: Office of War Information, Intelligence and Leaflet Unit, 1 September 1945): 8–9 (via CARL).

67. "Some Results of Recent Leaflet Operations," 3.

68. Bartholomew-Feis, *OSS and Ho Chi Minh*, 169; Mark Moyar, *Triumph Forsaken: The Vietnam War, 1954–1965* (New York: Cambridge University Press, 2006), 15.

69. "OWI Propaganda against Japan," 9, 14.

70. Bartholomew-Feis, *OSS and Ho Chi Minh*, 78.

71. Duiker, *Ho Chi Minh*, 279, 283, 308 (quote).

72. Ibid., 292 (quote); Bartholomew-Feis, *OSS and Ho Chi Minh*, 233.

73. Duiker, *Ho Chi Minh*, 311; Bartholomew-Feis, *OSS and Ho Chi Minh*, 233; "Vietnamese Declaration of Independence, 1945," www.fordham.edu/halsall/mod/1945vietnam.html.

74. Duiker, *Ho Chi Minh*, 286, 291; Logavall, *Embers of War*, 100–102; Bartholomew-Feis, *OSS and Ho Chi Minh*, 130 (quote).

75. Quinn-Judge, *Ho Chi Minh*, 247; Bui Tin, *Following Ho Chi Minh: The Memoirs of a North Vietnamese Colonel* (Honolulu: University of Hawaii Press, 2003),13; Duiker, *Ho Chi Minh*, 391 (quote).

76. Duiker, *Ho Chi Minh*, 326, 344 (quote).

77. Quinn-Judge, *Ho Chi Minh*, 3.

78. Duiker, *Ho Chi Minh*, 421.

79. Edward Miller, *Misalliance: Ngo Dinh Diem, the United States, and the Fate of South Vietnam* (Cambridge, MA: Harvard University Press, 2013), 24–25, 53 (quote).

80. Duiker, *Ho Chi Minh*, 348, 491; Miller, *Misalliance*, 32.

81. Paddock, *U.S. Army Special Warfare*, 20 (quote); Linebarger, *Psychological Warfare*, 168.

82. *Functions of the 5th Army Combat Propaganda Team*, 5th Mobile Radio Broadcasting Company, Fifth Army, 5 April 1944, www.psywar.org/psywar/reproductions/5ACPTPWB.pdf. This is the first example of a field manual that could be found. See also Linebarger, *Psychological Warfare*, 168.

2. Cold War Developments

1. William E. Daugherty, ed., *Psychological Warfare Casebook* (Baltimore: Johns Hopkins University Press, 1958), 29. The closest thing to a history of the psychological war was Paul M. A. Linebarger, *Psychological Warfare* (1948; reprint, New York: Arno Press, 1972), 68.

2. Daugherty, ed., *Psychological Warfare Casebook,* 29; Kenneth Osgood, *Total Cold War: Eisenhower's Secret Propaganda Battle at Home and Abroad* (Lawrence: University Press of Kansas, 2006), 51 (quote).

3. Robert J. Kodosky, *Psychological Operations American Style: The Joint United States Public Affairs Office, Vietnam and Beyond* (Plymouth, UK: Lexington Books, 2007), 68.

4. Scott Lucas, "Campaigns of Truth: The Psychological Strategy Board and American Ideology, 1951–1953," *International History Review* 18, no. 2 (May 1996): 284.

5. Charles H. Briscoe, "The 1st L&L in Korea, A Photographer's Record, 1952–53," *Veritas, Journal of Army Special Operations History* 3, no. 4 (2007): 14.

6. Department of the Army, *Field Manual 33-5, Psychological Warfare Operations* (Washington, DC: Department of the Army, March 1955), 59.

7. Charles H. Briscoe, "A Clearer View of Psywar at Fort Riley & Fort Bragg, 1951–1952," *Veritas, Journal of Army Special Operations History* 5, no. 4 (2009): 61–65.

8. George S. Pettee, *U.S. PSYWAR Operations in the Korean War* (Washington, DC: Operations Research Office, 1951), 29.

9. Linebarger, *Psychological Warfare*, 307.

10. Alfred H. Paddock, *U.S. Army Special Warfare: Its Origins*, rev. ed., Modern War Studies series (Lawrence: University Press of Kansas, 2002), 117.

11. Briscoe, "Clearer View of Psywar," 66; Department of the Army, *Field Manual 33-5* (1955), 32.

12. Paddock, *U.S. Army Special Warfare*, 56. Also Herbert Avedon, "Psychological Warfare Operational Deficiencies Noted in Korea—A Study," 28 August 1953, 57; and "Korean War After Action Reports Lessons Learned Documents, Battle Assessments," BACM Research, available at paperlessarchives.com.

13. Pettee, *U.S. PSYWAR Operations*, 37.

14. Chief of Army Field Forces, "Dissemination of Combat Information," 29 August 1952, 17, www.koreanwar-educator.org/topics/reports/after_action/combat_information_bulletins/combat_ information_bulletins_520829_350_05_56.pdf.

15. Chief of Army Field Forces, "Dissemination of Combat Information," 17 February 1953, 18, available at paperlessarchives.com.

16. Pettee, *U.S. PSYWAR Operations*, 36, 46 (quote).

17. Avedon, "Psychological Warfare," 5, 36.

18. Ibid., 50, attached memorandum with no page number (quote).

19. Michael E. Haas, *Apollo's Warriors: United States Air Force Special Operations during the Cold War* (Maxwell AFB: Air University Press, 1997), 38, 78 (quote), 80; Briscoe, "1st L&L in Korea," 14–24.

20. Haas, *Apollo's Warriors*, 106.

21. Ibid., 104.

22. Ibid., 93, 104 (quote), 108.

23. Paddock, *U.S. Army Special Warfare*, 133.

24. Ibid., 140.

25. Robert W. Jones Jr., "The OSS (Office of Strategic Services)," *Veritas, Journal of Army Special Operations History* 3, no. 4 (2007), 34–51.

26. Eugene G. Piasecki, "The A Team Numbering System," *Veritas, Journal of Army Special Operations History* 5, no. 4 (2009), 39–42.

27. Department of the Army, *Field Manual 33-5* (1955), 32–33.

28. Ibid., 15.

29. Lucas, "Campaigns of Truth," 281, 296.

30. Osgood, *Total Cold War*, 47.

31. Lucas, "Campaigns of Truth," 296 (quote); Osgood, *Total Cold War*, 55.

32. US Information Agency (USIA), *USIA Administrative History*, Lyndon B. Johnson Library (LBJL), USIA box 1, folder 1 (hereafter *USIA Administrative History)*; William Hoffer, "Broadcasting in an Insurgency Environment: USIA in Vietnam, 1965–1970" (PhD diss., University of Wisconsin, 1972), 285 (quote); Daugherty, ed., *Psychological Operations Casebook*, 31.

33. Michael G. Barger, "Psychological Operations Supporting Counterinsurgency: 4th PSYOP Group in Vietnam" (master's thesis, US Army Command and General Staff College, 2007), 15.

34. Hoffer, "Broadcasting in an Insurgency Environment," 75; Military Assistance Command, Vietnam (MACV), *Command History, United States Military Assistance Command, Vietnam 1964* (Saigon: MACV Military History Branch, 15 October 1965), 178, available at the Defense Technical Information Center (DTIC), AD0955106.

35. Central Intelligence Agency (CIA), "Memorandum on Pasternak's Dr. Zhivago," 12 December 1957, CIA Online Archive (www.cia.gov/library/readingroom) doc. (hereafter CIA doc.) 0005795615; CIA, "Memorandum on the Availability of Dr. Zhivago in Russian," 8 September 1958, CIA doc. 0005795664; Osgood, *Total Cold War*, 8.

36. CIA, From Chief, WH to Chief of Station, Lincoln, 8 June 1954, "Subject: KUGOWN Propaganda Guidance," CIA doc. 0000921974; Nicholas Cullather, *Operation PBSUCCESS: The United States and Guatemala, 1952–1954* (Washington, DC: CIA History Staff, Center for the Study of Intelligence, 1994), 1–3.

37. CIA, "Progress Report—PBSUCCESS for the 8–14 June 1954," 15 June 1954, CIA doc. 0000921914.

38. CIA, "CIA's Role in the Overthrow of Arbenz," 12 May 1975, CIA doc. 0000919933.

39. CIA, "Memorandum for Chief of Project, Subject PM Operation," 14 July 1954, 7, CIA doc. 0000920336. This document provides an after-action review of the Guatemalan operation.

40. CIA, LINCOLN to PBSUCCESS, Headquarters, "Foreign Policy and Propaganda Problems, Caused by Final Phase of PBSUCCESS," 1 July 1954, 2, CIA doc. 0000920823.

41. CIA, "Current Situation in Iran, 9 November 1949," CIA Historical Review, CIA doc. 0000258579.

42. William J. Duiker, *Ho Chi Minh: A Life* (New York: Hyperion, 2000), 424.

43. Thomas L. Ahern Jr., *CIA and the House of Ngo: Covert Action in South Vietnam, 1954–1963* (Washington, DC: Center for the Study of Intelligence, 2000), 15.

44. "Edward G. Lansdale, Lecture at Armed Forces Staff College," 7 January 1960, 3, Texas Tech Vietnam Archive (TTVA), Vladimir Lehovich Collection, box 1, folder 7.

45. "Organization and Operation of the 1st Psywar Battalion," Undated Briefing, 1963, TTVA, Douglas Pike Collection (DPC): Unit 3—Insurgency Warfare, box 15, folder 5.

46. FBIS, 14 July 1953, DD1, Saigon English Service.

47. Lien-Hang T. Nguyen, *Hanoi's War: An International History of the War for Peace in Vietnam* (Chapel Hill: University of North Carolina Press, 2012), 36.

48. Bui Tin, *Following Ho Chi Minh: The Memoirs of a North Vietnamese Colonel* (Honolulu: University of Hawaii Press, 1995), 15.

49. Duiker, *Ho Chi Minh*, 445 (quote); Jean-Louis Panné, *The Black Book of Communism: Crimes, Terror, Repression* (Cambridge, MA: Harvard University Press: 1999), 569–570.

50. Duiker, *Ho Chi Minh*, 423, 436–437.

51. David W. P. Elliott, *The Vietnamese War: Revolution and Social Change in the Mekong Delta, 1930–1975* (Armonk, NY: M. E. Sharpe, 2003), 141.

52. Duiker, *Ho Chi Minh*, 470; Edward Miller, *Misalliance: Ngo Dinh Diem, the United States, and the Fate of South Vietnam* (Cambridge, MA: Harvard University Press, 2013), 6; Edwin E. Moise, "First Indochina War," clemson.edu/coah/history/faculty/pages/edmoise/viet4.html; Mark Moyar, *Triumph Forsaken: The Vietnam War, 1954–1965* (New York: Cambridge University Press, 2006), 59 (quote).

53. Miller, *Misalliance*, 41.

54. Ahern, *CIA and the House of Ngo*, 34–37, 69–73; Moyar, *Triumph Forsaken*, 48.

55. Ahern, *CIA and the House of Ngo*, 15; Neil Sheehan, *The Pentagon Papers. As Published by the* New York Times (New York: Quadrangle Books, 1971), 54–66; Cecile B. Currey, *Edward Lansdale: The Unquiet American* (Boston: Houghton-Mifflin, 1988), 142.

56. Seth Jacobs, *Cold War Mandarin: Ngo Dinh Diem and the Origins of America's War in Vietnam, 1950–1963* (Lanham, MD: Rowman & Littlefield, 2006), 50, 52–53.

57. Miller, *Misalliance*, 99.

58. Ahern, *CIA and the House of Ngo*, 44.

59. Ibid., 45.

60. Jeffrey Race, *War Comes to Long An: Revolutionary Conflict in a Vietnamese Province* (1972; reprint, Berkeley: University of California Press, 2010), 96.

61. Edward G. Lansdale, "Memorandum for the Record—'Pacification' in Vietnam," 15 July 1958, TTVA, DPC: Unit 6—Democratic Republic of Vietnam, box 17, folder 26.

62. Rufus Phillips, *Why Vietnam Matters: An Eyewitness Account of Lessons Not Learned* (Annapolis, MD: Naval Institute Press, 2008), 18.

63. Ibid., 68 (quote), 80.

64. Miller, *Misalliance*, 141.

65. Ahern, *CIA and the House of Ngo,* 77; Moyar, *Triumph Forsaken,* 47, 58; Ronald H. Spector, *Advice and Support: The Early Years, 1941–1960* (Washington, DC: Center of Military History, 1983), 235–237.

66. Lansdale, "'Pacification' in Vietnam," n.p. (quote); James Lawton Collins, *The Development and Training of the South Vietnamese Army, 1950–1972* (Washington, DC: Department of the Army, 2002), 138.

67. Lansdale, "'Pacification' in Vietnam," n.p.

68. Frank Scotton, *Uphill Battle: Reflections on Viet Nam Counterinsurgency* (Lubbock: Texas Tech University Press, 2014), 4–5.

69. Elliott, *Vietnamese War*, 164; Lansdale, "'Pacification' in Vietnam," n.p.

70. Moyar, *Triumph Forsaken*, 36–37; Ahern, *CIA and the House of Ngo*, 11–12.

71. Hoffer, "Broadcasting in an Insurgency Environment," 74, 75 (quote).

72. Records of the Military Assistance Advisory Group Vietnam (MAAGV), *1950–1964, Adjutant General Division Security Classified Files*, "Weekly Report 11–18 April 1964" (Bethesda, MD: University Publications of America) (hereafter MAAGV Records).

73. Chief CATD Division Memorandum to Chief of Staff, MAAG, "Psychological Warfare Seminar- Okinawa," n.d., MAAGV Records.

74. Bradford Buttler, "MAAG to RVNAF General Staff," MAAGV, Saigon, Vietnam, 24 November 1958, MAAGV Records; "Psychological Operations Vietnam, Period 1 May 1958 to 31 October 1958," MAAGV Records; ARVN, "Organization and Operation of the 1st Psywar Battalion," 1, TTVA, DPC: Unit 3—Insurgency Warfare, box 15, folder 5.

75. Egerton L. Ballachey, "Psychological Operations: South Vietnam, Project PROSYMS" (Washington, DC: American University, Special Operations Research Office, April 1959), 1, DTIC, AD310382.

76. Miller, *Misalliance*, 59, 66; see Walt W. Rostow, *The Stages of Economic Growth: A Non-Communist Manifesto* (1960; reprint, Cambridge: Cambridge University Press, 1990).

77. Miller, *Misalliance*, 12.

78. Ibid., 13, 40 (quote), 123.

79. Ibid., 181 (quote), 253.

80. Ibid., 179, 188 (quote).

81. Ibid., 201.

82. Ibid., 236.

83. Race, *War Comes to Long An*, 100.

84. Nguyen, *Hanoi's War,* 45; Pierre Asselin, *Hanoi's Road to the Vietnam War* (Oakland: University of California Press, 2013), 59 (quote).

85. Race, *War Comes to Long An*, 98.

86. Ibid., 110.

87. Nguyen, *Hanoi's War*, 46; Spector, *Advice and Support*, 330, 331, 337 (quote).

88. Duiker, *Ho Chi Minh*, 508; Moyar, *Triumph Forsaken*, 61.

89. Military History Institute of Vietnam, *Victory in Vietnam: The Official History of the People's Army of Vietnam, 1954–1975*, trans. Merle L. Pribbenow (Lawrence: University Press of Kansas, 2002), 65 (quote), 67.

90. Asselin, *Hanoi's Road*, 88 (quote), 95; Eric Bergerud, *Dynamics of Defeat: The Vietnam War in Hau Nghia Province* (Boulder, CO: Westview Press, 1991), 20.

91. Nhu Tang Truong, *A Vietcong Memoir: An Inside Account of the Vietnam War and Its Aftermath* (New York: Random House, 1986), 63, 68.

92. Ibid., 84.

93. Moyar, *Triumph Forsaken*, 94 (quote).

94. Race, *War Comes to Long An*, 123.

95. Ibid., 122.

96. Richard A. Hunt, *Pacification: The American Struggle for Vietnam's Hearts and Minds* (Boulder, CO: Westview, 1995), 6 (quote); Moyar, *Triumph Forsaken*, 104; Truong, *Vietcong Memoir*, 60–80.

97. Elliott, *Vietnamese War*, 164 (quote), 214–216; Duiker, *Ho Chi Minh*, 528.

98. Spector, *Advice and Support*, 354–361; Moyar, *Triumph Forsaken*, 99.

99. US Information Service (USIS), *Media Survey of Teachers* (Saigon: Center for Vietnamese Studies, Bureau of Social Science Research, August 1960), 1, TTVA, DPC: Unit 11—Monographs, box 11, folder 5.

100. Ibid., 2, 13.

101. USIA, *USIA Administrative History*, LBJL, USIA box 1, folder 1, 1–8.

3. America's "Special War," 1960–1962

1. Michael G. Barger, "Psychological Operations Supporting Counterinsurgency: 4th PSYOP Group in Vietnam" (master's thesis, US Army Command and General Staff College, 2007), 24.

2. Memo, "Dictated by President-elect Kennedy, 19 January 1961," 2, www.paperlessarchives.com/FreeTitles/Eisenhower-JFKContact.pdf.

3. "Clifford to Kennedy, Memorandum of Conference," 24 January 1961, 3, www.paperlessarchives.com/FreeTitles/Eisenhower-JFKContact.pdf.

4. Thomas L. Ahern Jr., *The Way We Do Things: Black Entry Operations into North Vietnam, 1961–1964* (Washington, DC: Center for the Study of Intelligence, May 2005), 10 (quote), 11.

5. US Information Agency, *USIA Administrative History*, LBJL, USIA box 1, folder 1, 1–8.

6. Robert W. Chandler, "U.S. Psychological Operations in Vietnam, 1965–72" (PhD diss., George Washington University, 1974), 150; USIA, *USIA Administration History*, vol. 1, 5–44; FBIS, Broadcasting Stations of the World, part 1, *AM Broadcasting* (Washington, DC: US Government Printing Office, 1961), n.p.; "Proposal (probably by Lansdale) for a Presidential Directive for a Task Force," 19 April 1961, 27. TTVA, DPC: Unit 1—Assessment and Strategy, box 2, folder 1.

7. "Lansdale to McNamara, Memorandum—Vietnam," 17 January 1961, 4, TTVA, DPC: Unit 1—Assessment and Strategy, box 2, folder 1.

8. Ibid., 8.

9. Ibid., 11.

10. Lansdale, "Basic Counterinsurgency Plan for Viet-Nam," 10, TTVA, DPC: Unit 2—Military Operations, box 1, folder 12.

11. Ibid., 9 (quote), 11.

12. Orrin Schwab, *Defending the Free World: John F. Kennedy, Lyndon Johnson, and the Vietnam War, 1961–1965* (Westport, CT: Praeger, 1998), 9.

13. NASM No. 2, "Development of Counter-guerrilla Forces," 3 February, 1961, John F. Kennedy Library (JFKL), www.jfklibrary.org/Asset-Viewer/Archives/JFKNSF-328-003.aspx.

14. National Security Action Memorandum (NSAM) No. 9, "General Lansdale's Story of the Counter Guerilla Case," 6 February 1961, JFKL.

15. Ibid.; An American Officer [Lansdale], "The Report the President Wanted Published," *Saturday Evening Post*, 20 May 1961, 31, 69–71.

16. "Lansdale to President Diem," 30 January 1961, TTVA, DPC: Unit 2—Military Operations, box 1, folder 12.

17. Edward Miller, *Misalliance: Ngo Dinh Diem, the United States, and the Fate of South Vietnam* (Cambridge, MA: Harvard University Press, 2013), 196.

18. Ibid., 15, 44 (quote).

19. Thomas L. Ahern Jr., *CIA and the House of Ngo: Covert Action in South Vietnam, 1954–1963* (Washington: Center for the Study of Intelligence, 2000), 157, 161 (quote).

20. Tran Van Don, *Our Endless War: Inside Vietnam* (San Francisco: Presidio Press, 1978), 63.

21. NSAM No. 28, "Guerilla Operations in Vietnam Territory," 9 March 1961, JFKL.

22. "Proposal (probably by Lansdale)," 29 (quote), 30.

23. NSAM No. 56, "Evaluation of Paramilitary Requirements," 28 June 1961, JFKL.

24. NSAM No. 29, "Re: Southeast Asia," 9 March 1961, JFKL; Michael E. Haas, *Apollo's Warriors: United States Air Force Special Operations during the Cold War* (Maxwell AFB: Air University Press, 1997), 130–140, 286.

25. United States Air Force, "Operation Ranch Hand: The Air Force and Herbicides in Southeast Asia, 1961–1971," 33, www.afhso.af.mil/shared/media/document/AFD-100928-054.pdf.

26. Haas, *Apollo's Warriors*, 286.

27. NSAM No. 119, "Civic Action," 18 December 1961, JFKL.

28. John M. Carland, *Combat Operations: Stemming the Tide, May 1965 to October 1966*, United States Army in Vietnam series (Washington, DC: United States Army Center for Military History, 2000), 8; Barger, "Psychological Operations Supporting Counterinsurgency," 18.

29. Department of the Army, *Field Manual 33-5, Psychological Operations* (Washington, DC: Department of the Army, 16 January 1962), 5.

30. Ibid., 244.

31. Ibid., 48.

32. Ibid., 262–263.

33. United States Army Special Warfare School, "Program of Instruction for 33-G-F7: Counterinsurgency and Special Warfare Staff Officer," March 1962, TTVA, DPC: Unit 3—Insurgency Warfare, box 18, folder 3.

34. Ahern, *Way We Do Things*, 31.

35. FBIS, 9 August 1962, JJJ2–3, RH in English to Europe, "Strategic Hamlets Destroyed in Raids."

36. FBIS, 13 August 1962, JJJ4, in Rhade, "Warning Issued to Highland Traitor."

37. FBIS, 15 August 1962, JJJ2–3, RH to RVN, "U.S.-Diemists Persecute Highlanders."

38. Philip E. Catton, *Diem's Final Failure: Prelude to America's War in Vietnam* (Lawrence: University Press of Kansas, 2002), 118 (quote), 187.

39. Nhu Tang Truong, *A Vietcong Memoir: An Inside Account of the Vietnam War and Its Aftermath* (New York: Random House, 1986), 46, 47.

40. Catton, *Diem's Final Failure*, 138.

41. FBIS, 20 September 1962, JJJ3, RH in Rhade, "Highlanders Resist Diem Resettlement."

42. Duiker, *Ho Chi Minh*, 531 (quote); FBIS, 20 September 1962, JJJ3, RH in Rhade, "Highlanders Resist Diem Resettlement"; FBIS, 3 July 1962, JJJ4, Hanoi in French to SEA [Southeast Asia], "Strategic Hamlet Program a New Fiasco."

43. Jeffrey Race, *War Comes to Long An: Revolutionary Conflict in a Vietnamese Province* (1972; reprint, Berkeley: University of California Press, 2010), 128.

44. Rufus Phillips, *Why Vietnam Matters: An Eyewitness Account of Lessons Not Learned* (Annapolis, MD: Naval Institute Press, 2008), 131.

45. Douglas S. Blaufarb, *The Counterinsurgency Era: U.S. Doctrine and Performance, 1950 to the Present* (New York: Free Press, 1977), 212.

46. William Colby with James McCargar, *Lost Victory: A Firsthand Account of America's Sixteen-Year Involvement in Vietnam* (Chicago: Contemporary Books, 1989), 188.

47. Thomas L. Ahern Jr., *Undercover Armies: CIA and Surrogate Warfare in Laos, 1961–1973* (Washington, DC: Center for the Study of Intelligence, 2006), 39.

48. FBIS, 21 July 1962, JJJ2, RH to Vietnam, "NFLSV sees Laos as Example for Accord."

49. FBIS, "One Year of Arduous and Victorious Struggle by the Southern Compatriots against U.S. Aggressors and Their Lackeys," 4 January 1963, JJJ7; Pierre Asselin, *Hanoi's Road to the Vietnam War* (Oakland: University of California Press, 2013), 128 (quote).

50. CIA Task Force Vietnam, "Status Report on Covert Actions in Vietnam," 16 May 1962, 3, CIA doc. 0000530448.

51. FBIS, 23 July 1962, JJJl, RH to RVN, "The U.S.-Diemists Are Getting More and More Frightened."

52. Military History Institute of Vietnam, *Victory in Vietnam: The Official History of the People's Army of Vietnam, 1954–1975*, trans. Merle L. Pribbenow (Lawrence: University Press of Kansas, 2002), 65 (quote 1), 109; Mark Moyar, *Triumph Forsaken: The Vietnam War, 1954–1965* (New York: Cambridge University Press, 2006), xv, 155, 164 (quote 2).

53. Frank Scotton, *Uphill Battle: Reflections on Viet Nam Counterinsurgency* (Lubbock: Texas Tech University Press, 2014), 47.

54. CIA Task Force Vietnam, "Status Report," 16 May 1962, 1–3.

55. Richard H. Shultz, *The Secret War against Hanoi: Kennedy's and Johnson's Use of Spies, Saboteurs, and Covert Warriors in North Vietnam* (New York: HarperCollins, 1999), 151.

56. FBIS, 20 July 1962, JJJ5–7, RH in English to Europe, "Testimony Heard in Spy-commando Trial."

57. FBIS, 28 August 1962, JJJ2, "Greater Vigilance Needed in Struggle."

58. Asselin, *Hanoi's Road*, 99.

59. FBIS, 5 July 1962, JJJ3–4, RH in English, "American Mother Used to Seek Peace."

60. Richard A. Hunt, *Pacification: The American Struggle for Vietnam's Hearts and Minds* (Boulder, CO: Westview Press, 1995), 7.

61. Ibid., 22.

62. Elliott, *Vietnamese War*, 337.

63. FBIS, 20 September 1962, JJJ4, RH in English, "U.S. Adviser Sees Failure of Diem Plan."

64. FBIS, 4 January 1963, JJJ3–4, VNA Domestic Service [Vietnamese News Agency, the official DRV news agency], "One Year of Arduous and Victorious Struggle by the Southern Compatriots against U.S. Aggressors and Their Lackeys"; Moyar, *Triumph Forsaken,* 160–161, 185 (quote).

65. FBIS, 3 January 1963, JJJ2–3, "0511 GMT."

66. David Toczek, *The Battle of Ap Bac, Vietnam: They Did Everything but Learn from It* (Annapolis, MD: Naval Institute Press, 2007), 69–70; William P. Head, "The March to Oblivion: The Defeat at Ap Bac Hamlet and the Americanization of the Vietnam War," *Journal of Third World Studies* (Fall 2014): 57–81.

67. Neal Sheehan, *A Bright Shining Lie: John Paul Vann and America in Vietnam* (New York: Random House, 1988), 211–214; *Winnipeg Free Press*, "Viet Nam Defeats Vex U.S.," 9 January 1963, 45.

68. Sheehan, *Bright Shining Lie*, 211; Mai Viet Ha, *Steel and Blood: South Vietnamese Armor and the War for Southeast Asia* (Annapolis, MD: Naval Institute Press, 2008), 13 (quote).

69. Toczek, *Battle of Ap Bac*, 79; Lewis Sorley, ed., *The Vietnam War: An Assessment by South Vietnam's Generals*, Modern Southeast Asia series (Lubbock: Texas Tech University Press, 2010), 281; Moyar, *Triumph Forsaken*, 189–193.

70. FBIS, 7 January 1963, BBB23, Peking NCNA in English.

71. Moyar, *Triumph Forsaken*, 37.

72. Sheehan, *Bright Shining Lie* (see sketch after page 430); FBIS, 7 January 1963, JJJ1 (quote); David Toczek, *Battle of Ap Bac*, 79.

73. Miller, *Misalliance*, 251.

74. Ahern, *CIA and the House of Ngo*, 163 (quote); Moyar, *Triumph Forsaken*, 172.

75. Phillips, *Why Vietnam Matters*, 141; Moyar, *Triumph Forsaken*, 202.

76. Moyar, *Triumph Forsaken*, 205.

77. Ha, *Steel and Blood*, 15–16; Sheehan, *Bright Shining Lie*, 227; see organizations such as the Committee to Aid the NLF described below for uses of the battle as propaganda.

78. FBIS, 9 January 1963, JJJ1–2, VNA in English to Europe and Asia, "U.S. GIs Asked: Why Die in Vietnam."

79. PRP Administrative Committees Notes, "1963 Meeting of the Cadre of the Political Staff," 23 February 1963, TTVA, DPC: Unit 5—National Liberation Front, box 11, folder 9.

80. FBIS, 3 October 1963, KKK6, RH in English, "Victories in Anti-U.S. Campaign Noted."

81. Asselin, *Hanoi's Road*, 152 (quote); Vietnam War Casualty Figures, n.p., TTVA, DPC: Unit 3—Statistical Data, box 43, folder 2.

4. Buddhist Crisis and Coup, Spring–Fall 1963

1. Military Assistance Advisory Group Vietnam, *Lessons Learned Number 40: Corps Psywar/CA Operations Center,* enc 2. (Vietnam: MAAGV, 23 March 1964), DTIC, AD385117.

2. The People of Vietnam, *The Republic of Vietnam Armed Forces,* Monthly Publication of Psywar Directorate, no. 3 (June 1963), TTVA, Vladimir Lehovich Collection, box 2, folder 1.

3. FBIS, 23 May 1963, KKK1, "Large Scale Viet Cong Attacks Scored."

4. Army Concept Team in Vietnam (ACTIV), *The Heliborne Public Address System for Use in PSYWAR and Civic Action* (Saigon: ACTIV, 10 February 1964), 14 (hereafter ACTIV, *Heliborne Public Address System*).

5. FBIS, 8 May 1963, KKK1, RH in English, "Soviet Notes against Toxins Welcomed."

6. FBIS, 12 June 1963, JJJ1, Radio Hanoi English to Europe.

7. Richard H. Shultz, *The Secret War against Hanoi: Kennedy's and Johnson's use of Spies, Saboteurs, and Covert Warriors in North Vietnam* (New York: HarperCollins, 1999), 20; Thomas L. Ahern Jr., *The Way We Do Things: Black Entry Operations into North Vietnam, 1961–1964* (Washington, DC: Center for the Study of Intelligence, May 2005), 10 (quote).

8. Kenneth Conboy and Dale Andrade, *Spies and Commandos: How America Lost the Secret War in North Vietnam* (Lawrence: University Press of Kansas, 2000), 78–79; Ahern, *Way We Do Things*, 47 (quote).

9. William J. Duiker, *Ho Chi Minh: A Life* (New York: Hyperion, 2000), 261; Lien-Hang T. Nguyen, *Hanoi's War: An International History of the War for Peace in Vietnam* (Chapel Hill: University of North Carolina Press, 2012), 202; Conboy and Andrade, *Spies and Commandos*, 78–80.

10. Conboy and Andrade, *Spies and Commandos,* 110; Shultz, *Secret War against Hanoi*, 219 (quote); MACSOG Documentation Study, annex A to appendix C, Psychological Operations (Washington, DC: Office of the Secretary of Defense, 1970), 19–21.

11. FBIS, 27 August 1962, JJJ4, RH in English, "U.S. Publications Rap U.S.-Diem Tactics" (quote); Conboy and Andrade, *Spies and Commandos*, 218–219.

12. Ahern, *Way We Do Things*, 37.

13. Ibid., 49.

14. Mark Moyar, *Triumph Forsaken: The Vietnam War, 1954–1965* (New York: Cambridge University Press, 2006), 212.

15. Ibid., 214–218; Rufus Phillips, *Why Vietnam Matters: An Eyewitness Account of Lessons Not Learned* (Annapolis, MD: Naval Institute Press, 2008), 159.

16. Edward Miller, *Misalliance: Ngo Dinh Diem, the United States, and the Fate of South Vietnam* (Cambridge, MA: Harvard University Press, 2013), 262–263; Moyar, *Triumph Forsaken,* 216, 217.

17. Miller, *Misalliance*, 17.

18. FBIS, 12 June 1963, KKK1, Saigon, "Good Will Towards Buddhist Affirmed."

19. Thomas L. Ahern Jr., *CIA and the House of Ngo: Covert Action in South Vietnam, 1954–1963* (Washington, DC: Center for the Study of Intelligence, 2000), 166.

20. Moyar, *Triumph Forsaken,* xvi, 231 (quote), 233.

21. Nhu Tang Truong, *A Vietcong Memoir: An Inside Account of the Vietnam War and Its Aftermath* (New York: Vintage Books, 1985), 46–47; Moyar, *Triumph Forsaken*, 212–218; Ahern, *CIA and the House of Ngo*, 167; James McAllister, "'Only Religions Count in Vietnam': Thich Tri Quang and the Vietnam War," *Modern Asian Studies* 42, no. 4 (July 2008): 751–782.

22. Moyar, *Triumph Forsaken,* 223; FBIS, 23 July 1963, JJJ4, RH in English to Europe, "Americans' Letter Raps S. Vietnam War" (quote).

23. FBIS, 16 August 1963, JJJ1, Radio Hanoi to Europe, "U.S. Clergymen Hit Abuse of Buddhists."

24. "Clergymen Score U.S. Aid to Diem," *New York Times Magazine,* 15 August 1963.

25. Miller, *Misalliance,* 274; Nguyen, *Hanoi's War,* 300; FBIS, 8 July 1963, JJJ1–2, RH in Vietnamese to RVN, "Paper Comments on Nolting's Dismissal" (quote).

26. Moyar, *Triumph Forsaken,* 238, 241 (quote).

27. Miller, *Misalliance*, 297.

28. FBIS, 21 August 1963, KKK1–3, Saigon Domestic Service; Moyar, *Triumph Forsaken*, 232.

29. FBIS, 21 August 1963, KKK4–5, Saigon Domestic Service, "South Vietnam Army Statement."

30. FBIS, 29 August 1963, KKK8, Liberation Broadcast Service (LBS) to RVN, "U.S. Aggression Is the Principal Cause of All Bloodshed in South Vietnam"; Ahern, *CIA and the House of Ngo*, 172–173.

31. FBIS, 22 August 1963, KKKl, Saigon Domestic Service, "Search of Tu Dem Pagoda."

32. Ahern, *CIA and the House of Ngo*, 172.

33. FBIS, 22 August 1963, KKK5, Saigon Domestic Service, "Let Us Unmask Vietnamese Traitors Who Are Shamelessly Serving as Agents for the Communists."

34. FBIS, 26 August 1963, KKK8–10, "Vietnam Press."

35. FBIS, 27 August 1963, KKK1–5, Saigon Domestic Service.

36. Frank Scotton, *Uphill Battle: Reflections on Viet Nam Counterinsurgency* (Lubbock: Texas Tech University Press, 2014), 62–63; Phillips, *Why Vietnam Matters*, 159 (quote).

37. FBIS, 29 August 1963, KKK4, Vietnam Press Agency in English to Asia.

38. FBIS, 5 September 1963, KKK1, *Times of Vietnam,* "Viet Cong Plan to Profit by CIA Plot."

39. William Colby with James McCargar, *Lost Victory: A Firsthand Account of America's Sixteen-Year Involvement in Vietnam* (Chicago: Contemporary Books, 1989), 145.

40. ONE [Office of National Estimates] Memorandum, "South Vietnam's Leaders," 4 September 1963, 2, CIA doc. 0001166436.

41. FBIS, 4 September 1963, KKK2, Saigon Domestic Service, "What Has the Free World Done to Check the Red Wave in South Vietnam?"

42. FBIS, 6 September 1963, KKK1–2, "Vietnam Press Interview with Diem."

43. Ibid.

44. FBIS, 14 November 1963, KKK1, Liberation Press Agency [LPA] to RVN, "Ap Bac Emulation Drive to Be Stepped Up."

45. *Republic of Vietnam Armed Forces Magazine* (Saigon: PSYWAR Directorate, June and August 1963), TTVA, Vladimir Lehovich Collection, box 1, folder 1–2.

46. Moyar, *Triumph Forsaken,* 246, 254 (quote); Nguyen, *Hanoi's War*, n300. See Robert S. McNamara, "Report of McNamara 26 September 63 Interview with Professor Smith," Foreign Relations of the United States, 1961–1963, vol. 4, Vietnam, August to December 1963 (Washington, DC: US Government Printing Office, 1991), 293–295.

47. ACTIV, *Heliborne Public Address System, Weekly Headway Report*, 9 to 16 October 1963, C-6. This study contains reproductions in the annex of MAAGV operations reports for 1963.

48. Ibid.

49. Ibid., 4 to 11 September 1963, C4.

50. Ibid., 12, C5, C6.

51. Scotton, *Uphill Battle*, 53.

52. Moyar, *Triumph Forsaken*, 207 (quote); Miller, *Misalliance*, 245.

53. Moyar, *Triumph Forsaken*, 208.

54. Military History Institute, *Victory in Vietnam*, 115.

55. Moyar, *Triumph Forsaken*, 210.

56. Bui Tin, *Following Ho Chi Minh: The Memoirs of a North Vietnamese Colonel* (Honolulu: University of Hawaii Press, 1995), 60.

57. Pierre Asselin, *Hanoi's Road to the Vietnam War* (Oakland: University of California Press, 2013), 161.

58. Lam Quang Thi, *The Twenty-five Year Century: A South Vietnamese General Remembers the Indochina War to the Fall of Saigon* (Denton: University of North Texas Press, 2001), 129.

59. FBIS, 1 November 1963, KKKl-2, Saigon Domestic Service.

60. FBIS, 1 November 1963, KKKl-2, Saigon Domestic Service, "1027-GMT, Reports on Coup against Diem Government" (quote); Moyar, *Triumph Forsaken*, 281–282.

61. Moyar, *Triumph Forsaken*, 284 (quote).

62. Nguyen, *Hanoi's War*, 77.

63. Duiker, *Ho Chi Minh*, 536–537, 539; FBIS, 7 April 1964, JJJ6, RH in English, "Political Conference Resolution."

64. FBIS, 4 November 1963, JJJ2, "Voice of Vietnam."

65. FBIS, 4 November 1963, JJJ5, RH in English, "Liberation Radio Comment."

66. FBIS, 5 November 1963, JJJ1, "More Hanoi Comment on Saigon Coup d'État."

67. FBIS, 6 November 1963, JJJ1, RH in Vietnamese to RVN, "U.S. Responsible for Diem's Downfall."

68. FBIS, 27 November 1963, JJJ1, Radio Hanoi International in English, "Diem Death a Lesson for Vietnam Catholics."

69. FBIS, 10 January 1964, KKK1, Saigon in English, "Reds Trying to Dupe Catholics."

70. FBIS, 13 November 1963, JJJ6, RH in English, Nhan Dan: "NFLSV Scores New Successes."

71. FBIS, 18 November 1963, KKK8–11, LBS to RVN, "NFLSV Radio Reviews Post-Coup Events."

72. FBIS, 14 November 1963, KKK5–6, Radio Hanoi International in English, "LPA—Many Strategic Hamlets Destroyed."

73. USIS, "Long An Province Survey January 1964," 9, TTVA, John Donnell Collection, box 3, folder 6.

74. FBIS, 21 November 1963, JJJ2, Radio Hanoi International in English, "Nhan Dan: U.S. Will Leave S. Vietnam." Bernard Fall was a war correspondent, historian, and author of many books on the Vietnam War such as *Street without Joy: Indochina at War, 1946–54.*

75. FBIS, 13 November 1963, KKK2, Liberation Press Agency (LPA) to RVN, "Future of Coup Leaders Hazardous."

76. ACTIV, *Heliborne Public Address System, Headway Addenda,* II CORPS 6 to 13 November 1963, C9.

77. Military Assistance Command, Vietnam, *Command History, United States Military Assistance Command, Vietnam 1964* (Saigon: MACV Military History Branch, 15 October 1965), 71, 554, available at DTIC, AD0955106.

78. J. A. Koch, *The Chieu Hoi Program in South Vietnam, 1963–1971* (Washington, DC: Rand Corporation for ARPA, January 1973), 21, fig. 6.

79. ACTIV, *Heliborne Public Address System*, 19–22.

80. John J. Tolson, *Vietnam Studies: Air Mobility, 1961–1971* (Washington, DC: Department of the Army, 1989), 29.

81. Charles R. Shrader, *History of Operations Research in the U.S. Army,* vol. 2: *1961–1973* (Washington, DC: Office of the Deputy Under Secretary of the Army for Operations Research, US Army, 2006), 298.

82. ACTIV, *Heliborne Public Address System*, iii (quote), 5, 7.

83. MACV, *Command History, 1964*, 75.

84. "Infiltration of Military and Technical Personnel from North to South Vietnam," 3, CIA Research Reports, 3 December 1964, TTVA, Vietnam Archive Collection (VAC), box 4, folder 422; Moyar, *Triumph Forsaken*, 305–306.

85. Moyar, *Triumph Forsaken*, 305–306.

86. Ibid., 292.

87. Colby, *Lost Victory*, 165 (quote); Ahern, *Way We Do Things*, 49, 50.

88. CIA, "Cable from CIA to State, OSD, WH, Reference DEPTEL 570," 14 November 1963, 1, 2 (quote), CIA doc. 0000881593; Ahern, *Way We Do Things*, 50.

89. John L. Plaster, *SOG: The Secret Wars of America's Commandos in Vietnam* (New York: Simon & Schuster, 1997), 118–120.

90. FBIS, 22 November 1963, KKK4, LBS to RVN, "Hamlets Sprayed with Poison Chemicals."

91. FBIS, 10 December 1963, JJJ2, RH in English, "Letter Protests U.S. Chemical Spraying."

92. FBIS, 27 January 1964, JJJ4.

93. FBIS, 28 January 1964, KKK3–4, "NFLSV Sees Victory over U.S. Inevitable."

94. FBIS, 16 January 1964, JJJ2–4.

95. FBIS, 3 February 1964, KKK8, LBS to RVN, "U.S. Uses Poison Chemicals in Combat."

96. USIS, "Long An Province Survey," 1964, n.p. TTVA, DPC: Unit 5—National, box 16, folder 13 (To Mike Dunn From Earl Young), "Subject: Preliminary Survey Long An Province," 15 December 1963, TTVA, Rufus Phillips Collection, box 2, folder 10.

97. Young to Dunn, "Subject: Preliminary Survey Long An Province," 15 December 1963.

98. USIS, "National Liberation Front Communication 1964: Analysis of VC Propaganda Activities in Long An Province" (Saigon: n.p., n.d.), 2, TTVA, DPC: Unit 5—National, box 16, folder 13.

99. ACTIV, *Heliborne Public Address System,* 13; USIS, "NLF Communications in Long An Province," 4, 5 (quote), 7.

100. USIS, "NLF Communications in Long An Province," 7 (quote), 18.

101. USIS, "Long An Province Survey," January 1964, 37, 153 (quote), TTVA, DPC: Unit 5—National, box 16, folder 13.

102. USIS, "Long An Province Survey," 1964, 13 (quote); Scotton, *Uphill Battle*, 78–88.

103. FBIS, 3 December 1963, KKK6–7, LBS to RVN, "Hiep Hoa Revenge Mop-up Operation Fails."

104. FBIS, 9 December 1963, JJJ5, RH to RVN, "*Quan Doi Nhan* on Cha La victory."

105. FBIS, 26 November 1963, JJJ2, "Hiep Hoa U.S. Special Forces Camp Overrun."

106. FBIS, 3 December 1963, KKK6–7, LBS to RVN, "Hiep Hoa Revenge Mop-up Operation Fails."

107. FBIS, 6 December 1963, KKK5, "Viet Cong Propaganda."

108. Eric Bergerud, *Dynamics of Defeat: The Vietnam War in Hau Nghia Province* (Boulder, CO: Westview Press, 1991), 71; Garth S. Jowett and Victoria O'Donnell, *Propaganda and Persuasion,* 5th ed. (Thousand Oaks, CA: Sage Publications, 2012), 180.

109. Robert Manning, "Memorandum for the President from Assistant Secretary of State. Information Program for Southeast Asia," n.d., 1–2, LBJL, box 54, Southeast Asia Memos, Information Program.

5. "Only Hard Ones Lie Ahead," 1964

1. "PSYCH OPS Printing Support for MACV," 7 January 1964, digital copy in possession of the author.

2. MACV, *Command History, United States Military Assistance Command, Vietnam 1964* (Saigon: MACV Military History Branch, 15 October 1965), 75, available at DTIC, AD0955106.

3. MAAGV Records, *1950–1964, Adjutant General Division Security Classified Files,* MAAGV Lessons Learned Number 40: Corps Psywar/CA Operations Center, 23 August 1958 (Bethesda, MD: University Publications of America), 6.

4. "Organization and Operation of the 1st PSYWAR BN (ARVN)," 1, TTVA, DPC: Unit 3—Insurgency Warfare, box 15, folder 5, 1–7.

5. FBIS, 2 January 1964, KKK1, Saigon, Vietnam Press in English.

6. MACV, *Command History, 1964*, 52.

7. "Psywarriors Gathered for Sessions in Saigon," n.p., 9 January 1964, unattributed newspaper clipping, TTVA, DPC: Unit 3—US Economy and US Mission (Saigon), box 42, folder 03.

8. Robert Manning, "Memorandum for the President from Assistant Secretary of State.

Information Program for Southeast Asia," n.d., 14, LBJL, box 54, Southeast Asia Memos, Information Program.

9. FBIS, 2 January 1964, KKK8, Radio Hanoi International, "Army Officer Joins Liberation Front."

10. Pierre Asselin, *Hanoi's Road to the Vietnam War* (Oakland: University of California Press, 2013), 165; "Lao Dong Ninth Plenum Decision on War in South," December 1963, TTVA, DPC: Unit 1—Assessment and Strategy, box 3, folder 2.

11. Asselin, *Hanoi's Road*, 171.

12. Lien-Hang T. Nguyen, *Hanoi's War: An International History of the War for Peace in Vietnam* (Chapel Hill: University of North Carolina Press, 2012), 55; Asselin, *Hanoi's Road*, 173.

13. Nguyen, *Hanoi's War*, 50.

14. Ilya V. Gaiduk, *The Soviet Union and the Vietnam War* (Chicago: Ivan R. Dee Publisher, 1996), 10.

15. Gaiduk, *Soviet Union,* 5, 6 (quote), 11, 16.

16. FBIS, 6 January 1964, KKK6, Radio Hanoi International, "LPA Hails 1963 Struggle Successes."

17. FBIS, 16 January 1964, JJJ2, Radio Hanoi, "Newsweek Voices Pessimism on Vietnam War."

18. FBIS, 16 January 1964, JJJ2–4.

19. Manning, "Information Program for Southeast Asia," 1–2.

20. USIS, "Long An Province Survey," January 1964, 29, TTVA, DPC: Unit 5—National, box 16, folder 13; FBIS, 3 February 1964, KKK3–4.

21. Ibid., 35.

22. Ibid., 38.

23. Ibid., 154.

24. Ibid., 24, 31.

25. Ibid., 32.

26. Ibid., 102.

27. Ibid., 159–161, 159 (quote).

28. FBIS, 3 February 1964, KKK7, Saigon Domestic Service, "Neutralization Plans for Vietnam Scored"; Rufus Phillips, *Why Vietnam Matters: An Eyewitness Account of Lessons Not Learned* (Annapolis, MD: Naval Institute Press, 2008), 227.

29. FBIS, 6 January 1964, KKKl, Saigon, "Government Raps Neutralization Solution."

30. FBIS, 26 February 1964, JJJ2–3, Hanoi to RVN, "Vietnamese Warned of U.S. Associations."

31. United States and KTO Press, *The Cumulated Indexes to the Public Papers of the Presidents of the United States, Lyndon B. Johnson, 1963–1969,* book 1: *1963–64* (Millwood, NY: KTO Press, 1978), 257–260.

32. FBIS, 12 February 1964, JJJl, Hanoi VNA in English.

33. Avedon, *Psychological Warfare Operational Deficiencies,* "Field Reports 6 November 1952," n.p.

34. John L. Plaster, *SOG: The Secret Wars of America's Commandos in Vietnam* (New York: Simon & Schuster, 1997), 346; Thomas L. Ahern Jr., *The Way We Do Things: Black Entry Operations into North Vietnam, 1961–1964* (Washington, DC: Center for the Study of Intelligence, 2005), 52, 56; Tom G. Abbey, *Southeast Asia Report: The Role of the USAF in Support of Special Activities in SEA* (Hickam AFB, HI: CHECO/Corona Harvest Division, Operations Analysis Office, Department of the Air Force Headquarters Pacific Air Forces, 1 July 1976), 48 (quote).

35. Ahern, *Way We Do Things*, 54.

36. Herbert Weisshart interview by Richard H. Shultz, in *Oral History Interviews with Military & CIA Officers, Who Served in MACVSOG OP 39 Covert Psychological Operations Against North Vietnam,* July 1997, 7. TTVA (hereafter *MACV-SOG OP 39*).

37. Fred Stables, *MACV-SOG OP 39,* 72, 76 (quote).

38. FBIS, 23 March 1964, JJJ2, RH to RVN, "The Failure of a Plot."

39. FBIS, 8 April 1964, JJJ4, RH in English, "Military Court Sentences Spy Commandos."

40. Plaster, *SOG,* 120, 123; MACSOG Documentation Study, annex A to appendix C, Psychological Operations (Washington, DC: Office of the Secretary of Defense, 1970), 19–21.

41. FBIS, 13 April 1964, KKK1, Saigon, "*Dan Quyen.*"

42. United States Senate, Select Committee on Intelligence, *Vietnamese Commandos,* testimony of John Singlaub before the Committee, 104th Cong., 2nd sess., 19 June 1996, 15–18.

43. FBIS, 20 April 1964, JJJl RH to RVN, "Military Court Holds Ranger Spy Trial."

44. FBIS, 9 March 1964, JJJ1, RH in English, "Johnson Charge of DRV Aid to NFLSV Scored."

45. FBIS, 16 March 1964, JJJ4, RH to RVN, "It Is True That McNamara Is Absent Minded."

46. Phillips, *Why Vietnam Matters*, 230.

47. FBIS, 8 April 1964, JJJ1, RH to RVN, "Despite Their Increased Espionage Activities the U.S. Aggressors Will Surely Be Defeated."

48. FBIS, 8 April 1964, KKK3, LBS to RVN, "Liberation Radio Hits Pacification Plan."

49. FBIS, 20 March 1964, JJJ1–3 Radio Hanoi Domestic, "Hoc Tap Editorial on Party Goals."

50. NSAM No. 325, "Informational and Psychological Warfare Programs in South Vietnam," 17 March 1964, 1, LBJL, www.lbjlib.utexas.edu/johnson /archives.hom/NSAMs/nsamhom.asp.

51. FBIS, 8 April 1964, KKK1, Saigon Domestic Service, "Hoan Speech on Safeguarding Freedoms."

52. FBIS, 14 April 1964, KKK1, Saigon Domestic Service, "Khanh Initiates Weekly Radio Addresses."

53. MAAGV, Lessons Learned no. 40: "Corps Psywar/CA Operations Center," 23 March 1964, 3.

54. Ibid., 6.

55. FBIS, 13 April 1964, JJJ6–7, RH in English, "Journalists Welcome Burchett to Hanoi."

56. FBIS, 12 May 1964, JJJ5, "*Nhan Dan* published letters"; "Pilot's Despair over Vietnam: Last Letters of a U.S. Flyer Killed in Action," *Life*, 8 May 1964, 34–40.

57. FBIS, 25 May 1964, JJJl, RH in English, "Ho Chi Minh Letter to American Paper."

58. Foreign Language Publishing House published the *Vietnam Courier* weekly (May 1964–1973) in Hanoi, Vietnam.

59. *Vietnam Courier*, WorldCat. Accessed 11 December 2017. www.worldcat.org.

60. Based on perusal of issues at the University of North Texas Library. Hanoi began publishing *Vietnam* around 1957.

61. FBIS, 30 March 1964, BBB9, Peking NCNA in English, "U.S. Student Condemnation."

62. MACV Records, "Weekly Report, 3 May to 9 May Weekly Report," Records of the Military Assistance Command, Vietnam, *The War in Vietnam, 1964–1973,* reel 2, part 1, MACV Historical Office Documentary Collection (Bethesda, MD: University Publications of America) (hereafter MACV Records, Weekly Report).

63. Ibid.

64. MACV Records, Weekly Report, 25 April–2 May.

65. Records of the Military Assistance Advisory Group Vietnam, 1950–1964, Adjutant General Division Security Classified Files, 3 (quote), 32.

66. MACV Records, Weekly Report, 13–20 June 1964, 22; MACV Records, Weekly Report, 5–11 April 1964, 36 (quote); Memorandum for Bundy, "Subject: First Status Report on the Forty-one Non-military Actions in Vietnam," 28 April 1964; NSAM No. 328, "Presidential Decisions with Respect to Vietnam," LBJL.

67. Memorandum, "Improvement of Informational-Psychological Program in South Viet-Nam," undated memo following May 1964 Honolulu meeting. LBJL, box 5, Memos, May Improvement.

68. Smith-Mundt Act (United States Information and Education Exchange Act of 1948), 27 January 1948, www.state.gov/documents/organization/177574.pdf.

69. Memorandum, "Improvement of Informational-Psychological Program in South Viet-Nam," 2 (quote 1), 3, 4 (quote 2).

70. Ibid., 3.

71. United States Army Special Warfare School, "Program of Instruction for Military Assistance Training Advisor Course," May 1964, I-1 (quote), IV-2–3. US Army Special Operations Command Archives (USASOC), PSYOP Papers, 1961–1968, box 3; MACV Records, Weekly Report, 13–20 June 1964, 35.

72. Thomas L. Ahern Jr., *CIA and Rural Pacification in Vietnam* (Washington, DC: Center for the Study of Intelligence, 2001), 162; Frank Scotton, *Uphill Battle: Reflections on Viet Nam Counterinsurgency* (Lubbock: Texas Tech University Press, 2014), 91.

73. Douglas S. Blaufarb, *The Counterinsurgency Era: U.S. Doctrine and Performance, 1950 to the Present* (New York: Free Press, 1977), 227.

74. Ibid., 228.

75. Scotton, *Uphill Battle*, 91.

76. Memo from USIA Director Rowan to President, "Information Program for Southeast Asia," 26 May 1964, LBJL, NSF Country File Vietnam box 5, Memos, May Improvement (quote 1); William M. Hammond, *Public Affairs: The Military and the Media, 1962–1968,* US Army in Vietnam series (Washington, DC: US Army Center for Military History, 1988), 73, 81 (quote 2).

77. Memorandum, "Improvement of Informational-Psychological Program in South Viet-Nam."

78. Memo from USIA Director Rowan to President, 26 May 1964.

79. MACV Records, Weekly Report, 13–20 June 1964, 35; MACV, *Command History, 1964*, 41 (quote).

80. MACV Records, Weekly Report, 30 May–6 June 1964, 39; MACV Records, Weekly Report, 13–20 June 1964, 35.

81. MACV Records, Weekly Report, 9–16 May 1964, 45.

82. MACV Records, Weekly Report, 13–20 June 1964, TTVA, Vietnam Archive Collection at Texas Tech [VAC], box 2, folder 733; Richard A. Hunt, *Pacification: The American Struggle for Vietnam's Hearts and Minds* (Boulder, CO: Westview Press, 1995), 23.

83. MACV Records, Weekly Report, 13–20 June 1964.

84. "National Board of Estimates Chairman Sherman Kent to the Director," 9 June 1964, LBJL, NSF Country File Vietnam, box 54, Southeast Asia Memo, Domino.

85. Walt Rostow/Jack Valenti, "Memo to the President," 7 June 1964, LBJL, NSF Country File Vietnam, box 54, Southeast Asia Memo, Rostow.

86. Hammond, *Public Affairs*, 78.

87. "Memo for Secretary from William P. Bundy Subject Highlights of Honolulu Conference," 3 June 1964, 4, LBJL, NSF Country File Vietnam, box 53.

88. MACV Records, Weekly Report, 3–9 May 1964.

89. FBIS, 1 June 1964, JJJ5, RH in English, "VPA Protests U.S. Violations to ISCC."

90. "Outgoing DOS Telegram to Saigon Embassy," 28 July 1964, LBJL, NSF Country File Vietnam, box 6, Cables.

91. "Outgoing DOS Telegram to Saigon Embassy," 29 July 1964, LBJL, NSF Country File Vietnam, box 6, Cables.

92. MACV Records, Weekly Report, 13–20 June 1964, 35–36, TTVA, VAC, box 2, folder 733.

93. MACV Records, Weekly Report, 18–25 July 1964, 56.

94. Ibid.

95. Ibid., 55.

96. "Department of State Tel. Saigon to SECSTATE," 24 July 1964, 2, LBJL, NSF Country File Vietnam, box 6, Cables.

97. Ibid., 2.

98. CIA, "Situation in South Vietnam Weekly Report," 29 July 1964, TTVA, VAC, box 3, folder 682; "Dept of State Telegram from Saigon to SECSTATE, Flash, 25 July 1964 6:03 AM Actual" (quote), LBJL, NSF Country File Vietnam, box 6, Cables.

99. Robert K. Brigham, *Guerrilla Diplomacy: The NLF's Foreign Relations and the Viet Nam War* (Ithaca, NY: Cornell University Press, 1999), 36; MACV Records, Weekly Report 18–25 July 1964 (quote), TTVA.

100. MACV Records, Weekly Report, 27 June–4 July 1964.

101. CIA, "Situation in South Vietnam Weekly Report," 29 July 1964.

102. MACV Records, Weekly Report, 18–25 July 1964, 57; CIA, "Situation in South Vietnam Weekly Report," 29 July 1964 (quote); FBIS, 1 June 1964, JJJ6, RH in English, "S. Vietnamese Boats Violate DRV Waters."

103. FBIS, 10 June 1964, JJJ1, RH in English, "U.S. Trains Agents to Sabotage the DRV."

104. Franklin J. Yurco, "5th Special Forces Unit History, 1962–1970, Part 1," TTVA, Bud Harton Collection, box 00, folder 1.

105. "Southeast Asia Resolution Memo," 11 June 1964, LBJL, NSF Country File Vietnam, box 54, Southeast Asia Memo, Resolution.

106. Mark Moyar, *Triumph Forsaken: The Vietnam War, 1954–1965* (New York: Cambridge University Press, 2006), 310–316.

107. Weapons Systems Evaluation Group, *The Flaming Dart Airstrikes against North Vietnam: 7 to 11 February 1965*, 2 August 1967, CI no. 9, Working Paper No. 4, 1, 11, www.dod.mil/pubs/foi/Reading_Room/International_Security_Affairs/550.pdf; Charles Tustin Kamps, "The JCS 94—Target List: A Vietnam Myth That Still Distorts Military Thought," *Air and Space Power Journal* (Spring 2001): 68.

108. FBIS, 6 August 1964, JJJ2–3, Radio Hanoi Domestic, "VPA Press Conference 5 August."

109. FBIS, 10 August 1964, JJJ6, Radio Hanoi Domestic, "Ho Citation for VPA Units."

110. MACV Records, Weekly Report, 8–13 August 1964.

111. FBIS, 19 December 1964, JJJ2, RH in English, "Commentaries to U.S. Servicemen in South Vietnam."

112. Philip Shenon, "Hanoi Hannah Looks Back, with Few Regrets," *New York Times*, 26 November 1994, section 1, 4.

113. Don North, "The Search for Hanoi Hannah," www.psywarrior.com/hannah.html.

114. Ibid.

6. Organizational Problems, Fall 1964–Winter 1965

1. MACV Records, Weekly Report, 8–15 August 1964, 47.

2. Ibid., 42.

3. Ibid., 47.

4. "Viet Cong Morale and Vulnerabilities," 27 January 1965, 8. TTVA, DPC: Unit 3—Insurgency Warfare, box 13, folder 6.

5. MACV Records, Weekly Report, 20–26 September 1964, 43; MACV Records, Weekly Report, 22–28 August 1964, 3, 33–34; "The Virtual Wall," www.VirtualWall.org/dh/HarrimanAB01a.htm.

6. Mark Moyar, *Triumph Forsaken: The Vietnam War, 1954–1965* (New York: Cambridge University Press, 2006), 320; MACV Records, Weekly Report, 8–15 August 1964 (quote); Charles Tustin Kamps, "The JCS 94—Target List: A Vietnam Myth That Still Distorts Military Thought," *Air and Space Power Journal* (Spring 2001): 68.

7. USIS, "Results of Long An Province Survey," 15 August 1964, 3, 37, TTVA, John Donnell Collection, box 3, folder 6; Frank Scotton, *Uphill Battle: Reflections on Viet Nam Counterinsurgency* (Lubbock: Texas Tech University Press, 2014), 78–88.

8. Scotton, *Uphill Battle*, 47.

9. USIS, "Results of Long An Province Survey," 15 August 1964, 4, 15 (quote).

10. Ibid., 17–18.

11. Ibid., 21.

12. Ibid., 22.

13. Ibid., 27 (quote), 29, 38.

14. Ibid., 62.

15. Ibid., 5.

16. Ibid., 44.

17. FBIS, 27 August 1964, RH in English to American Servicemen.

18. FBIS, 17 July 1964, JJJ5, RH to RVN, "Special feature for American Servicemen in South Vietnam, U.S. Policy Puts U.S. Troops in Peril."

19. MACV Records, Weekly Report, 5–12 September 1964, 33.

20. MACV Records, Weekly Report, 26 September–3 October 1964, 32–33.

21. MACV Records, Weekly Report, 8–15 August 1964, 49–50; MACV Records, Weekly Report, 26 September–3 October 1964, 32–33.

22. NSAM No. 314, "Resumption of Operations in SEA," 10 September 1964, 1, LBJL, www.lbjlib.utexas.edu/johnson/archives.hom/NSAMs/nsamhom.asp.

23. US Joint Chiefs of Staff, "Memo for Record, October Schedule for OPLAN 34A, Draft for Cyrus Vance," TTVA, David Lifton Collection, box 1, folder 4.

24. MACSOG Documentation Study, annex A to appendix C, Psychological Operations (Washington, DC: Office of the Secretary of Defense, 1970), 55–58; Richard H. Shultz, *The Secret War against Hanoi: Kennedy's and Johnson's Use of Spies, Saboteurs, and Covert Warriors in North Vietnam* (New York: HarperCollins, 1999), 113.

25. Cyrus Vance, Memo for Record, "October Schedule for OPLAN 34A," 1 October 1964, 5.

26. Ibid., 6.

27. Office of Special Assistant for Counterinsurgency and Special Activities Programs and Review Division (OSACSAPRD), Major General Anthis, "Subject: Report on Forrestal Group Trip to SVN," TTVA, David Lifton Collection, box 1, folder 5.

28. MACV Records, Weekly Report, 3–10 October 1964, 36; Military Assistance Command, Vietnam, *Command History, United States Military Assistance Command, Vietnam 1964* (Saigon: MACV Military History Branch, 15 October 1965), 75, available at DTIC, AD0955106.

29. James L. Collins, *The Development and Training of the South Vietnamese Army, 1950–1972* (US Department of the Army, 1975), 138.

30. FBIS, 22 September 1964, JJJ5, Radio Hanoi Domestic, "Southern People Seize Two Radio Stations."

31. FBIS, 22 September 1964, JJJ5, RH in English, "Qui Nhon Station."

32. FBIS, 30 September 1964, JJJ5, RH in Jarai to RVN, "Appeal to Montagnards."

33. MACV Records, Weekly Report, 26 September–3 October 1964, 30.

34. MACV Records, Weekly Report, 5–12 December 1964, 32; FBIS, 12 November 1964, JJJ5, RH to RVN in Rhade (quote), "Letter to the Revolutionaries in Ban Me Thuot and Pleiku."

35. FBIS, 17 November 1964, JJJ3, RH in Jarai, "U.S.-Khanh Treachery."

36. MACV Records, Weekly Report, 26 September–3 October 1964, 29.

37. MACV, *Command History, 1964*, 176.

38. William Hoffer, "Broadcasting in an Insurgency Environment: USIA in Vietnam, 1965–1970" (PhD diss., University of Wisconsin, 1972), 376–384.

39. MACV Records, Weekly Report, 26 September–3 October 1964, 29. MACV Records, Weekly Report, 15–21 November 1964, 37–38.

40. MACV Records, Weekly Report, 17–24 October 1964, 35; MACV, *Command History, 1964*, 157.

41. Joseph B. Knotts interview with Allen B. Richardson, United States Marine Corps History and Museums Division Oral History Project, n.d., 1, 3 (quote), 5, TTVA.

42. FBIS, 7 October 1964, JJJ3, RH in English, "U.S. Faced with Impasse in Vietnam Situation."

43. Pierre Asselin, *Hanoi's Road to the Vietnam War* (Oakland: University of California Press, 2013), 204.

44. FBIS, 2 November 1964, JJJ3, RH to RVN, "Let Us Acclaim the New Resounding Feat of the South Vietnamese People's Armed Forces."

45. Ibid., JJJ3–4.

46. William M. Hammond, *Public Affairs: The Military and the Media, 1962–1968*, US Army in Vietnam series (Washington, DC: US Army Center for Military History, 1988), 113; Moyar, *Triumph Forsaken,* 331; H. R. McMaster, *Dereliction of Duty: Lyndon Johnson, Robert McNamara, the Joint Chiefs of Staff, and the Lies That Led to Vietnam* (New York: HarperCollins, 1997), 175.

47. FBIS, 27 October 1964, JJJ6, RH to RVN, "Talk to Southern Air Force Men about the Battle of Xeo Ro in Rach Gia."

48. MACV Records, Weekly Report, 24–31 October 1964, 29.

49. Ibid., 35 (quote); Collins, *Development and Training,* 58; MACV Records, Weekly Report, 31 October–7 November 1964, 30.

50. MACV Records, Weekly Report, 20–26 September 1964, 42; MACV Records, Weekly Report, 8–14 November 1964, 37; Weekly Report, 9–15 December 1964 (quote).

51. MACV Records, Weekly Report, 15–21 November 1964, 37–38.

52. MACV Records, Weekly Report, 5–12 December 1964, 38; FBIS, 16 November 1964, JJJ5, RH to RVN, "Developing Their Determination to Fight and Gain Success, the Southern Compatriots Firmly Resist Natural Calamities" (quote).

53. FBIS, 18 November 1964, JJJ1, RH in English, "U.S.-Khanh Exploit Flood to Repress People"; FBIS, 4 December 1964, KKK4, LBS to RVN, "Liberation Army Statement on Flood Relief" (quote).

54. MACV Records, Weekly Report, 15–21 November 1964, 37–38.

55. MACV Records, Weekly Report, 21–28 November 1964, 35.

56. Memorandum for the Chairman, Joint Chiefs of Staff (JCS), "Report on Col. Knight's visit to RVN," 21 December 1964, tab 3, page 5, TTVA, David Lifton Collection, box 1, folder 5.

57. Ibid., tab 2, page 1; MACSOG Documentation Study, 58, 62.

58. Ibid., tab 3, page 5.

59. Ibid., cover letter to Anthis.

60. Robert M. Gillespie, *Black Ops, Vietnam: An Operational History of MACVSOG* (Annapolis, MD: Naval Institute Press, 2011), 103; MACSOG Documentation Study, 76, 119.

61. Memorandum for the Chairman, JCS, "Report on Col. Knight's visit to RVN," tab 2, page 2.

62. Ibid., Memo, Russell to Chief of Staff, "Consideration OPLAN 34-A Surfacing," cover letter; MACSOG Documentation Study, 7–10.

63. Gillespie, *Black Ops*, 29.

64. Memorandum for the Chairman, JCS, Memo, Russell to Chief of Staff, "Consideration OPLAN 34-A Surfacing," 3–5.

65. MACSOG Documentation Study, 13.

66. MACV, *Command History, 1964*, 72.

67. MACV Records, Weekly Report, 9–15 December 1964, 34–35, 37.

68. MACV, *Command History, 1964*, 77.

69. MACV Records, Weekly Report, 5–12 December 1964, 37–40, annex B, table 6.

70. Ibid.

71. J. A. Koch, *The Chieu Hoi Program in South Vietnam, 1963–1971* (Washington, DC: Rand Corporation for ARPA, January 1973), 24.

72. MACV, *Command History, 1964*, 69–70; Douglas S. Blaufarb, *The Counterinsurgency Era: U.S. Doctrine and Performance, 1950 to the Present* (New York: Free Press, 1977), 221 (quote).

73. MACV, *Command History, 1964,* 78 (quote), 123.

74. MACV, *Command History, 1964,* 47; John M. Carland, *Combat Operations: Stemming the Tide, May 1965 to October 1966,* US Army in Vietnam series (Washington, DC: US Army Center for Military History, 2000), 12; MACV Records, Weekly Report, 9–15 December 1964, 35.

75. Weapons Systems Evaluation Group, *Flaming Dart Airstrikes,* 21; FBIS, 28 December 1964 JJJ1, Radio Hanoi, "Bomb Destroys U.S. Officers Club in Saigon"; Moyar, *Triumph Forsaken*, 348 (quote).

76. MACV Records, Weekly Report, 2–9 May 1964, 43.

77. Message from USIS to USIA, "Plans and Special Projects for 1965 in Psychological Operations Field," 26 January 1965, 3, TTVA, Larry Berman Collection, box 1, folder 15.

78. FBIS, 29 December 1964, JJJ2–3, Radio Hanoi, "English for U.S. Servicemen."

79. FBIS, 29 January 1965, JJJ3, RH to RVN, "Open Letter to Johnson."

80. USIS, "Plans and Special Projects for 1965," 1, 8 (quote).

81. FBIS, 14 January 1965, JJJ4, RH to RVN, "The Greater the Victories We Score in the South Vietnam, the More Divided the Americans Become."

82. USIS, "Plans and Special Projects for 1965," 4.

83. MACV, *Command History, 1964*, 75.

84. USIS, "Plans and Special Projects for 1965," 4, 2.

85. Ibid., 3.

86. Ibid., 5.

87. Ibid., 6.

88. Hoffer, "Broadcasting in an Insurgency," 117.

89. Message from USIS to USIA, "Effectiveness of Field Representatives at the Field Level," 3–4, LBJL, NSF box 15, memorandums 2.

90. FBIS, 4 January 1965, JJJ2, RH in Rhade, "U.S. Khanh Step up Recruiting of Youth."

91. FBIS, 12 January 1965, JJJ4–5, RH in Rhade, "Highlands People Should Join People's Side."

92. Weapons Systems Evaluation Group, *Flaming Dart Airstrikes*, 9.

93. McMaster, *Dereliction of Duty*, 99, 160.

94. Weapons Systems Evaluation Group, *Flaming Dart Airstrikes*, 17.

95. Weapons Systems Evaluation Group, *Flaming Dart Airstrikes*, 21; McMaster, *Dereliction of Duty*, 62, 100–101.

96. Carland, *Combat Operations*, 24.

97. Weapons Systems Evaluation Group, *Flaming Dart Airstrikes,* 3 (quote); Caroline Page, *U.S. Official Propaganda during the Vietnam War, 1965–1973: The Limits of Persuasion* (London: Leicester University Press, 1996), 89, 132.

98. FBIS, 9 February 1965, JJJ2, "Activities in Washington" (quote); Carland, *Combat Operations*, 14.

99. Weapons Systems Evaluation Group, *Flaming Dart Airstrikes,* 5; McMaster, *Dereliction of Duty*, 62, 63, 229.

100. FBIS, 11 February 1965, JJJ17, RH in English to American Servicemen.

101. FBIS, 15 February 1965, JJJ16, Radio Hanoi International Service, "Shumaker at Press Conference."

102. Moyar, *Triumph Forsaken,* 367, 416.

103. William J. Duiker, *Ho Chi Minh: A Life* (New York: Hyperion, 2000), 549.

104. "Text of U.S. White Paper on North Vietnam's Growing Role in War in the South," *New York Times,* 27 February 1965, 30.

105. Hammond, *Public Affairs,* 133.

106. FBIS, 18 March 1965, JJJ4–5, "U.S. Public Opinion Reveals the True Face of the U.S. Aggressors" (quote); William L. Lunch and Peter W. Sperlich, "American Public Opinion and the War in Vietnam," *Western Public Opinion Quarterly* 32, no. 1 (March 1979): 21–23.

107. FBIS, 3 March 1965, JJJ10, RH to RVN, "A Typical Example of the U.S. Imperialists' Trick of Calling for Help while Committing Robbery."

7. Expansion, Spring and Summer 1965

1. Memorandum for the President from Carl Rowan, 26 February 1965, LBJL, box 14, memorandums 2.

2. US Information Agency, *USIA Administrative History*, LBJL, USIA box 1, folder 1 (hereafter *USIA Administrative History*), 5–53.

3. Department of State, Research Reference Service, "Facts and Attitudes: Long An Province," February 1965, 10, LBJL, NSF Country File Vietnam box 14, memorandums 2.

4. Ibid.

5. Ibid., 16.

6. FBIS, 12 February 1965, JJJ9, RH in English to American Servicemen, "Thu Huong: Commentary for U.S. Servicemen."

7. FBIS, 23 February 1965, JJJ5, RH to American Servicemen, "Comment for U.S. Servicemen."

8. FBIS, 25 February 1965, JJJ5 RH to RVN, "The Development of the Anti-U.S. Movement among Soldiers and Offensive of the Southern Armed Forces."

9. MACV, "Orientation Material: Viet Cong," 10 February 1965, 5, CARL.

10. Dong Van Khuyen, *RVNAF History*, part 4—"Political Warfare," 323, Indochina Monographs, Washington, DC: US Army Center for Military History, 1978.

11. MACV Memo, "Orientation Material: Viet Cong," 10 February 1965, 5.

12. Ibid., 4, 9 (quote).

13. Leon Goure (Rand Study), "Some Impressions of the Effects of Military Operations on Viet Cong Behavior," March 1965, LBJL, NSF Vietnam, box 15, memorandums 1.

14. CIA, Weekly Report, Republic of Vietnam, "The Situation in South Vietnam," CIA Research, 6, TTVA, VAC, box 5, folder 561.

15. Goure, "Some Impressions," 3 (quote), 6.

16. USIS Saigon to USIA, "Transmittal of Memorandum of Conversation," 9 March 1965, LBJL, NSF Vietnam, box 15, memorandums 2.

17. "Nonlethal Gas Warfare Used against Viet Cong," *Dallas Morning News,* 23 March 1965, 2.

18. Department of State, Outgoing Telegram to Amembassy Saigon, 24 March 1965, LBJL, box 15, cables 21.

19. Department of State, Incoming Telegram from Saigon, "From Taylor Subj Crop Destruction," LBJL, NSF Vietnam, box 15, cables 21.

20. FBIS, Special Memorandum: Foreign Reaction to the Use of Tear and Nausea Gases in South Vietnam, 26 March 1965, LBJL, box 194, file "Gas," vol. 1, FBIS; United Nations Office for Disarmament Affairs, "United States of America: Ratification of 1925 Geneva Protocol," disarmament.un.org/treaties/a/1925/unitedstatesofamerica/rat/paris. When East Germany began this campaign, refering to West Germans as the Vietnam Legion, West Germany began an extensive counterpropaganda effort to immunize the population. See Peter Busch, "The 'Vietnam Legion' West German Psychological Warfare against East German Propaganda in the 1960s," *Journal of Cold War Studies* 16, no. 3 (Summer 2014): 164–189.

21. Greenfield from Zorthian, Incoming State Telegram, From Saigon Embassy, 28 March 1965, LBJL, NSF Vietnam, box 15, cables 21.

22. Memo from General Westmoreland, "Riot Control Munitions," LBJL, box 194, file "Gas," vol. 1, Westmoreland.

23. John McCone to Senator Stennis, 9 April 1965, "Letter and Special DCI Briefing for Senator Stennis," 18 (quote), TTVA, Central Intelligence Agency Collection, box 1, folder 11.

24. Ibid., 1–6.

25. Ibid., 16 (quote).

26. NSAM No. 330, "Intensified and Expanded Psychological Operation Activities in Vietnam," From Bundy to Sec State/Def Dir USIA/CI, Administrator USAID, 9 April 1965, 1, LBJL (quote); NSAM No. 328, "Presidential Decisions with Respect to Vietnam," 4 April 1965, 2, LBJL; Mark Moyar, *Triumph Forsaken: The Vietnam War, 1954–1965* (New York: Cambridge University Press, 2006), 355.

27. FBIS, 26 May 1965, JJJ8, RH to American Servicemen, "Robert Williams Notes U.S. Crimes in Vietnam."

28. FBIS, 2 March 1965, JJJ8, Radio Hanoi International in English, "Malcolm X Murder Shows Imperialists' Cruelty."

29. FBIS, 18 March 1965, JJJ4–5, RH to American Servicemen, "Robert Williams' Remarks for US Servicemen."

30. FBIS, 17 March 1965, JJJ5, "NHAN DAN Condemns US Repression of Negroes."

31. Richard A. Hunt, *Pacification: The American Struggle for Vietnam's Hearts and Minds* (Boulder, CO: Westview Press, 1995), 31.

32. FBIS, 1 March 1965, JJJ9, "Remark on Worldwide Protests against the US War in Vietnam."

33. Committee to Aid the NLF (CANLF), www.redandgreen.org/CANLF/Index.htm (site by Walter Teague, who maintains archive of committee documents).

34. *CANLF Bulletin,* 20 May 1966, 19.

35. *CANLF Bulletin,* 4 February 1967, "Progress Report," 13.

36. FBIS, 10 March 1965, JJJ5, RH to American Servicemen, "12 Americans Seek Repatriation from South."

37. Commander in Chief Pacific/Commander US MACV, CINCPAC/COMUSMACV, *Command Report on the War in Vietnam (to May 1968),* 20 (n.p., n.d.), BACM Research, paperless archives.com.

38. Department of State, Incoming State Telegram from Taylor, 26 March 1965, LBJL, box 15, cables 21, leaflet (quote); Department of State, Outgoing State Telegram, 9 April 1965, "Air Strikes against North Viet Nam," LBJL, box 16, cables 3, 9 April.

39. Robert W. Chandler, *War of Ideas: The U.S. Propaganda Campaign in Vietnam* (Boulder, CO: Westview Press, 1981), 77.

40. CINCPAC/COMUSMACV, "Command Report on the War in Vietnam," 20–21; Chandler, *War of Ideas*, 113.

41. Jeanne S. Mintz, et al., *A Short Guide to Psychological Operations in the Republic of Vietnam, 1966* (Washington, DC: Special Operations Research Office, 1965), 14.

42. Ibid., 109, 119.

43. United States Broadcasting and Visual Support Activity Pacific, *Special Orders Number 134—Extract* (Headquarters, USAB&VAP, Okinawa, 22 July 1965), 1–2, www.timyoho.com/BVAPage/HistoryPsyWar/1st-PsyOps.htm; Naval Advisory Group, "Historical Review," 4–5, 6 (quote), DTIC, AD954004.

44. Joseph B. Knotts interview with Allen B. Richardson, USMC0028, United States Marine Corps, History and Museums Division, Oral History Collection, n.d., 5, Vietnam Center and Archive, Texas Tech University.

45. Ibid. (quote 1); Dong Van Khuyen, *RVNAF History,* Indochina Monographs, part 4—"Political Warfare," 299, TTVA, Michael Sloniker Collection, box 5, folder 2 (quote 2).

46. FBIS, 14 May 1965, JJJ9, RH in English, "U.S. Takes People's Land for Troop Barracks."

47. FBIS, 11 May 1965, JJJ6, RH to RVN, "The U.S. Aggressors Are Daydreaming with Bombs in Both Hands."

48. FBIS, 11 May 1965, JJJ12, RH to American Servicemen, "U.S. Troops Cannot Rely on Aroused Vietnamese."

49. FBIS, 11 May 1965, JJJ11, RH to American Servicemen, "U.S. Paratroopers Oppose Duty in Vietnam."

50. FBIS, 14 May 1965, JJJ5, RH to American Servicemen, "U.S. Youth Oppose Johnson's Vietnam Policy."

51. FBIS, 12 May 1965, JJJ10, RH to RVN, "Saigon Troops Urged to Turn Guns on Americans."

52. Rowan Memorandum for the President, 26 May 1965, 3, LBJL, NSF Vietnam, box 5, Rowan.

53. Ibid., 4.

54. William Hoffer, "Broadcasting in an Insurgency Environment: USIA in Vietnam, 1965–1970" (PhD diss., University of Wisconsin, 1972), 123.

55. FBIS, 20 May 1965, JJJ6–7, Radio Hanoi International, "U.S. Marines Stage Bloody Brawl in Da Nang."

56. FBIS, 26 May 1965, JJJ8, RH to American Servicemen, "Robert Williams Notes U.S. Crimes in Vietnam."

57. FBIS, 3 May 1965, JJJ6, Radio Hanoi International, "DRV Protests Dispatch of Australian Battalion."

58. John M. Carland, *Combat Operations: Stemming the Tide, May 1965 to October 1966,* US Army in Vietnam series (Washington, DC: US Army Center for Military History, 2000), 26. Ian McNeill, *To Long Tan: The Australian Army and the Vietnam War, 1950–1966* (Sidney: Allen and Unwin, 1993), 293. Derrill De Heer, "Psychological Operations Conducted by the Australian Army in Phuoc Tuy Province South Vietnam: 1965–1971" (Master of Arts [Research] History, University of New South Wales School of Humanities and Social Science, Australian Defence Force Academy, 2009), 68.

59. Military Assistance Command, Vietnam, *Command History, United States Military Assistance Command, Vietnam 1965* (Saigon: USMACV Military History Branch, 20 April 1966), 448.

60. Ibid., 253.

61. Ibid., 253, 458.

62. Joint United States Public Affairs Office (JUSPAO), *JUSPAO Guidance 2: Missions and Functions of JUSPAO Planning Staff* (Saigon: JUSPAO, 4 August 1965), 2 (quote); MACV, *Command History, 1965,* 254, 449.

63. Joint United States Public Affairs Office, *JUSPAO Guidance 9: Exploitation of VC Vulnerabilities* (Saigon: JUSPAO, 5 August 1965), 2.

64. Ibid., 3.

65. Ginger Davis, "The Visible Persuaders: A Study of America's Use of Psychological Warfare Using the Chieu Hoi Program as a Case Study" (master's thesis, Texas Tech University, 1998), 71; Joseph J. McDonough, "Analysis of Official U.S. Military Psychological Warfare Efforts in the Vietnam conflict" (MPIA thesis, Boston University School of Public Communications, August 1968), 18.

66. J. M. Carrier and C. A. H. Thomson, *Viet Cong Motivation and Morale: The Special Case of Chieu Hoi* (Santa Monica, CA: Rand Corporation, 1966), 42.

67. Ibid., xvii (quote), 44.

68. McDonough, "Analysis of Official U.S. Military Psychological Warfare," 45 (quote), 51.

69. JUSPAO, *Exploitation of VC Vulnerabilities*, 4.

70. Ibid., 5.

71. Carrier and Thomson, *Viet Cong Motivation,* 50; Duiker, *Ho Chi Minh*, 553.

72. Carrier and Thomson, *Viet Cong Motivation,* xiii, 22.

73. David W. P. Elliott, *The Vietnamese War: Revolution and Social Change in the Mekong Delta, 1930–1975* (Armonk, NY: M. E. Sharpe, 2003), 153, 783.

74. FBIS, 15 June 1965, JJJ8, RH to American Servicemen, "Thu Huong Introduces Anna Louise Strong."

75. FBIS, 29 June 1965, JJJ10, RH to American Servicemen, "U.S. Officer Refuses to Go to Vietnam"; "Steinke on Duty," *Pacific Stars and Stripes,* 30 June 1965, page 2 (?).

76. Hoffer, "Broadcasting in an Insurgency," 368, 369 (quote).

77. Ibid., 369.

78. Ibid., 368.

79. Ibid., 365.

80. Ibid., 209, 215.

81. MACV, *Command History, 1965*, 450 (quote); James Lawton Collins, *The Development and Training of the South Vietnamese Army, 1950–1972* (Washington, DC: Department of the Army, 2002), 138.

82. Khuyen, *RVNAF History,* 294, 314 (quote).

83. MACV, *Command History, 1965,* 454, 456 (quote).

84. Headquarters, 173rd Airborne Bde. (SEP), "Commander's Combat Note Number 73, Duty in Vietnam," 5 August 1965, USASOC, Vietnam 1966–2, box 7, 1.

85. Ibid.; US Air Force, "5th Special Operations Squadron," www.919sow.afrc.af.mil/library/factsheets/factsheet.asp?id=16212.

86. Headquarters, 173rd Airborne Bde. (SEP), "Commander's Combat Note Number 73," 6.

87. Ibid., 39.

88. Ibid., 1.

89. Ibid.

90. FBIS, 11 August 1965, JJJ5, RH in English, "Liberation Forces Attack Post at Duc Co."

91. 5th Special Forces Group, "Quarterly Command Report for the Period Ending 31 December 1965," 10, inclosures 25 and 27, DTIC, AD391693.

92. Headquarters, 173rd Airborne Bde. (SEP), "Commander's Combat Note Number 78," 2, 3.

93. FBIS, 25 August 1965, JJJ11, RH to RVN, "Bacteriological Warfare Unit Sent to South."

94. Ibid.

95. FBIS, 9 September 1965, JJJ3, Radio Hanoi International in English, "In a Story about U.S. Use of Poison Gas."

96. FBIS, 10 September 1965, JJJ3, Radio Hanoi Domestic, "Dispatch of Chemical Warfare Team Protested."

97. FBIS, 13 September 1965, JJJ2, Radio Hanoi Domestic, "DRV Denounces U.S. Use of Gas in the South."

98. FBIS, 29 September 1965, JJJ5, RH to RVN, "U.S. Imperialist Crimes Must Be Stopped."

99. FBIS, 15 September 1965, JJJ4, Radio Hanoi, *Quan Doi Nhan Dan,* "Psychological Warfare Is Part of the U.S. Imperialists' Aggressive War."

100. Ibid.

101. Combined Document Exploitation Center (CDEC), "Against the United States' 'Psychological Warfare,'" Hoc Tap 9 (September 1965) 3–6, 3 (quote), TTVA, VAC, box 163, folder 396.

102. JUSPAO, "Exploitation of VC Vulnerabilities," 8, DTIC ADA032189.

103. Carrier and Thomson, *Viet Cong Motivation,* 136.

104. Ibid., 138.

105. Goure, "Some Impressions," 77.

106. JUSPAO, *Exploitation of VC Vulnerabilities,* 11 (quote); MACV, *Command History, 1965*, 14.

107. FBIS, 27 September 1965, KKK1, LBS to RVN, "Liberation Forces Execute Two Americans."

108. JUSPAO Field Memorandum No. 2, "Review of Current Viet Cong Propaganda Efforts" (Saigon: Vietnam JUSPAO Planning Staff, 30 September 1965), 1–5.

109. *Newsweek* clipping, "PSYWAR," about Barry Zorthian, 4 October 1965, TTVA, DPC: Unit 3—US Economy and US Mission (Saigon), box 42, folder 3.

110. US Operations Mission (USOM), "Reports of JUSPAO Field Representatives, 30 November 1965," 4; Herb Friedman, "24th PSYOP Detachment," psywarrior.com/24thPsyopDet.html; MACV, *Command History, 1965*, 254.

111. USOM, "Reports of JUSPAO Field Representatives, 30 November 1965," 4; MACV, *Command History, 1965,* 254; US Air Force, "Factsheet: 5th Special Operations Squadron,"

www.919sow.afrc.af.mil/library/factsheets/factsheet.asp?id=16212; John Schlight, *A War Too Long: The USAF in Southeast Asia, 1961–1975* (Washington, DC: Air Force History and Museums Program, 1996), 160.

112. 1st PSYOP Field Detachment (Prov), "Leaflet Catalog," 18 October 1965, U.S. Army Education and Historical Center (AEHC), 1ST PSYOP box, leaflets; 5th Special Forces Group, "Quarterly Command Report for the Period Ending 31 December 1965," inclosure 2, DTIC, AD391693.

113. 1st Cavalry Division, "Quarterly Command Report," December 1965, 1–15, 28 (quote), 33, DTIC, AD390501; 5th Special Forces Group, "Quarterly Command Report for the Period Ending 31 December 1965," inclosure 2.

114. Mark Moyar and Harry G. Summers, *Phoenix and the Birds of Prey: The CIA's Secret Campaign to Destroy the Viet Cong* (Annapolis, MD: Naval Institute Press, 1997), 303.

115. 1st Cavalry Division, "Quarterly Command Report," December 1965, 1–15, 15 (quote); 5th Special Forces Group, "Quarterly Command Report for the Period Ending 31 December 1965," 9.

116. 5th Special Forces Group, "Quarterly Command Report for the Period Ending 31 December 1965," 8.

117. 1st Cavalry Division, "Quarterly Command Report," 29.

118. Ibid., 33.

119. USOM, "Reports of JUSPAO Field Representatives, 30 November 1965," 24, TTVA, Rufus Phillips Collection, box 1, folder 3; 1st Cavalry Division, "Quarterly Command Report," 28.

120. USOM, "Reports of USOM Provincial Representatives for Month Ending 31 December 1965," TTVA, Rufus Phillips Collection, box 1, folder 4.

8. Refined and Expanded Targeting, Fall 1965

1. FBIS, 12 October 1965, KKK1, Radio Hanoi International in English, "NFSLV Calls for Support in Hate U.S. Month."

2. FBIS, 19 February 1965, JJJ9, VNA International Service, "Nhan Dan Publishes USSR-DPRK Statement Text."

3. FBIS, 12 October 1965, KKK4, "Editors Note."

4. FBIS, 15 October 1965, KKKl-2, LBS to RVN, "Peace Committee Letter to American People."

5. FBIS, 4 November 1965, JJJ2, Radio Hanoi International in English, "Morrison Immolation Hailed by People, Press"; FBIS, 5 November 1965, KKK2, Radio Hanoi International in English, "Vietnam Martyr's Wife Writes Mrs. Morrison"; FBIS, 5 November 1965, JJJl, RH to RVN, "Norman Morrison Torch Will Blaze Forever" (quote).

6. FBIS, 9 November 1965, JJJ5, RH in English, "Yale Youth Protests Policy with Hunger Strike" (quote); Jane Fonda, *My Life So Far* (New York: Random House, 2005), 342; Robert F. Turner, "Triumph Forsaken as a Path to Setting the Record Straight," in Andrew Wiest and Michael Doidge, eds., *Triumph Revisited: Historians Battle for the Vietnam War* (New York: Routledge, 2010), 101. Bui Tin later acknowledged the psychological value of actions such as Jane Fonda's.

7. FBIS, 19 October 1965, JJJ1–2, Radio Hanoi International, "Americans Demonstrate against War in Vietnam."

8. FBIS, 22 October 1965, JJJ3, Radio Hanoi International, "DRV Groups Hail Antiwar Struggle in U.S."

9. FBIS, 27 October 1965, JJJ9, RH in English to RVN, "Progressive Americans to Stage Peace March."

10. *CANLF Bulletin,* October 1965, 26, www.redandgreen.org/CANLF/Index.htm.

11. *CANLF Bulletin,* November 1965, 68; Fred Halstead, *Out Now! A Participant's Account of the American Movement against the Vietnam War* (New York: Monad Press, 1978) 142–143. The author knew Teague and discusses the flag controversy.

12. FBIS, 16 November 1965, BB3, Moscow in Mandarin to China, "Socialists Must Spur Antiwar Struggle."

13. FBIS, 30 November 1965, BB7, USSR to North America, "Moscow Radio Debates Pros, Cons of Viet War,"

14. FBIS, 22 November 1965, JJJ1, Radio Hanoi International in English, "President Ho Chi Minh Writes U.S. Professor" (quote); FBIS, 22 November 1965, JJJ3–4, "The Month of High Spirited Struggle among Americans Supporting Vietnam and Protesting the War of Aggression of the Johnson Clique."

15. William J. Duiker, *Ho Chi Minh: A Life* (New York: Hyperion, 2000), 552.

16. FBIS, 14 October 1965, KKK3, Radio Hanoi in Korean to RVN, "Appeal to the Officers and Men of the South Korean Army Dispatched to South Vietnam."

17. 5th Special Forces Group, "Quarterly Command Report for the Period Ending 31 December 1965," inclosure 26, 155–156, DTIC, AD391693.

18. Ibid.

19. 5th Special Forces Group, "Quarterly Command Report for the Period Ending 31 December 1965," inclosure 26, 155.

20. Ibid., 153–158.

21. 1st Cavalry Division, "Quarterly Command Report," Period Ending 1 December 1965, 28.

22. USOM, "Reports of JUSPAO Field Representatives," 30 November 1965, 17. TTVA, Rufus Phillips Collection, box 1, folder 3.

23. 5th Special Forces Group, "Quarterly Command Report for the Period Ending 31 December 1965," 166.

24. Ibid., 12.

25. Ibid., inclosure 25.

26. 5th Special Forces Group, "Quarterly Command Report for the Period Ending 31 December 1965," inclosure 25.

27. FBIS, 24 November 1965, JJJ10, Radio Hanoi Domestic, "U.S. 'Running' Cavalry."

28. USOM, "Reports of JUSPAO Field Representatives," 30 November 1965, 12 (quote); Gregory Daddis, *No Sure Victory: Measuring U.S. Army Effectiveness and Progress in the Vietnam War* (Oxford, UK: Oxford University Press, 2011), 85, 91.

29. FBIS, 9 November 1965, JJJ7, RH to RVN, "Use of More U.S. Troops Cannot Alter Outcome."

30. FBIS, 9 November 1965, KKK1, LBS to RVN, "NFLSV Appeals to Saigon-Gia Dinh Compatriots."

31. USOM, "Reports of JUSPAO Field Representatives," 30 November 1965, 9.

32. Military Assistance Command, Vietnam, *Command History, United States Military Assistance Command, Vietnam 1965* (Saigon: USMACV Military History Branch, 20 April 1966), 253.

33. U.S. Army Special Warfare School, "An Evaluation of Psychological Operations Officer Training Program," 7 December 1965, 2.

34. Department of the Army, "5th SF Group MTOE 31–106E," n.p., USASOC, 1966–3, box 10; 5th Special Forces Group, "Quarterly Command Report for the Period Ending 31 December 1965," 16 (quote).

35. 5th Special Forces Group, "Quarterly Command Report for the Period Ending 31 December 1965," 13–14.

36. USAB&VAP, "Special Orders Number 134—Extract," 1–2; USOM, "Reports of JUSPAO Field Representatives, 30 November 1965," 45, TTVA, Rufus Phillips Collection, box 1, folder 4.

37. USOM, "Reports of JUSPAO Field Representatives, 30 November 1965," 6.

38. Ibid., 1.

39. Ibid., 2 (quote), 29.

40. Ibid., 9, 15 (quote).

41. Ibid., 21–22.

42. Ibid., 11.

43. Ibid., 37.

44. Stuart I. Rochester, *The Battle behind Bars: Navy and Marine POWs in the Vietnam War*, US Navy and the Vietnam War series (Washington, DC: Naval History and Heritage Command, Department of the Navy, 2010), 10.

45. FBIS, 2 December 1965, JJJ4, Radio Hanoi International in English, "American Troops Receive Antiwar Leaflets" (quote); Angelo Collura, *The Case of Pvt. Robert R. Garwood, USMC*, vol. 1 (Report to the Assistant Secretary of Defense for Command, Control, Communication and Intelligence, June 1993), 20.

46. FBIS, 3 December 1965, JJJ6, Radio Hanoi International in English, "*NHAN DAN* Discusses Statement of U.S. POWs."

47. FBIS, 14 December 1965, JJJ3, RH to American Servicemen, "Account of Ceremony Releasing U.S. POWs."

48. John M. Carland, *Combat Operations: Stemming the Tide, May 1965 to October 1966*, US Army in Vietnam series (Washington, DC: US Army Center for Military History, 2000), 92.

49. Thomas L. Ahern Jr., *Vietnam Declassified: The CIA and Counterinsurgency* (Lexington: University Press of Kentucky, 2010), 179.

50. Mark Moyar and Harry G. Summers, *Phoenix and the Birds of Prey: The CIA's Secret Campaign to Destroy the Viet Cong* (Annapolis, MD: Naval Institute Press, 1997), 150; Ahern, *CIA and Rural Pacification*, 193.

51. Ahern, *Vietnam Declassified*, 169.

52. Robert W. Chandler, "U.S. Psychological Operations in Vietnam, 1965–72" (PhD diss., George Washington University, 1974), 86.

53. JUSPAO, *National Psychological Operations Plan Proposed for Vietnam* (Saigon: JUSPAO Office of the Director, December 1965), 1, USASOC Archives, PSYOP Papers 1961–1968, box 3.

54. Ibid., 13–14.

55. Ibid., 3.

56. Ibid., 3–4.

57. J. A. Koch, *The Chieu Hoi Program in South Vietnam, 1963–1971* (Washington, DC: Rand Corporation for ARPA, January 1973), vi, 6 (quote), 11, 26, 37.

58. United States Military Assistance Command, *Command History Vietnam: 1967* (Saigon: USMACV Military History Branch, 16 September 1968), 949, 959, 965.

59. Robert Fantina, *Desertion and the American Soldier* (New York: Algora Publishing, 2006), 116; United States Senate, "Treatment of Military Deserters," Report of the Armed Forces Committee (Washington, DC: US Government Printing Office, March 1969), 24; *The Big Picture*, "The Unique War," 24 min. (US Army Pictorial Center, 1966).

60. William Hoffer, "Broadcasting in an Insurgency Environment: USIA in Vietnam, 1965–1970" (PhD diss., University of Wisconsin, 1972), 209, 215, 383 (quote).

61. Nguyen Cao Ky and Marvin J. Wolf, *Buddha's Child: My Fight to Save Vietnam* (New York: St Martin's Press, 2002), 187 (quote); MACV, *Command History, 1964*, 77.

62. JUSPAO, "General Briefing Book," n.p., n.d., 35; Robert M. Gillespie, *Black Ops, Vietnam: An Operational History of MACVSOG* (Annapolis, MD: Naval Institute Press, 2011), 101; Hoffer, "Broadcasting in an Insurgency," 209, 215; MACSOG Documentation Study, annex A to appendix C, Psychological Operations (Washington, DC: Office of the Secretary of Defense, 1970), 34, 40.

63. Hoffer, "Broadcasting in an Insurgency," 215, 376; Bruce Wahl, "History of Project Jenny," www.afvn.tv/ProjectJenny/history.html.

64. Hoffer, "Broadcasting in an Insurgency," 368, 376 (quote).

65. Ibid., 369, 383.

66. Ibid., 363 (quote), 369, 615.

67. Oliver Quayle and Co., "A Survey of Public Opinion in South Viet Nam Study 245," unpublished study, vol. 1, January 1966, 3, 7 (quote), LBJL, NSF Country File Vietnam, box 213.

68. Ibid., 8, 16, 21–28.

69. FBIS, 20 January 1966, KKK3, RH in English, "NFLSV Lists Military Achievements for 1965."

70. J. M. Carrier and C. A. H. Thomson, *Viet Cong Motivation and Morale: The Special Case of Chieu Hoi* (Santa Monica, CA: Rand Corporation, 1966), 142.

71. JUSPAO Planning Office, "PSYOPS in Vietnam: Indications of Effectiveness" (Saigon: JUSPAO, 1967), 33.

72. Ibid., 47.

73. Ibid.

74. FBIS, 6 January 1966, JJJ4–5, RH in English, "Speech by U.S. Visitor Herbert Aptheker."

75. FBIS, 26 January 1966, JJJ6, RH to American Servicemen, "Aptheker Article Hails Morrison's Martyrdom."

76. FBIS, 28 January 1966, JJJ12, RH in English to American Servicemen, "Staughton Lynd Gives Talk on Visit to DRV"; Tom Hayden, *Reunion: A Memoir* (New York: Random House, 1988), 222; Herbert Aptheker, *Mission to Hanoi* (New York: International Publishers, 1966), 51. Aptheker also describes (page 46) being on the receiving end of American leaflet and transistor radio drops.

77. "Donovan Unofficial Site," donovan-unofficial.com/music/singles/universal_soldier.html; FBIS, 4 January 1966, JJJ8–15, RH to American Servicemen, "Radio Stateside Program to U.S. Servicemen."

78. Ibid.

79. William D. Parker, *U.S. Marine Corps Civil Affairs in I Corps, Republic of Vietnam: April 1966 to April 1967* (Washington, DC: Historical Division, Headquarters, US Marine Corps, 1970), 13.

80. Ibid., 9.

81. Headquarters, 173rd Airborne Bde., "Commander's Combat Note Number 91," 22 January 1966, USASOC, Vietnam 1966–3, box 10, 4.

82. Ibid.

83. MACV, *Command History, United States Military Assistance Command, Vietnam: 1966* (Saigon: MACV Military History Branch, 19 April 1967), 559.

84. Ibid., 558–559.

85. Parker, *U.S. Marine Corps,* 13 (quote), 14–15; MACV, "Counterinsurgency Lessons Learned No. 61: Salient Lessons Learned," 27 January 1967, 12, TTVA, Bud Harton Collection, box 00, folder 1.

86. Parker, *U.S. Marine Corps*, 14.

87. Eric Bergerud, *Dynamics of Defeat: The Vietnam War in Hau Nghia Province* (Boulder, CO: Westview Press, 1991), 152–153 (quote); Frank Scotton, *Uphill Battle: Reflections on Viet Nam Counterinsurgency* (Lubbock: Texas Tech University Press, 2014), 190.

88. Parker, *U.S. Marine Corps*, 9.

89. MACV, *Command History, 1965*, 254 (quote); 6th Psychological Operations Battalion, Status Report, 21 April 1966, 4–5, AEHC, 6th Psychological Operations Battalion, Status Report.

90. MACV, *Command History, 1966*, 555.

91. Headquarters I Field Force Vietnam, OR for 31 July 1966, 25 August 1966, 21–23, USASOC, Vietnam 1966–2, box 7.

92. MACV, *Command History, 1966*, 555.

93. "Briefing on the 6th Psychological Operations Battalion," n.d., n.p., USASOC, PSYOP Papers 0000–1944, box 1; 6th Psychological Operations Battalion, Status Report, 21 April 1966, 1, AEHC, 6th PSYOP Battalion, Status Report.

94. Howard Holiday to Commanding Officer, 2nd Psychological Operations Group, 24 October 1966, USASOC, Avedon Collection, box 9, 245th Loudspeaker Study.

95. Headquarters, I Field Force Vietnam, Operational Report for 31 July 1966, 22–23.

96. MACV, *Command History, 1966*, 576.

97. Headquarters, I Field Force, "Operational Report, Period Ending 31 July 1966," 22; John B. Dwyer, U.S. Navy Beach Jumpers Association, "Vietnam War" www.beachjumpers.com/History/1960sJBD.htm.

98. MACV, *Command History, 1966*, 569.

99. MACV, *Command History, 1965*, 459 (quote); MACV, *Command History, 1966,* 571; Tom G. Abbey, *Southeast Asia Report: The Role of the USAF in Support of Special Activities in SEA* (Hickam AFB, HI: CHECO/Corona Harvest Division, Operations Analysis Office, Department of the Air Force Headquarters Pacific Air Forces, 1 July 1976), 31; MACSOG Documentation Study, annex A to appendix C, Psychological Operations (Washington, DC: Office of the Secretary of Defense, 1970), 160, 163.

9. Second Republic Surge, Spring and Summer 1966

1. Garth S. Jowett and Victoria O'Donnell, *Propaganda and Persuasion,* 5th ed. (Thousand Oaks, CA: Sage Publications, 2012), 263.

2. USIA, *USIA Administrative History*, 1–19; Franklin D. Roosevelt, Executive Order No. 918, 13 June 1942, www.cia.gov/library/readingroom/docs/Executive_Order_9182_13_Jun_1942.pdf.

3. JUSPAO, *General Briefing Book,* n.p., n.d., 22.

4. Ibid., 25; William Hoffer, "Broadcasting in an Insurgency Environment: USIA in Vietnam 1965–1970" (PhD diss., University of Wisconsin, 1972), annex C-10.

5. JUSPAO, *General Briefing Book*, 11, 12 (quote).

6. Ibid., 27 (quote), 28.

7. MACV, *Command History, United States Military Assistance Command, Vietnam: 1966* (Saigon: MACV Military History Branch, 19 April 1967), 571.

8. FBIS, 14 January 1966, JJJ1, RH to RVN, "Comment on Johnson State of Union Message" (quote); President Lyndon Johnson, "President Johnson's State of the Union Address, 1966," www.lbjlib.utexas.edu/johnson/archives.hom/speeches.hom/660112.asp.

9. Hoffer, "Broadcasting in an Insurgency," 325.

10. FBIS, 17 January 1966, JJJ7–10, RH to RVN, "From the Medley of Peace Baskets to the Masterly Speech of Johnson."

11. FBIS, 6 January 1966, KKK9, LBS to RVN, "Radio Comments on Cease-fire Significance."

12. John Schlight, *The War in South Vietnam: The Years of the Offensive, 1965–1968*, US Air Force in Southeast Asia series (Washington, DC: Office of Air Force History, 1988), 195.

13. JUSPAO Planning Office, *PSYOPS in Vietnam: Indications of Effectiveness* (Saigon: n.p., 1967), 59, 60.

14. Stanley Sandler, *"Cease Resistance: It's Good for You!": A History of U.S. Army Combat Psychological Operations*, 2nd ed. (Washington, DC: US Army Special Operations Command, Directorate of History and Museums, 1999), 318.

15. 5th Special Forces Group, "Operational Report Lessons Learned for Period Ending 30 April 1966" (Nha Trang, 10 May 1966), 10 (quote), 11, DTIC, AD391694.

16. Headquarters 1st Cavalry Division, "Combat Operations After Action Report, Operation Jim Bowie," 6, USASOC, Vietnam 1966–3, box 10.

17. Ibid., 6 (quote), 16.

18. Ibid., 13, 17.

19. CDEC, "Personal History Statement," Enemy Document Bulletin 1260, 15 October 1966, TTVA, Garnett Bell Collection, box 16, folder 28.

20. FBIS, 19 April 1966, KKK1, Radio Hanoi International, "LPA Quarterly Review of S. Vietnam Situation" (quote); MACV, *Command History, 1966*, 559.

21. FBIS, 23 March 1966, KKK1–2, Hue Domestic Service, "The Failures of the Military"; FBIS, 28 March 1966, KKK1, Hue to RVN, "Protests Greet U.S. Servicemen's Acts in Hue" (quote).

22. FBIS, 5 April 1966, KKK6 Saigon Domestic, "Thieu: Measures to Be Taken to Restore Order"; FBIS, 6 April 1966, JJJ5, RH to RVN, "Clearly Identifying the Enemy and Uniting in the Struggle" (quote).

23. FBIS, 4 March 1966, KKK1, LBS Radio to RVN, "Let Everybody Rise up and Ask for Salary Increase."

24. FBIS, 6 April 1966, KKKl, LBS to RVN, "Step up Further the Task of Revolutionary Proselyting among the Soldiers of the Puppet Armed Forces."

25. FBIS, 7 April 1966, JJJ6, Radio Hanoi International, "Fatherland Front Denounces U.S. Use of Gas."

26. 5th Special Forces Group, "Operational Report Lessons Learned for Period Ending 30 April 1966," 10, DTIC, AD391694.

27. Ibid., 12.

28. 5th Special Forces Group, "CIDG Civic Action/PSYOP Team Manual" (Nha Trang, 30 March 1966,) 67–69, USASOC Vietnam 1966–2, box 7; 5th Special Forces Group, "Operational Report Lessons Learned for Period Ending 30 April 1966," 11 (quote). 5th Special Forces Group, "Quarterly Command Report for the Period Ending 31 December 1965," inclosure 26, 155–156, DTIC, AD391693.

29. 5th Special Forces Group, "Operational Report Lessons Learned for Period Ending 30 April 1966," 12.

30. Ibid., inclosure 14, 8.

31. Ibid., 9.

32. Ibid.

33. Ibid., inclosure 23 to Section II.

34. Ibid., inclosure 24.

35. Andrew J. Birtle, "PROVN, Westmoreland, and the Historians: A Reappraisal," *Journal of Military History* 72, no. 4 (October 2008): 1213–1247; Eric Bergerud, *Dynamics of Defeat: The Vietnam War in Hau Nghia Province* (Boulder, CO: Westview Press, 1991), 112, 166; Office of the Deputy Chief of Staff for Military Operations, *A Program for the Pacification and*

Long-Term Development of South Vietnam (PROVN), US Army, 1 March 1966, vol. 1, 104 (quote), DTIC, AD3777743 (hereafter PROVN).

36. PROVN, 3–52.

37. Ibid., 3–53.

38. Directive Number 515-1, USMACV, Psychological Operations Political Warfare, 26 April 1966, 1 (quote); MACPD, "PSYOP Guide," 6 August 1967, 2, AEHC Library.

39. Bergerud, *Dynamics of Defeat*, 129.

40. Ibid., 112, 166 (quote); MACV, "Battlefield Reports—A Summary of Lessons Learned," 30 June 1966, 171, USASOC, Vietnam 1966–3, box 10.

41. Derrill De Heer, "Psychological Operations Conducted by the Australian Army in Phuoc Tuy Province South Vietnam: 1965–1971" (Master of Arts [Research] History, University of New South Wales School of Humanities and Social Science, Australian Defence Force Academy, 2009), 209.

42. Gregory Daddis, *No Sure Victory: Measuring U.S. Army Effectiveness and Progress in the Vietnam War* (Oxford, UK: Oxford University Press, 2011), 97.

43. Bergerud, *Dynamics of Defeat*, 168.

44. MACV, "Battlefield Reports—A Summary of Lessons Learned," 30 June 1966, 171.

45. FBIS, 11 May 1966, JJJ2, RH in English, "Nhan Dan on 'Mutiny' of U.S. Battalion."

46. FBIS, 2 March 1966, JJJ9, RH to American Servicemen, "Antiwar Rally speeches by Duncan, Scheer"; FBIS, 4 March 1966, JJJ7, RH to American Servicemen, "American POW urges Troops to Stop Fighting."

47. FBIS, 4 May 1966, JJJ5, RH to American Servicemen, "U.S. Pilot Message to American Troops."

48. FBIS, 13 May 1966, JJJ7, Tokyo *Akahata* (Red Flag) in Japanese, "Taped Statement by Captured U.S. Pilot."

49. FBIS, 17 May 1966, JJJ3–4, RH to American Servicemen, "Excerpts from Gen. Vuong Thua Vu Article."

50. Headquarters (HQ), 1st Cavalry Division, "Combat Operations After Action Report, Operation Davy Crockett," 25 May 1966, 2, USASOC, Vietnam 1966–3, box 10; Sophie Quinn-Judge, *Ho Chi Minh: The Missing Years, 1919–1941* (Berkeley: University of California Press, 2002), 22.

51. HQ, 1st Cavalry Division, "Critique of Operation Crazy Horse," 27 June 1966, 1, USASOC, 1966–3, box 10.

52. MACV, *Command History, 1966*, 569.

53. 1st Cavalry Division, "Operation Davy Crockett," 4, 14 (quote); 1st Cavalry Division, "Operation Crazy Horse," 12; MACV, "Battlefield Reports—A Summary of Lessons Learned," 30 June 1966, 169, USASOC, Vietnam 1966–3, box 10.

54. Headquarters I Field Force Vietnam, "Operational Report for Quarterly Period Ending 31 July 1966," 25 August 1966, Vietnam 1966–2, box 7, 4 (quote); MACV, *Command History, 1966*, 556.

55. MACV, *Command History, 1966*, 555.

56. "VC Document on Antiwar Movement," US State Department release of Viet Cong Foreign Affairs and Propaganda Activities captured document, 4 April 1967, 9, 10, 11, 17, 19, TTVA, DPC: Unit 3—Antiwar Activities, box 5, folder 14; TTVA, DPC: Unit 3—Antiwar Activities; *CANLF Bulletin*, 9 May 1967, 21.

57. "VC Document on Antiwar Movement," 10, 12.

58. Tran Cong Tuong and Phan Thanh Vinh, *The Failure of the Special War, 1961–65* (Hanoi: Foreign Languages Publishing House, 1967), 63; *The N.F.L. Symbol of Independence,*

Democracy and Peace in South Viet Nam (Hanoi: Foreign Languages Publishing House, 1967), 20.

59. *CANLF Bulletin*, 9 June 1966, 18.

60. *CANLF Bulletin*, 19 March 1966, 18.

61. FBIS, 5 July 1966, KKK2, Radio Hanoi International, "War Crimes Dossier"; Harish C. Mehta, "North Vietnam's Informal Diplomacy with Bertrand Russell: Peace Activism and the International War Crimes Tribunal," *Peace & Change* 37, no. 1 (January 201): 64–94.

62. FBIS, 14 June 1966, JJJ7, RH in English, "Speech by Bertrand Russell to American Servicemen in South Vietnam."

63. FBIS, 21 June 1966, JJJ1–4, RH in English, "Appeal to U.S. Servicemen" (quote); Bertrand Russell, *Autobiography of Bertrand Russell* (Boston: Little, Brown, 1967), 666.

64. FBIS, 9 June 1966, JJJ3–4, "Viet Cong POWs Said to Have Been Tortured."

65. FBIS, 8 June 1966, JJJ4, Radio Hanoi International, "DRV Protests W. German Aid for Vietnam War."

66. FBIS, 7 June 1966, KKK2, Hue Voice of Buddhist Salvation, "Thich Tri Quang Appeal."

67. FBIS, 10 June 1966, KKK4, Hue VOBS, "U.S.-Thieu-Ky 'Lackeys' Destroy Buddha Altars."

68. FBIS, 14 June 1966, KKK4, LBS, "Disturbances in Hue."

69. FBIS, 19 July 1966, KKK2, Voice of Anticommunist Buddhist Force (clandestine), "Political Motives Harm Buddhists' Cause."

70. FBIS, 21 June 1966, KKK1, Radio Hanoi International, "NFLSV Flays Attacks on North Vietnam Dikes."

71. FBIS, 30 June 1966, JJJ2–3, Radio Hanoi, "U.S. Pilot's Recorded Statement, Cpt. Murphy Neal Jones."

72. FBIS, 6 July 1966, JJJ2, Peking NCNA International, "Atmosphere of Preparedness Prevails in Hanoi."

73. FBIS, 7 July 1966, JJJ2, Radio Hanoi International, "Population's Anger."

74. JUSPAO, "Impact of Allied PSYOP on North Vietnam," Saigon, 17 August 1966, TTVA, DPC: Unit 3—Insurgency Warfare, box 13, folder 8.

75. MACV, *Command History, 1966*, 570.

76. JUSPAO, "Impact of Allied PSYOP on North Vietnam," 5.

77. Ibid., 3.

78. Ibid., 4.

79. CDEC, "PSYWAR Tricks of the American Aggressors," *Nhan Dan*, 18 July 1966, 2, TTVA, VAC, box 136, folder 1616.

80. Ibid., 2.

81. CDEC, "Chief Justice of the Supreme People's Court," 4 May 1966 (quote), TTVA, VAC, box 124, folder 2263; Martin Grossheim, "Fraternal Support: The East German 'Stasi' and the Democratic Republic of Vietnam during the Vietnam War," Cold War International History Project, CWIHP Working Paper No. 71 (September 2014).

82. Duke University Study Team, "Report on the Psychological Operations Department of the Special Warfare School," Army Research Office, Durham, NC, 1 July 1966, 17, USASOC, Avedon Papers, box 9.

83. Headquarters, 6th Psychological Operations Battalion, "Operational Report Lessons Learned," 15 August 1966, 2; Holley, "Report on the Psychological Operations Department of the Special Warfare School," 17.

84. Headquarters, 6th Psychological Operations Battalion, "Operational Report for Quarterly Period Ending 31 October 1966," 17 January 1967, 3; 6th Psychological Operations Battalion, "Operational Report Lessons Learned," 15 August 1966, 2–5, DTIC, AD394030.

85. Department of the Army, *Field Manual 33-5, Psychological Operations, Techniques and Procedures* (Washington, DC: Department of the Army, October 1966), 137–140.

86. De Heer, "Psychological Operations Conducted by the Australian Army in Phuoc Tuy Province South Vietnam: 1965–1971," 50.

87. 6th Psychological Operations Battalion, "Operational Report Lessons Learned, Period 1 May–31 Jul 1966," 5.

88. Headquarters I Field Force Vietnam, "Operational Report for Quarterly Period Ending 31 July 1966," 22; Kenneth Sams, *Operation Hawthorne (Project CHECO Southeast Asia Report)* (Hickam AFB, HI: Department of the Air Force, Headquarters Pacific Air Forces, CHECO/Corona Harvest Division, Operations Analysis Office, 8 September 1966), 20.

89. Headquarters, 1st Bde., 101st Abn., "Combat Operations After Action Report, Operation Beauregard," 10 August 1966, 23, DTIC, AD394091.

90. Headquarters, 1st Bde., 101st Abn., "Combat Operations AAR, Operation John Paul Jones," 28, DTIC, AD388882.

91. 1st Infantry Division, "Annex J (PsyOp/Civil Affairs) to Field SOP," 31 July 1966, J-1, USASOC, PSYOP Papers 1961–1968, box 3.

92. Ibid.

93. Ibid.; Headquarters I Field Force Vietnam, "Operational Report for Quarterly Period Ending 31 July 1966," 21.

94. Ibid., 55.

95. Ibid., 52.

96. Ibid., 50 (quote 1), 56; MACV, *Command History, 1966*, 556; Headquarters 6th Psychological Operations Battalion, "Operational Report for Quarterly Period Ending 31 October 1966," 17 January 1967, 3 (quote 2).

97. Headquarters 6th Psychological Operations Battalion, "Operational Report for Quarterly Period Ending 31 October 1966," 17 January 1967, 1, DTIC 393030; "Briefing on the 6th Psychological Operations Battalion," n.d., n.p., USASOC, PSYOP Papers 0000–1944, box 1.

98. MACV, *Command History, 1966*, 562 (quote), 563.

99. 5th Special Forces Group, *Monthly Operational Summary for Period 1–31 August 1966,* 15 September 1966, CIDG program, 31, 39 (quote), 61, 62, USASOC, Vietnam 1966–3, box 10.

100. Ibid., 6 (quote), 11, 24.

101. William D. Parker, *U.S. Marine Corps Civil Affairs in I Corps, Republic of Vietnam: April 1966 to April 1967* (Washington, DC: Historical Division, Headquarters, US Marine Corps, 1970), 46.

102. FBIS, 20 July 1966, KKK2–3, LBS to RVN, "Let Us Overcome All Obstacles and Positively Produce Food."

103. Headquarters I Field Force Vietnam, "Operational Report for 31 July 1966," 24.

104. FBIS, 8 August 1966, JJJ9–10, RH to American Servicemen, "Petition of Two American Pilots to U.S. Government"; The POW Network, "Rivers, Wendell Burke," www.pownetwork.org/bios/r/r040.htm.

105. FBIS, 17 August 1966, JJJ5–7, RH in English to American Servicemen on 14 August.

106. FBIS, 15 August 1966, JJJ5, RH to RVN, "The Americans Escalated the War and Practiced Deception."

107. CDEC, Translation Report, 30 September 1966, "Entertainment Plan of Tay Ninh Province from 20 July to December 1966," TTVA, Garnett Bell Collection, box 10, folder 2.

108. CDEC, Translation Worksheet: "Notebook of a VC PSYWAR Entertainment Cadre," 13 September 1966, TTVA, VAC, box 12, folder 2.

109. CDEC, "Rallier Interrogation Report, Huyen Thi Phen," 1 March 1966, TTVA, VAC, box 24, folder 2346.

110. Andrew R. Molnar, *Human Factors Considerations of Undergrounds in Insurgencies* (Washington, DC: Department of the Army, September 1966), 84, 85 (quote).

111. Bergerud, *Dynamics of Defeat,* 158; MACV, *Command History, 1966*, 564.

112. Simulmatics Corporation, "Improving the Effectiveness of the Chieu Hoi Program: A Summary," September 1967, 15, 21, LBJL, box 239, Simulmatics, Chieu Hoi.

113. CDEC, Translation Report: "Resolution Adopted by Quang Tri, Thua Thien Military Region Party Committee," 19 June 1967, 1–12, 11 (quote), TTVA, Garnett Bell Collection, box 26, folder 13.

114. FBIS, 5 August 1966, KKK4, Saigon Domestic, "Ky Opens Psychological Warfare Conference" (quote 1); Rufus Phillips, *Why Vietnam Matters: An Eyewitness Account of Lessons Not Learned* (Annapolis, MD: Naval Institute Press, 2008), 264 (quote 2); 5th Special Forces Group, *Operational Report Lessons Learned for the Period Ending 31 October 1966*, 19, DTIC 393837.

115. FBIS, 18 August 1966, KKK4, Radio Hanoi, "Other Comment"; Phillips, *Why Vietnam Matters*, 264.

116. FBIS, 12 September 1966, KKK3, Saigon Domestic, "Election Turnout Heavy Throughout Nation."

117. HQ, II Field Force Vietnam, *Operational Report for Quarterly Period Ending 31 October 1966*, 10, DTIC, AD512784; MACV, *Command History, 1966*, 570.

118. FBIS, 13 September 1966, JJJl, RH to RVN, "In Response Claim People Forced to Vote."

119. FBIS, 13 September 1966, KKK5, LBS to RVN, "Shameful Failure of Fraud," and HHHH1, Havana Domestic, "S. Vietnamese Balloting Called Farce."

120. FBIS, 16 September 1966, JJJ2–3, RH to RVN, "Determination to Exterminate the U.S. Aggressors Who Are Stepping up the War with Toxic Gas."

10. Growing Pains, Fall 1966–May 1967

1. Headquarters, II Field Force Vietnam, "Operational Report for Quarterly Period Ending 31 October 1966," 12.

2. Frank Newport and Joseph Carroll, "Iraq Versus Vietnam: A Comparison of Public Opinion," www.gallup.com/poll/18097/Iraq-Versus-Vietnam-Comparison-Public-Opinion.aspx; Paul Burstein and William Freudenburg, "Changing Public Policy: The Impact of Public Opinion, Antiwar Demonstrations, and War Costs on Senate Voting on Vietnam War Motions," *American Journal of Sociology* 84, no. 1 (July 1978): 9–12; E. M. Schreiber, "Anti-War Demonstrations and American Public Opinion on the War in Vietnam," *British Journal of Sociology* 27, no. 2 (June 1976): 225–236, 227 (quote); American Institute for Political Communications, "The Vietnam Issue: University Student Attitudes and the 1966 Congressional Elections" (Washington, DC: AIPC, 1966).

3. Department of the Army, *Field Manual 33-5, Psychological Operations, Techniques and Procedures, 1966, Operations* (Washington, DC: Department of the Army, 20 October 1966), 109–110.

4. Headquarters, 6th Psychological Operations Battalion, "Operational Report for Quarterly Period Ending 31 October 1966," 1, AEHC, MACV Command Histories Collection, box 32; MACV, *MACV Directive Number 10-1: Organization and Functions, Psychological Operations* (Saigon: MACPD, 29 September 1966), 3, 410, USASOC, PSYOP Papers 1961–1968, box 3.

5. 15th Psychological Operations Det. (STRAT), 7th Psychological Operations Group, "Report on Psychological Operations Intelligence in Vietnam," 17 November 1967, 2, TTVA, DPC: Unit 3—Insurgency Warfare, box 14, folder 1.

6. Ibid., 1 (quote); HQ, 6th Psychological Operations Battalion, "Operational Report for Quarterly Period Ending 31 October 1966," 5.

7. 15th Psychological Operations Det., "Report on Psychological Operations Intelligence in Vietnam," 5.

8. Ibid., 13; 6th Psychological Operations Battalion, "Operational Report for Quarterly Period Ending 17 January 1967," 6.

9. 6th Psychological Operations Battalion, "Operational Report for Quarterly Period Ending 31 October 1966," 5 (quote), 6.

10. 6th Psychological Operations Battalion, "Operational Report for Quarterly Period Ending 31 January 1967," 2–5, 3 (quote); AEHC, MACV Command Histories, 6th PSYOP Battalion.

11. 6th Psychological Operations Battalion, "Operational Report for Quarterly Period Ending 31 October 1966," 9, 11 (quote); II Field Force Vietnam, "Operational Report for Quarterly Period Ending 31 October 1966," 12.

12. 6th Psychological Operations Battalion, "Operational Report for Quarterly Period Ending 31 October 1966," 11.

13. JUSPAO Planning Office, "Long An Province Survey—1966," October 1966, 1–3, TTVA, John B. O'Donnell Collection, box 1, folder 19.

14. Ibid., 5, 22.

15. Ibid., 5.

16. 6th Psychological Operations Battalion, "Operational Report for Quarterly Period Ending 31 October 1966," 5.

17. Ibid.; II Field Force Vietnam, "Operational Report for the Period Ending 31 October 1966," 27. Rallier figures provided in the I Field Force report for the period contain much higher numbers.

18. 6th Psychological Operations Battalion, "Operational Report for Quarterly Period Ending 31 October 1966," 5, 8 (quote).

19. JUSPAO Planning Office, *PSYOPS in Vietnam: Indications of Effectiveness* (Saigon: n.p., 1967), 75.

20. Ibid., 76.

21. "Report of the Internal Defense/Development Psychological Operations Instructors Conference," Fort Bragg, 31 October–4 November 1966, 252, USASOC, Avedon Collection, box 10.

22. Richard A. Hunt, *Pacification: The American Struggle for Vietnam's Hearts and Minds* (Boulder, CO: Westview, 1995), 61.

23. CDEC, "Subject: COSVN Survey on Counter Propaganda, 3 March 1967," TTVA, VAC, box 63, folder 1981.

24. Headquarters, 6th Psychological Operations Battalion, "Operational Report for Quarterly Period Ending 31 January 1967," 7, AEHC, MACV Command History box, folder 6th PSYOP Battalion.

25. Ibid.

26. FBIS, 21 November 1966, JJJ2, RH to American Servicemen, "Bertrand Russell to American Negro Soldiers in Vietnam."

27. FBIS, 25 November 1966, KKK1, Radio Hanoi International in English, "Liberation Radio Hails War Crimes Tribunal."

28. FBIS, 21 October 1966, JJJ6, RH to RVN, "U.S. Pilots Refuse to Go on Bombing Missions."

29. 6th Psychological Operations Battalion, "Operational Report for Quarterly Period Ending 31 January 1967," 7.

30. FBIS, 21 October 1966, JJJ5, RH in English, "Denton Confesses to Bombing Innocent People."

31. CDEC, Translation, "Notes on COSVN Current Affairs Committee Briefing," taken from VC Cadre notebook, 1, TTVA, DPC: Unit 2—Military Operations, box 8, folder 11.

32. Ibid., 2.

33. FBIS, 9 December 1966, KKK4–5, LBS to RVN, "Winter Spring Production Conference Held."

34. MACV, *Command History, United States Military Assistance Command, Vietnam: 1966* (Saigon: MACV Military History Branch, 19 April 1967), 559; J. A. Koch, *The Chieu Hoi Program in South Vietnam, 1963–1971* (Washington, DC: Rand Corporation for ARPA, January 1973), 28.

35. Koch, *Chieu Hoi Program*, 65.

36. William Hoffer, "Broadcasting in an Insurgency Environment: USIA in Vietnam, 1965–1970" (PhD diss., University of Wisconsin, 1972), 362.

37. MACV, *Command History, 1966*, 572.

38. Headquarters, 6th Psychological Operations Battalion, "Operational Report for Quarterly Period Ending 31 January 1967," 3; "Briefing on the 6th Psychological Operations Battalion," n.d., n.p., USASOC, PSYOP Papers 0000–1944, box 1.

39. *Ψ[psy] Observer* 1, 1 November 1966, 1, USASOC, PSYOP Papers 1961–1968, box 3 (newspaper of the 6th Psychological Operations Battalion).

40. FBIS, 16 December 1966, KKKl, LBS to SVN, "LAF Command Hails Attack on Tan Son Nhut."

41. "Battalion Hit by VC Terrorist Attack," *Ψ[psy] Observer* 2, 1 January 1967, 1; 6th Psychological Operations Battalion, "Operational Report," 27 February 1967, 1; 6th Psychological Operations Battalion, "Operational Report for Quarterly Period Ending 31 January 1967," 3.

42. Briefing on the 6th Psychological Operations Battalion, n.p.

43. MACV, *Counterinsurgency Lessons Learned No. 61: Salient Lessons Learned,* 27 January 1967, 12, BACM Research, paperlessarchives.com.

44. JUSPAO, *PSYOPS in Vietnam*, v.

45. Ibid., 16.

46. Ibid., 32.

47. Ibid., 36.

48. H. R. McMaster, *Dereliction of Duty: Lyndon Johnson, Robert McNamara, the Joint Chiefs of Staff, and the Lies That Led to Vietnam* (New York: HarperCollins, 1997), 235; Wesley R. C. Melyan and Lee Bonettii, *Rolling Thunder: July 1965–December 1966* (Hickam AFB, HI: CHECO/Corona Harvest Division, Operations Analysis Office, Department of the Air Force, Headquarters, Pacific Air Forces, 15 July 1967), 15–16, DTIC, ADA486572. Ball and McNamara had been on the Strategic Bombing Survey team that found increased civilian resolve from World War II strategic bombing.

49. United States Military Assistance Command, *Command History Vietnam: 1967* (Saigon: USMACV Military History Branch, 16 September 1968), 642.

50. Melyan and Bonettii, *Rolling Thunder*, 3–4 (quote); MACV, *Command History, 1967*, 657.

51. Ilya V. Gaiduk, *The Soviet Union and the Vietnam War* (Chicago: Ivan R. Dee, 1996), 79.

52. MACV, *Command History, 1967*, vol. 2, 657.

53. FBIS, 26 January 1967, JJJ1, Radio Hanoi International, "Report on Theft of USSR Missiles Is Slander."

54. Richard H. Shultz, *The Secret War against Hanoi: Kennedy's and Johnson's Use of Spies, Saboteurs, and Covert Warriors in North Vietnam* (New York: HarperCollins, 1999), 118, MACSOG Documentation Study, annex A to appendix C, Psychological Operations (Washington, DC: Office of the Secretary of Defense, 1970), 20–25.

55. Shultz, *The Secret War,* 114 (quote); John L. Plaster, *SOG: The Secret Wars of America's Commandos in Vietnam* (New York: Simon & Schuster, 1997), 118.

56. Robert M. Gillespie, *Black Ops, Vietnam: An Operational History of MACVSOG* (Annapolis, MD: Naval Institute Press, 2011), 68.

57. FBIS, 1 February 1967, JJJ7, Radio Hanoi International, "American Woman Says Vietnamese Morale High."

58. FBIS 28 February 1967, JJJ4, Radio Hanoi International, "More Americans Condemn U.S. Vietnam Policy."

59. D. J. Kinney, "Selling Greenland: The Big Picture Television Series and the Army's Bid for Relevance during the Early Cold War," *Centaurus* 55 (2013): 346 (quote), 349–350.

60. Based on analysis of TV listings for the decade from the *Dallas Morning News.*

61. The Army Pictorial Center website, "Catalog of the Big Picture," www.armypictorialcenter.com/Catalog_of_the_Big_Picture.htm.

62. USIS, Vietnam Feature Service, *Operation Cedar Falls . . . Out of the Iron Triangle,* 1, TTVA, DPC: Unit 2—Military Operations, box 8, folder 7.

63. Ibid., 3, 4 (quote).

64. Ibid., 5–16.

65. MACV, *Command History, 1967,* 656; FBIS, 20 February 1967, KKK2, LBS to RVN, "NLF Central Committee Statement on Tet Truce" (quote).

66. William D. Parker, *U.S. Marine Corps Civil Affairs in I Corps, Republic of Vietnam: April 1966 to April 1967* (Washington, DC: Historical Division, Headquarters, US Marine Corps, 1970), 68.

67. Headquarters, 6th Psychological Operations Battalion, "Operational Report for Quarterly Period Ending 30 April 1967," 19 May 1967, 22.

68. JUSPAO, Field Development Division, "Communicating with Vietnamese through Leaflets," Saigon: RVN, March 1967, 1, 5, 32; 6th Psychological Operations Battalion, "Operational Report for Quarterly Period Ending 30 April 1967," 1, 6 (quote), TTVA, DPC: Unit 3—Insurgency Warfare, box 13, folder 11.

69. William B. Sellen to Commander, 5th Air Commando Squadron, "SUBJECT: Favorable Times for Loudspeaker Employment," 6th Psychological Operations Battalion, 3 February 1967, USASOC, Avedon Collection, box 9.

70. 6th Psychological Operations Battalion, "Operational Report for Quarterly Period Ending 31 January 1967," 3.

71. Ibid., 5–7, 7 (quote).

72. Ibid., "Operational Report for Quarterly Period Ending 30 April 1967," 7.

73. Parker, *U.S. Marine Corps Civil Affairs in I Corps, Republic of Vietnam*, 94.

74. 6th Psychological Operations Battalion, "Operational Report for Quarterly Period Ending 30 April 1967," 6.

75. FBIS, 11 January 1967, KKK1, Seoul Domestic, "ROK Plans Radio Stations in South Vietnam"; 6th Psychological Operations Battalion, "Operational Report for Quarterly Period Ending 30 April 1967," 6.

76. Michael E. Haas, *Apollo's Warriors: United States Air Force Special Operations during the Cold War* (Maxwell AFB, AL: Air University Press, 1997), 288; US Air Force, "9th Special Operations Squadron," www.hurlburt.af.mil/AboutUs/FactSheets/FactSheets/tabid/140/Article/204527/9th-special-operations-squadron.aspx; 4th Psychological Operations Group,

"Operational Report for Quarterly Period Ending 31 January 1968," 3 April 1968, 26, AEHC, MACV Command History, 6th PSYOP Battalion; James N. Peale, "Victory over Viet Cong Terror," 73, unpublished manuscript, AEHC, Peale Papers.

77. MACV, *Command History, 1967*, 664–667, 668–669 (quote).

78. MACV, *Command History, 1967*, 688.

79. Interview with William Shelton conducted by Steve Maxner, Vietnam Archive Oral History Project, 5 June 2000, 105–106, TTVA.

80. 6th Psychological Operations Battalion, "Weekly Activity Report (16–22 June 1967)," 22 June 1967, USASOC, PSYOP Papers, 61–68, box 3–3, folder 54.

81. JUSPAO Planning Office, Policy Number 36, "The Use of Superstitions in Psychological Operations in Vietnam," Saigon, RVN, 10 May 1967, 75–76.

82. CDEC, "Translation of Document Captured April 2, 1967 in Operation Lam Son 67, Countering of 'Open Arms' Activities to Be Increased," 1, TTVA, DPC: Unit 2—Military Operations, box 8, folder 5.

83. Ibid., 2.

84. Ibid., 3.

85. US Mission, Saigon, "VC Increase Action Against Chieu Hoi Program," Press Release, 4 May 1967, TTVA, DPC: Unit 1—Assessment and Strategy, box 8, folder 4.

86. MACV, *Command History, 1967*, 656.

87. Simulmatics Corporation, "Improving the Effectiveness of the Chieu Hoi Program: A Summary," September 1967, 2, 4 (quote), 5, 6, LBJL, box 239, Simulmatics, Chieu Hoi.

88. Ibid., 2, 4, 5, 6 (quote).

89. David W. P. Elliott, *The Vietnamese War: Revolution and Social Change in the Mekong Delta, 1930–1975* (Armonk, NY: M. E. Sharpe, 2003), 788.

90. Lansdale to McDonald, "Subject: VC Taxes," 2 July 1967, LBJL, NSF Country File Vietnam, box 107, Lansdale.

91. CDEC, "Counter Measures against Enemy PSYWAR and Chieu Hoi Activity," Liberation Armed Forces Political Department, n.d., TTVA, VAC, box 63, folder 1974.

92. Combined Intelligence Center Vietnam (CIVC), "VC Propaganda Factbook," May 1967, 14, TTVA, VAC, box 24, folder 779.

93. Ibid., 25.

11. Measuring Effects, June–Fall 1967

1. United States Military Assistance Command, *Command History Vietnam: 1967* (Saigon: USMACV Military History Branch, 16 September 1968), 669–670.

2. William Hoffer, "Broadcasting in an Insurgency Environment: USIA in Vietnam, 1965–1970" (PhD diss., University of Wisconsin, 1972), 327–334 (quote), 334; James N. Peale, "Victory over Viet Cong Terror," 73, unpublished manuscript, AEHC, Peale Papers, 95–97.

3. Peter Arnett, "Map Error Makes Saigon Psywar Faces Red," *Washington Post*, 22 April 1967, TTVA, DPC: Unit 2—Military Operations, box 8, folder 14.

4. MACV, *Command History, 1967*, 670–672.

5. Hoffer, "Broadcasting in an Insurgency," 327.

6. Opinion Research Corporation, "The People of South Vietnam: How They Feel about the War," CBS Public Opinion Survey, March 1967, 1–15, TTVA, DPC: Unit 11—Monographs, box 8, folder 15.

7. JUSPAO, Field Memorandum 42, "Lessons Learned from Evaluation of Allied PSYOP Media in Viet Nam" (Saigon: JUSPAO, 13 December 1967), 8, TTVA, DPC: Unit 3—Insurgency Warfare, box 13, folder 14.

8. CDEC, CMIC interview of Nguyen Van Duc, March 1967, 3, TTVA, VAC, box 166, folder 2098.

9. William D. Parker, *U.S. Marine Corps Civil Affairs in I Corps, Republic of Vietnam: April 1966 to April 1967* (Washington, DC: Historical Division, Headquarters, US Marine Corps, 1970), 93.

10. 6th Psychological Operations Battalion, "Operational Report for Quarterly Period Ending 30 April 1967," 7, DTIC, AD387091.

11. Ibid., 15. Production time comparison based on the author's experience in Afghanistan.

12. MACV, *Command History, 1967*, 643.

13. Ibid.

14. 6th Psychological Operations Battalion, "Operational Report for Quarterly Period Ending 30 April 1967," 23.

15. Ibid., 7.

16. Ibid., 22.

17. JUSPAO, Guidance 38, "GVN National Reconciliation Policy (Doan Ket)," Saigon, 19 April 1967, 85–86.

18. CDEC, "VC Desertion as a Result of Chieu Hoi Program in MR IV," 27 June 1967, TTVA, VAC, box 160, folder 236.

19. MACV, *Command History, 1967*, 687; Parker, *U.S. Marine Corps*, 107.

20. Combined Intelligence Center Vietnam (CIVC), "VC Propaganda Factbook," May 1967, 3, TTVA, VAC, box 24, folder 779.

21. FBIS, 17 April 1967, KKK1 LBS to RVN, "Plot to Sabotage Economy Must Be Smashed"; Advanced Battle Analysis, "Operation Junction City Vietnam 1967," 28, US Army Command and General Staff College, 1983, DTIC, AD139612.

22. FBIS, 5 April 1967, KKK1 LBS to RVN, "LAF Urged to Destroy Pacification Groups."

23. CICV, "VC Propaganda Factbook," 28.

24. *CANLF Bulletin*, 4 February 1967, "Progress Report," 13, www.redandgreen.org/CANLF/Index.htm.

25. *CANLF Bulletin*, 9 May 1967, "Progress Report," 21.

26. Ibid.

27. FBIS, 13 April 1967, JJJ2, Radio Hanoi International, "Pham Van Dong Thanks Americans for Telegram."

28. FBIS, 17 April 1967, JJJ2, Radio Hanoi International, "Nhan Dan Hails Antiwar Demonstrations in U.S."

29. 6th Psychological Operations Battalion, "Operational Report for Period Ending 31 January 1967," 7 (quote); 4th Infantry Division, "ORLL (RCS CSFOR-65) for Quarterly Period Ending 30 April 1967," 15 June 1967, 5, DTIC, AD385851; CICV, "VC Propaganda Factbook," 28.

30. CICV, "VC Propaganda Factbook," 25.

31. Ibid., 26–27.

32. FBIS, 5 May 1967, Radio Hanoi International, "GI's Maim Kill S. Vietnamese Troops."

33. FBIS, 9 May 1967, JJJ8, Radio Hanoi International, "U.S. Crimes Sadden GI in South Vietnam."

34. FBIS, 10 May 1967, JJJ6, "Captured Pilot Hails Antiwar Struggle in U.S."

35. CICV, "VC Propaganda Factbook," 3, 5.

36. Ibid., 9, 10 (quote).

37. Ibid., 17.

38. Ibid., 19; MACSOG Documentation Study, annex A to appendix C, Psychological Operations (Washington, DC: Office of the Secretary of Defense, 1970), 65, 66.

39. Rostow to the President, "Project Compatriot," 22 May 1967, 2, LBJL, box 88, Compatriot, folder 6.

40. Ibid., 1–12, 10 (quote).

41. George S. Eckhardt, *Vietnam Studies: Command and Control, 1950–1969* (Washington, DC: Department of the Army, 1991), 69–70; BDM Corporation, *Study of Strategic Lessons Learned in Vietnam*, vol. 6: *Conduct of the War*, book 2: *Functional Analysis* (McLean, VA: BDM Corporation, 2 May 1980), 13–9; Richard A. Hunt, *Pacification: The American Struggle for Vietnam's Hearts and Minds* (Boulder, CO: Westview, 1995), 75; JUSPAO, *General Briefing Book*, 41 (quote), 44, 45; MACV, *Command History, 1967*, 635.

42. 6th Psychological Operations Battalion, "Operational Report for Period Ending 30 April 1967," 9.

43. Ibid., 4.

44. Ibid., 9–10, 20.

45. Ibid., 5.

46. Ibid., 11.

47. Ibid., 9–10, 20; 4th Infantry Division, "ORLL (RCS CSFOR-65) for Quarterly Period Ending 30 April 1967," 9 (quote), DTIC, AD383351.

48. 6th Psychological Operations Battalion, "Operational Report for Period Ending 30 April 1967," 12–13; CDEC, "Results of PSYWAR Activities Conducted by RAF troops," 16 June 1967 (quote), TTVA, VAC, box 142, folder 984.

49. CDEC, "Subject: Ba Bien Prov. Youth Group," TTVA, VAC, box 161, folder 1224.

50. FBIS, 6 June 1967, KKK1, Radio Hanoi International, "Australian Troops Denounced for Using Gas."

51. CDEC, "Subject: District in Long An Province," 20 June 1967, TTVA, VAC, box 161, folder 2104.

52. CDEC, "Subject: Tuy Hoa Dist., Phu Yen Province," 14 July 1967, TTVA, VAC, box 167, folder 175.

53. 4th Infantry Division, "ORLL (RCS CSFOR-65) for Quarterly Period Ending 30 April 1967," 8.

54. Ibid., 17, 19, 41 (quote).

55. Ibid., 2, 4, 19.

56. Thomas L. Ahern Jr., *Vietnam Declassified: The CIA and Counterinsurgency* (Lexington: University Press of Kentucky, 2010), 248.

57. Ibid., 249.

58. CDEC, "Subject: Thu Dau Mot Prov Unit, Circular 001/CTBV," 8 July 1967, TTVA, VAC, box 163, folder 1336.

59. CDEC, "Subject: Thu Dau Mot Prov Unit, Circular 40/CV-U," 8 July 1967, TTVA, VAC, box 163, folder 1341.

60. Walt Rostow to President Johnson, "South Vietnam Nationwide Hamlet Public Reaction Survey," 30 November 1967, 1–2, TTVA, Veteran Members of the 109th Quartermaster Company (Air Delivery) Collection, box 2, folder 25.

61. Ibid., 4–5.

62. USIA, Memorandum for Arthur MacCaferty, White House, "Nationwide Hamlet Public Reaction Survey," 5 December 1967, 4–8, 7 (quote), LBJL, box 151, HES Categories, Survey 1967.

63. Rostow to Johnson, "South Vietnam Nationwide Hamlet Public Reaction Survey," 6.

64. MACV, *Command History, 1967*, 680.

65. Mark Moyar and Harry G. Summers, *Phoenix and the Birds of Prey: The CIA's Secret Campaign to Destroy the Viet Cong* (Annapolis, MD: Naval Institute Press, 1997), 308.

66. Ibid., 309.

67. MACV, *Command History, 1967*, 660.

68. FBIS, 15 June 1967, KKK1, "NFLSV to Try Americans If Patriots Executed."

69. FBIS, 18 July 1967, JJJ2, Radio Hanoi International, "Foreign Visitors Inspect Bombed Dike Section."

70. FBIS, 5 July 1967, JJJ1, Radio Hanoi International, "Conservancy Ministry Flays Raids on Dikes."

71. FBIS, 24 July 1967, JJJ4, Radio Hanoi International, "U.S. Faces More Difficulties in S. Vietnam."

72. FBIS, 27 June 1967, JJJ2, RH to RVN, "Large Numbers of Puppet Troops Annihilated and Disintegrated during 1966–67."

73. JUSPAO, Consolidation of JUSPAO Guidance 1–22, "Excerpts from *Tuyen Huan,* June 1967," Saigon, RVN, June 1967, 82–84.

74. Ibid.

75. Lien-Hang T. Nguyen, *Hanoi's War: An International History of the War for Peace in Vietnam* (Chapel Hill: University of North Carolina Press, 2012), 106.

76. Ibid., 102.

77. JUSPAO, "Excerpts from *Tuyen Huan,* June 1967," 82–84, 86 (quote).

78. John L. Plaster, *SOG: The Secret Wars of America's Commandos in Vietnam* (New York: Simon & Schuster, 1997), 125; Robert M. Gillespie, *Black Ops, Vietnam: An Operational History of MACVSOG* (Annapolis, MD: Naval Institute Press, 2011), 103 (quote). The Psywarrior website has excellent coverage of the use of banknote leaflets: "Propaganda Banknotes of the Vietnam War," www.psywarrior.com/Vietnambanknote.html.

79. Plaster, *SOG*, 125; Gillespie, *Black Ops, Vietnam*, 104 (quote), 107. Interview with William Shelton conducted by Steve Maxner, Vietnam Archive Oral History Project, 5 June 2000, TTVA; MACSOG Documentation Study, 91–93.

80. Richard H. Shultz, *The Secret War against Hanoi: Kennedy's and Johnson's Use of Spies, Saboteurs, and Covert Warriors in North Vietnam* (New York: HarperCollins, 1999), 158 (quote).

81. MACV-CORDS Civil Operations and Revolutionary Development Support, "SUBJECT: Psychological Operations Division," 29 June 1967, 3–4, TTVA, US Marine Corps History Division Vietnam War Documents Collection, folder 65.

82. 6th Psychological Operations Battalion, "Weekly Activities Report (16–22 June 1967)," USASOC, PSYOP Papers 1961–1968, box 2, folder 2.

83. 6th Psychological Operations Battalion, "Operational Report for Period Ending 31 October 1967," 15 November 1967, 3–4, AEHC.

84. 6th Psychological Operations Battalion, "Weekly Activities Report (16–22 June 1967)."

85. Government Accountability Office, Henry Eschwege, "Use of Agent Orange in Vietnam," 16 August 1978, www.gao.gov/products/CED-78-158 ; JUSPAO, Policy Guidance 31, "PSYOP Aspects of Defoliation," Saigon, RVN, February 1967.

86. JUSPAO, Policy Guidance 39, "Viet Cong Cadre Vulnerabilities," Saigon, RVN, 30 June 1967, 1–7.

12. The 4th PSYOP Group and the Tet Offensive, Fall 1967

1. FBIS, 30 August 1967, KKK4, Radio Hanoi International, "U.S., Puppet Troops Clash in Quang Tri."

2. FBIS, 10 August 1967, JJJ3, RH to RVN, "Shifted to Pacification Work, the Puppet Troops Have Disintegrated More Rapidly"; "Indochina War Statistics, Dollars and Deaths—Congressional Record, May 15, 1975," TTVA, DPC: Unit 3—Statistical Data, box 44, folder 9.

3. Thomas C. Thayar, *A Systems Analysis View of the Vietnam War, 1965–1972*, vol. 7: *The Republic of Vietnam Armed Forces*, n.p., 18–19, DTIC, AD051612.

4. Robert Ginsburgh, Memorandum for Mr. Rostow, "Some Miscellaneous Items of Interest from Recent DIA Briefing," 19 September 1967, LBJL, NSF Vietnam File, box 90, Viet Cong Offensive.

5. CIA, "Intelligence Information Cable, Summary of Viet Cong Activities in Phong Dinh Province," 4 (quote), 5, LBJL, NSF Vietnam Files, box 90, Viet Cong Offensive.

6. Ginsburgh to Rostow, "Some Miscellaneous Items," 1–2.

7. Konrad Kellen, *A View of the VC: Elements of Cohesion in the Enemy Camp* (Santa Monica, CA: Rand Corporation, October 1968), v, LBJL, NSF Vietnam, box 89, NLF Political.

8. FBIS, 10 August 1967, KKK4, Radio Hanoi International, "Students' Message to SNCC Flays US policy."

9. FBIS, 30 August 1967, JJJ1, Radio Hanoi Domestic, "Stokely Carmichael Arrives in North Vietnam."

10. FBIS, 21 September 1967, JJJ2–3, RH to American Servicemen, "Carmichael Praises Cassius Clay's Defiance."

11. FBIS, 18 August 1967, KKK1, LBS to RVN, "Liberation Peasants Appeal to Compatriots."

12. FBIS, 5 September 1967, KKK1, LBS to RVN, "The Rigged Results of a Deceitful Election Farce."

13. CIA, "Intelligence Information Cable, Summary of Viet Cong Activities in Kien Hoa Province during September 1967," 19 October 1967, LBJL, NSF Vietnam, box 90, Viet Cong Offensive.

14. FBIS, 5 September 1967, JJJ8–9, Radio Hanoi International, "NFLSV Congress Adopts Political Program."

15. Message to Walt Rostow from Ambassador Locke, 24 November 1967, n.p., LBJL, NSF Vietnam, box 89, NLF Political, Director.

16. CIA, "Intelligence Memorandum, Communist Counteraction against the Revolutionary Development Program in South Vietnam," 20 November 1967, 6 (quote), 8, LBJL, box 90, Viet Cong Offensive.

17. *CANLF Bulletin*, 9 May 1967, "Progress Report," 34, www.redandgreen.org/CANLF/Index.htm.

18. FBIS, 17 October 1967, JJJ4, RH to RVN, "U.S. Paper Says War Drains Nation's Strength."

19. FBIS, 24 October 1967, JJJ3, RH to SEA in English, "Dellinger Statement."

20. CDEC, "Agency in Thu Dau Mot Province," November 1967, Document 12-1390-67, TTVA, DPC: Unit 2—Military Operations, box 10, folder 1.

21. 6th Psychological Operations Battalion, "Operational Report for Period Ending 31 October 1967," 15 November 1967, 7, AEHC, MACV Command History, 6th Psychological Operations Battalion.

22. Ibid., 8.

23. Ibid., 19; 6th Psychological Operations Battalion, "Weekly Operational Report 10–16 Nov 1967," 20 November 1967, 4 (quote), USASOC, PSYOP Papers 1961–1968 box 3; Headquarters, 6th Psychological Operations Battalion, "Weekly Operational Report 3 to 9 November 1967," 15 November 1967, USASOC, PSYOP Papers 1961–1968, box 3, 4.

24. CIA, "Intelligence Information Cable, Summary of Viet Cong Activities in Phong Dinh Province," 31 October 1967, 3–4, LBJL, NSF Vietnam, box 90, Viet Cong Offensive.

25. United States Military Assistance Command, *Command History Vietnam: 1967* (Saigon: USMACV Military History Branch, 16 September 1968), 647.

26. MACV, *Command History, 1967*, 649.

27. FBIS, 31 October 1967, JJJ4, RH to American Servicemen, 30 October Report, "From the Pacific to Truc Bach Lake."

28. MACV, *Command History, 1967*, 643.

29. Ibid., 663.

30. Ibid., 664.

31. Ibid.

32. MACV, Psychological Operations Directorate, "PSYOP Effort and Effectiveness: An Objective Look," October 1967, 1, Joint Forces Staff College Library.

33. Ibid., 10, 11.

34. Ibid., 13.

35. Ibid., 16.

36. Ibid., 27.

37. MACV, *Command History, 1967*, 644.

38. 15th PSYOP Detachment, "Report on Psychological Operations Intelligence in Vietnam," 13–2, AEHC, Vietnam War Psychological Operations Forces, Intelligence, folder 2.

39. 6th Psychological Operations Battalion, "S2 ORLL Input, 3rd Quarter 1967," 5 November 1967, USASOC, PSYOP Papers 1961–1968, box 3.

40. Ibid., 2.

41. 6th Psychological Operations Battalion, "Weekly Operational Report 27 October to 2 November 1967," 6 November 1967, 2, USASOC, PSYOP Papers 1961–1968, box 3.

42. Ibid., 10 to 16 November 1967 and 20 November 1967, 2.

43. Ibid., 3 to 9 November 1967, 15 November 1967, 2.

44. 15th PSYOP Detachment, "Report on Psychological Operations Intelligence in Vietnam," 5-F, 6-F, 13-F.

45. Ibid., 8-D, 16-C.

46. Ibid.

47. 6th Psychological Operations Battalion, "Weekly Operational Report 27 October–2 November 1967," 6 November 1967, 2, USASOC, PSYOP Papers 1961–1968, box 3; Psychological Operations Battalion, "Weekly Operational Report 3 to 9 November 1967," 2; Psychological Operations Battalion, "Weekly Operational Report 10–16 November 1967," 2.

48. 5th Special Forces Group, "Subject: After Action Report, 1 December 1967," 6, USASOC, PSYOP Papers, 61–68, box 3-3, folder 54.

49. Ibid., 1–5, 6 (quote).

50. Ibid., 8.

51. Robert M. Gillespie, *Black Ops, Vietnam: An Operational History of MACVSOG* (Annapolis, MD: Naval Institute Press, 2011), 102; MACSOG Documentation Study, appendix C, MACSOG Operations against North Vietnam (Washington, DC: Office of the Secretary of Defense, 1970), 51.

52. MACV, *Command History, 1967*, 624.

53. MACV, "Monthly Pacification Status Report," 26 December 1967, 1–6, 4–1. TTVA, VAC box 1, folder 601.

54. "Guidance for Evaluating Hamlet Evaluation System Factors," annex E, TTVA, DPC: Unit 3—Statistical Data, box 43, folder 12.

55. 6th Psychological Operations Battalion, "Weekly Operational Report 3 to 9 November 1967," 3–6.

56. James N. Peale, "Victory over Viet Cong Terror," 73, unpublished manuscript, AEHC, Peale Papers, 18; 15th PSYOP Detachment, "Report on Psychological Operations Intelligence in Vietnam," 10-F.

57. Ibid., 22.

58. MACV, *Command History, 1967*, 634.

59. Ibid., 641–642.

60. 4th PSYOP Group, "Handbook," 4, TTVA, John Cheney Collection, box 1, folder 3; 4th PSYOP Group, "Operational Report for Period Ending 31 January 1968," 3 April 1968, 3, AEHC, MACV Command History, 6th Psychological Operations Battalion.

61. Ibid., 35.

62. JUSPAO, Field Memorandum 42, "Lessons Learned from Evaluation of Allied PSYOP Media in Viet Nam" (Saigon: JUSPAO, 13 December 1967), 7, TTVA, DPC: Unit 3—Insurgency Warfare, box 13, folder 13.

63. Ibid.

64. Ibid.

65. 6th Psychological Operations Battalion, "Quarterly Operational Report," 17 November 1967, 2, AEHC, MACV Command Histories Collection, box 32.

66. 4th PSYOP Group, "Operational Report for Period Ending 31 January 1968," 24.

67. CDEC, Bulletin No. 9006, "Enemy Documents," 18 January 1968, TTVA, Garnett Bell Collection, box 26, folder 13.

68. CIA, Intelligence Information Cable, "Viet Cong Claims in Long An Province," 11 December 1967, 2 (quote), 3, LBJL, box 89, NLF Political, IIC.

69. CIA, Intelligence Information Cable, "Summary of Viet Cong Activities in Phuong Dinh Province," 17 January 1968, 1–7, 2 (quote), LBJL, NSF Vietnam, box 90, Viet Cong Offensive.

70. FBIS, 16 November 1967, JJJ2, RH International, "U.S. Peace Activist Receives American POW's."

71. William Westmoreland, "In 1965 the Enemy was Winning, Today He Is Certainly Losing," speech at the National Press Club, 21 November 1967, in American Foreign Policy Current Documents, 1967, Historical Office, Bureau of Public Affairs, Department of State (Washington, DC: US Government Printing Office, 1969).

72. MACV, *Command History 1967*, 655.

73. National Military Command Center, "Subject: Possible U.S. defector to Viet Cong," 19 January 1968, LBJL, box 89, NLF Political 2; Richard Linnett, "The Curious Case of Private McKinley Nolan," *Vietnam* (June 2011): 46–51.

74. United States Senate. "Treatment of Military Deserters," Report of the Armed Forces Committee (Washington, DC: US Government Printing Office, March 1969), 11, 22 (quote), 24.

75. JUSPAO, JUSPAO Policy Guidance 51, "Priorities in the PSYOP Effort," Saigon, RVN, 28 December 1968.

76. MACSOG Documentation Study, annex A to appendix C, Psychological Operations, 17, 81–83.

77. JUSPAO, *PSYOPS in Vietnam: Indications of Effectiveness,* III-4, H-13, TTVA, DPC: Unit 3—Insurgency Warfare, box 13, folder 12.

78. ACTIV, Final Report, "Employment of U.S. Army Psychological Operations Units in Vietnam," 7 June 1969, H-6, H-7, DTIC, AD855595.

79. MACV, *Command History, United States Military Assistance Command, Vietnam: 1968*, vol. 2 (Saigon: USMACV Military History Branch, 1969), 604–605.

Conclusion

1. *Saigon: Target Zero,* 19 min. (Saigon: Political Warfare Directorate, 1968). This film contains footage of the events described in the paragraphs below and activities of the ARVN political warfare units.

2. John Prados, *Vietnam: The History of an Unwinnable War, 1945–1975* (Lawrence: University Press of Kansas, 2009), 239. See the discussion on planning at pages 190–196.

3. 4th PSYOP Group, "Semi-Monthly Operations Report 1 February–15 February 1968," USASOC, Avedon Collection, Vietnam 1968–69, box 6. Charge made in the film *Saigon: Target Zero.*

4. 4th PSYOP Group, "Semi-Monthly Operations Report 1 February–15 February 1968."

5. Prados, *Vietnam,* 224; Eric Bergerud, *Dynamics of Defeat: The Vietnam War in Hau Nghia Province* (Boulder, CO: Westview Press, 1991), 197–203; James H. Willbanks: *The Tet Offensive, A Concise History* (New York: Columbia University Press, 2007), 14–17.

6. Prados, *Vietnam*, 223–225.

7. Ang Cheng Guan, "Khe Sanh—From the Perspective of the North Vietnamese Communists," *War in History* 8, no. 1 (2008): 87–98; Bergerud, *Dynamics of Defeat*, 216. Bergerud argues that in Hau Nghia Province the ARVN and militia fought well.

8. Prados, *Vietnam*, 229, 231, 233, 243; Ang Cheng Guan, "Decision-Making Leading to the Tet Offensive (1968—The Vietnamese Communist Perspective)," *Journal of Contemporary History* 33, no. 3 (July 1998): 343–346.

9. Bui Tin, *Following Ho Chi Minh: The Memoirs of a North Vietnamese Colonel* (Honolulu: University of Hawaii Press, 1995), 63.

10. ACTIV, "Employment of U.S. Army Psychological Operations Units in Vietnam," 7 June 1969, III-6, DTIC, AD855595.

11. Prados, *Vietnam*, 255.

12. James N. Peale, "Victory over Viet Cong Terror," 73, unpublished manuscript, AEHC, Peale Papers, 75–76; William Hoffer, "Broadcasting in an Insurgency Environment: USIA in Vietnam 1965–1970" (PhD diss., University of Wisconsin, 1972), 307; ACTIV, "Employment of U.S. Army Psychological Operations Units in Vietnam," v (quote).

13. ACTIV, "Employment of U.S. Army Psychological Operations Units in Vietnam," II-13, II-15.

14. Hoffer, "Broadcasting in an Insurgency," 310.

15. MACV, MACPD Directive 10-1, 11 December 1967, 7, TTVA, DPC: Unit 3—Insurgency Warfare, box 13, folder 17.

16. ACTIV, "Employment of U.S. Army Psychological Operations Units in Vietnam," III-6.

17. Benjamin Franklin, *Autobiography of Benjamin Franklin with Introduction and Notes* (1901; reprint, London: MacMillan Company, 1921), 87 (quote); Garth S. Jowett and Victoria O'Donnell, *Propaganda and Persuasion,* 5th ed. (Thousand Oaks, CA: Sage Publications, 2012), 267.

18. Richard H. Shultz, *The Secret War against Hanoi: Kennedy's and Johnson's Use of Spies, Saboteurs, and Covert Warriors in North Vietnam* (New York: HarperCollins, 1999), 169, xvii (quote).

19. Robert M. Gillespie, *Black Ops, Vietnam: An Operational History of MACVSOG* (Annapolis, MD: Naval Institute Press, 2011), 104 (quote); Shultz, *Secret War against Hanoi*, 172.

20. Thomas L. Ahern Jr., *Vietnam Declassified: The CIA and Counterinsurgency* (Lexington: University Press of Kentucky, 2010), 251.

21. FBIS, 27 December 1967, JJJ2; James B. Overton, *Rolling Thunder: January 1967–November 1968* (Hickam AFB, HI: CHECO/Corona Harvest Division, Operations Analysis Office, Department of the Air Force, Headquarters, Pacific Air Forces, 1967), figure 6, DTIC, ADA486572.

22. FBIS, 23 July 1962, JJJ3, Radio Hanoi, "Let the Dog Bark, the Caravan Goes."

23. *The Fog of War: Eleven Lessons from the Life of Robert S. McNamara*, 138 min. (Sony Pictures, 2003); Prados, *Vietnam*, 210–211, provides additional indications of this influence.

24. Shultz, *Secret War against Hanoi*, 169, 172 (quote).

25. Gregory Seese and Paul N. Smith, "Measuring PSYOP Effectiveness," *Special Warfare* (November–December 2008): 31 (31–34).

26. J. A. Koch, *The Chieu Hoi Program in South Vietnam, 1963–1971* (Washington, DC: Rand Corporation for ARPA, January 1973), 11.

27. ACTIV, "Employment of U.S. Army Psychological Operations Units in Vietnam," IV-3, II-17.

28. Frank Newport and Joseph Carroll, "Iraq Versus Vietnam: A Comparison of Public Opinion," www.gallup.com/poll/18097/Iraq-Versus-Vietnam-Comparison-Public-Opinion.aspx?utm_source=vietnam%20war%201967&utm_medium= search&utm_campaign=tiles.

29. See Frank Scotton, *Uphill Battle: Reflections on Viet Nam Counterinsurgency* (Lubbock: Texas Tech University Press, 2014), and Jeffrey Race, *War Comes to Long An: Revolutionary Conflict in a Vietnamese Province* (1972; reprint, Berkeley: University of California Press, 2010).

30. Lam Quang Thi, *The Twenty-five Year Century: A South Vietnamese General Remembers the Indochina War to the Fall of Saigon* (Denton: University of North Texas Press, 2001), 111.

31. See Kevin Michael Boylan, "The Red Queen's Race: The 173rd Airborne Brigade and Pacification in Binh Dinh Province, 1969–1970" (PhD diss., Temple University, 1994); Keith W. Taylor, "The Vietnamese War of 1955–1975 in Historical Perspective," in *Triumph Revisited: Historians Battle for the Vietnam War*, ed. Andrew Wiest and Michael Doidge (New York: Routledge, 2010), 18, 27.

32. Koch, *Chieu Hoi Study*, v–vii, 40–44.

Bibliography

Primary Sources

Army Heritage and Education Center, Carlisle, PA (AHEC)
4th Psychological Operations Group
James N. Peale Papers
MACV Command Histories Collection
Vietnam War Psychological Operations Forces
BACM Research, Paperless Archives (paperlessarchives.com)
Central Intelligence Agency Archives, Freedom of Information Act Electronic Reading Room. www.foia.cia.gov (CIA)
Cold War International History Project, Woodrow Wilson International Center for Scholars, Washington, DC (CWIHP)
Grossheim, Martin. "Fraternal Support: The East German 'Stasi' and the Democratic Republic of Vietnam during the Vietnam War." CWIHP Working Paper 71. September 2014.
Report of Pham No Mach [Pham Ngoc Thach] to the Soviet Envoy in Switzerland, A. G. Kulazhenkov, 20 September 1947.
Vietnamese Politburo to COSVN. North Vietnam Politburo Cable No. 320, 10 October 1968.
Combined Arms Research Library, Fort Leavenworth, KS (CARL)
Defense Technical Information Center (DTIC) (dtic.mil/dtic)
5th Special Forces Group. "Quarterly Command Report for the Period Ending 31 December 1965." DTIC, AD391693.
Army Concept Team in Vietnam (ACTIV). "Army Concept Team in Vietnam: The Heliborne Public Address System for Use in PSYWAR and Civic Action." Saigon: ACTIV, 10 February 1964. DTIC, AD348732.
———. "Final Report: Employment of U.S. Army Psychological Operations Units in Vietnam." Saigon: ACTIV, June 1969. DTIC, AD855955.
Ballachey, Egerton L. *Psychological Operations: South Vietnam, Project PROSYMS.* Washington, DC: Special Operations Research Office, American University, April 1959. DTIC, AD310382.
II Field Force Vietnam. "Operational Report for Quarterly Period Ending 31 October 1966." DTIC, AD512784.
Headquarters, 1st Bde., 101st Abn. "Combat Operations After Action Report, Operation Beauregard, 10 August 1966." DTIC, AD394091.
Headquarters, 1st Bde., 101st Abn. "Combat Operations After Action Report, John Paul Jones." DTIC, AD388882.

Headquarters, 1st Cavalry Division. "Quarterly Command Report, Period Ending 1 December 1965." DTIC, AD390501.

Headquarters, 4th Psychological Operations Group, "Operational Report—Lessons Learned, Period Ending 31 July 1969." Office of the Adjutant General, Washington, DC. 26 November 1969. DTIC AD502157.

Headquarters, 6th Psychological Operations Battalion. "Operational Report for Quarterly Period Ending 31 October 1966." 17 January 1967, DTIC AD393030.

Military Assistance Advisory Group Vietnam (MAAGV). *Lessons Learned Number 40: Corps Psywar/CA Operations Center,* enc 2. Vietnam: MAAGV, 23 March 1964, DTIC AD385117.

Office of the Deputy Chief of Staff for Military Operations. *A Program for the Pacification and Long-Term Development of South Vietnam* (PROVN). US Army, 1 March 1966, volume 1. DTIC AD3777743.

Thayar, Thomas C. *A Systems Analysis View of the Vietnam War 1965–1972.* vol. 7 *The Republic of Vietnam Armed Forces.* n.p. DTIC ADA051611.

US Joint Staff. "Operation Junction City Vietnam, 1967." DTIC AD139612.

John F. Kennedy Presidential Library (JFKL), Boston, MA. www.jfklibrary.org

National Security Action Memoranda

Lyndon B. Johnson Presidential Library (LBJL), Austin, TX

National Security Action Memoranda

National Security File (NSF), Vietnam Country File

US Information Agency (USIA). *USIA Administrative History*

National Archives and Records Administration, Washington, DC (NARA)

The following datasets found at the National Archives provide details about operations and data for geographic information systems (GIS) analysis:

RG 330, Enemy Base Area Files (BASFA), documenting the period July 1966–July 1971.

RG 330, Hamlet Evaluation System (HES) Files, 1967–1971.

RG 330, Southeast Asia Friendly Forces File (SEAFA), October 1966–July 1972.

RG 330, Viet Cong Initiated Incidents Files (VCIIA), January 1965–December 1968.

RG 472, Terrorist Incident Reporting System (TIRSA) Files, October 1967–February 1973.

US Army Special Operations Command Archives. Fort Bragg, NC (USASOC)

Herbert Avedon Collection

PSYOP Papers

Vietnam, 1966 Vietnam Archives. Texas Tech University, Lubbock, TX (TTVA)

Americal Division Association Collection

Bud Harton Collection

Central Intelligence Agency Collection

David Lifton Collection

Douglas Pike Collection

Unit 1—Assessment and Strategy

Unit 2—Military Operations

Unit 3—Antiwar Activities

Unit 3—Insurgency Warfare

Unit 3—Statistical Data

Unit 3—US Economy and US Mission (Saigon)

Unit 5—National Liberation Front

Unit 6—Democratic Republic of Vietnam
Unit 11—Monographs
Garnett Bell Collection
John Cheney Collection
John Donnell Collection
John B. O'Donnell Collection
Michael Sloniker Collection
Rufus Phillips Collection
US Marine Corps History Division Collection
Vietnam Archive Collection
Vietnam Members of the 109th Quartermaster Company
Vladimir Lehovich Collection

US Government Publications

Abbey, Tom G. *Southeast Asia Report: The Role of the USAF in Support of Special Activities in SEA.* Hickam AFB, HI: CHECO/Corona Harvest Division, Operations Analysis Office, Department of the Air Force, Headquarters, Pacific Air Forces, 1 July 1976.

Collura, Angelo. *The Case of Pvt. Robert R. Garwood, USMC,* vol. 1. June 1993. Report to the Assistant Secretary of Defense for Command, Control, Communication and Intelligence.

Committee on Public Information. "Four Minute Men News." Washington, DC: Division of Four Minute Men, n.d.

Department of the Army. *Field Manual 33-5: Psychological Warfare Operations.* Washington, DC: Department of the Army, March 1955.

———. *Field Manual 33-5: Psychological Operations.* Washington, DC: Department of the Army, 16 January 1962.

———. *Field Manual 33-5: Psychological Operations, Techniques and Procedures.* Washington, DC: Department of the Army, October 1966.

———. *Field Manual 33-1, Psychological Operations,* Washington, DC: Department of the Army, 21 June 1968.

The Joint Staff. *MACSOG Documentation Study.* Washington, DC: Office of the Secretary of Defense, 1970.

Journals of the Continental Congress, 1774–1789. Ed. Worthington C. Ford et al. Washington, DC: 1904–37, 5 (1776): 653–656.

Khuyen, Dong Van. *RVNAF History.* Indochina Monographs, part 4: "Political Warfare." Washington, DC: US Army Center for Military History, 1978.

Memorandum to the Joint Chiefs of Staff. "PSYCH OPS Printing support for COMUSMACV." 7 January 1964. Digital document in possession of the author.

Sams, Kenneth. *Southeast Asia Report, Operation Hawthorne.* Hickam AFB, HI: CHECO/Corona Harvest Division, Operations Analysis Office, Department of the Air Force, Headquarters, Pacific Air Forces, 8 September 1966.

United States. *Public Papers of the Presidents of the United States: Lyndon B. Johnson, 1963–64.* Containing the Public Messages, Speeches, and Statements of the President, 1967. Volume 1, January 1 to June 30, 1967. Washington, DC: US Government Printing Office, 1968.

United States and KTO Press. *The Cumulated Indexes to the Public Papers of the Presidents of the United States, Lyndon B. Johnson, 1963–1969,* book 1: *1963–64.* Millwood, NY: KTO Press, 1978.

US Congress. Senate Select Committee on Intelligence. *Vietnamese Commandos.* Hearing before the Select Committee on Intelligence. Testimony of John Singlaub. 104th Cong., 2nd sess., 19 June, 1996.

US Foreign Broadcast Information Service. *Daily Report.* Arlington, VA: Joint Publications Research Service, 1960–1972.

US Military Assistance Advisory Group Vietnam. *Records of the Military Assistance Advisory Group Vietnam, 1950–1964.* Adjutant General Division Security Classified Files. Bethesda, MD: University Publications of America, 2001.

US Military Assistance Command, Vietnam. *Command History, United States Military Assistance Command, Vietnam: 1964.* Saigon: USMACV Military History Branch, 15 October 1965.

———. *Command History, United States Military Assistance Command, Vietnam: 1965.* Saigon: USMACV Military History Branch, 20 April 1966.

———. *Command History, United States Military Assistance Command, Vietnam: 1966.* Saigon: USMACV Military History Branch, 19 April 1967.

———. *Command History, United States Military Assistance Command, Vietnam: 1967.* Annex G. Saigon: USMACV Military History Branch, 1968.

———. *Command History, United States Military Assistance Command, Vietnam: 1967.* Saigon: USMACV Military History Branch, 16 September 1968.

———. *Command History, United States Military Assistance Command, Vietnam: 1968.* Annex F. Saigon: USMACV Military History Branch, 1968.

———. Records of the Military Assistance Command Vietnam. *The War in Vietnam, 1964–1973.* MACV Historical Office Documentary Collection. Bethesda, MD: University Publications of America, [1988].

US Senate. "Treatment of Military Deserters." Report of the Armed Forces Committee. Washington, DC: US Government Printing Office, March 1969.

Weapons Systems Evaluation Group. *The Flaming Dart Airstrikes against North Vietnam: 7 to 11 February 1965.* CI no. 9, Working Paper No. 4. 2 August 1967.

Secondary Sources

Audio/Visual

The Big Picture. 28 min. US Army Audiovisual Center. 1953–1969. archive.org/search.php?query=Big%20Picture%20AND%20collection%3Anewsandpublicaffairs&and[]=mediatype%3A%22movies%22.

The Fog of War: Eleven Lessons from the Life of Robert S. McNamara. 138 min. Sony Pictures, 2003.

The Unique War. 24 min. New York: Army Pictorial Center, 1966.

Saigon: Target Zero. 19 min. Saigon: Political Warfare Directorate, 1968. archive.org/details/gov.archives.arc.32416

Staff Film Report 67–2: Chanh Luu Vietnam. 27 min. New York: Army Pictorial Center, 1967. archive.org/details/gov.dod.dimoc.27015.

Books

Ahern Jr., Thomas L. *CIA and the House of Ngo: Covert Action in South Vietnam, 1954–1963.* Washington, DC: Center for the Study of Intelligence, 2000.

———. *CIA and Rural Pacification in Vietnam.* Washington, DC: Center for the Study of Intelligence, 2001.

———. *The Way We Do Things: Black Entry Operations into North Vietnam, 1961–1964.* Washington, DC: Center for the Study of Intelligence, 2005.

———. *Undercover Armies, CIA and Surrogate Warfare in Laos, 1961–1973.* Washington, DC: Center for the Study of Intelligence, 2006.

———. *Vietnam Declassified: The CIA and Counterinsurgency.* Lexington: University Press of Kentucky, 2010.

American Institute for Political Communication. *The Vietnam Issue, University Students' Attitudes in the 1966 Congressional Elections: A Comparative Analysis of Group and Individual Attitudes Toward Viet Nam Issue Based on a Study of Persons Attending George Washington and Maryland Universities Respectively.* Washington: AIPC, 1966.

Aptheker, Herbert. *Mission to Hanoi.* New York: International Publisher, 1966.

Asselin, Pierre. *Hanoi's Road to the Vietnam War.* Oakland: University of California Press, 2013.

Bartholomew-Feis, Dixee R. *The OSS and Ho Chi Minh: Unexpected Allies in the War against Japan.* Lawrence: University Press of Kansas, 2006.

BDM Corporation. *A Study of Strategic Lessons Learned in Vietnam. Vol. 6, Conduct of the War. Book 2, Functional Analyses.* McLean, VA: BDM Corp., 1980.

Bergerud, Eric M. *Dynamics of Defeat: The Vietnam War in Hau Nghia Province.* Boulder, CO: Westview Press, 1991.

Blaufarb, Douglas S. *The Counterinsurgency Era: U.S. Doctrine and Performance, 1950 to the Present.* New York: Free Press, 1977.

Brigham, Robert K. *Guerrilla Diplomacy: The NLF's Foreign Relations and the Viet Nam War.* Ithaca, NY: Cornell University Press, 1999.

Bruntz, George G. *Allied Propaganda and the Collapse of the German Empire in 1918.* Redwood City, CA: Stanford University Press, 1938.

Bui Tin. *Following Ho Chi Minh: The Memoirs of a North Vietnamese Colonel.* Honolulu: University of Hawai'i Press, 2003.

Carland, John M. *Combat Operations: Stemming the Tide: May 1965 to October 1966.* Washington, DC: Center of Military History, 2000.

Carrier, J. M., and C. A. H. Thomson. *Vietcong Motivation and Morale: The Special Case of Chieu Hoi.* Santa Monica, CA: Rand Corporation, 1966.

Catton, Philip E. *Diem's Final Failure: Prelude to America's War in Vietnam.* Lawrence: University Press of Kansas, 2002.

Chandler, Robert W. *War of Ideas: The U.S. Propaganda Campaign in Vietnam.* Boulder, CO: Westview Press, 1981.

Colby, William, with James McCargar. *Lost Victory: A Firsthand Account of America's Sixteen-Year Involvement in Vietnam.* Chicago: Contemporary Books, 1989.

Collins, James Lawton. *The Development and Training of the South Vietnamese Army, 1950–1972.* Washington, DC: Department of the Army, 2002.

Conboy, Kenneth, and Dale Andrade. *Spies and Commandos: How America Lost the Secret War in North Vietnam.* Lawrence: University Press of Kansas, 2000.

Currey, Cecile B. *Edward Lansdale: The Unquiet American.* Boston: Houghton-Mifflin, 1988.

Daddis, Gregory A. *No Sure Victory: Measuring U.S. Army Effectiveness and Progress in the Vietnam War.* Oxford, UK: Oxford University Press, 2011.

Daugherty, William E. *A Psychological Warfare Casebook.* Baltimore: Published for the Operations Research Office by Johns Hopkins University Press, 1958.

Davidson, Philip G. *Propaganda and the American Revolution, 1763–1783.* Chapel Hill: University of North Carolina Press, 1941.

Don, Tran Van. *Our Endless War: Inside Vietnam.* Novato, CA: Presidio, 1978.

Duiker, William J. *Ho Chi Minh: A Life.* New York: Hyperion, 2000.

Eckhardt, George S. *Vietnam Studies: Command and Control, 1950–1969.* Washington, DC: Department of the Army, 1991.

Elliott, David W. P. *The Vietnamese War: Revolution and Social Change in the Mekong Delta, 1930–1975.* Armonk, NY: M. E. Sharpe, 2003.

Fantina, Robert. *Desertion and the American Soldier.* New York: Algora Publishing, 2006.

Feilitzsch, Heinrich Carl Philipp von and Christian Friedrich Bartholomai. *Diaries of Two Ansbach Jaegers.* Translated by Bruce E. Burgoyne. Bowie, MD: Heritage Books, 1997.

Fonda, Jane. *My Life So Far.* New York: Random House, 2005.

Foreign Broadcast Information Service. *Broadcasting Stations of the World,* part 1: *AM Broadcasting.* Washington, DC: US Government Printing Office, 1961.

Franklin, Benjamin. *Autobiography of Benjamin Franklin, with Introduction and Notes.* 1901. London: Macmillan Company, 1921.

Gaiduk, Ilya V. *The Soviet Union and the Vietnam War.* Chicago: Ivan R. Dee, 1996.

Gates, John M. *Schoolbooks and Krags: The United States Army in the Philippines, 1898–1902.* Westport, CT: Greenwood Press, 1973.

Gillespie, Robert M. *Black Ops, Vietnam: An Operational History of MACVSOG.* Annapolis, MD: Naval Institute Press, 2011.

Ha, Mai Viet. *Steel and Blood: South Vietnamese Armor and the War for Southeast Asia.* Annapolis, MD: Naval Institute Press, 2008.

Haas, Michael E. *Apollo's Warriors: United States Air Force Special Operations during the Cold War.* Maxwell AFB, AL: Air University Press, 1997.

Halstead, Fred. *Out Now! A Participant's Account of the American Movement against the Vietnam War.* New York: Monad Press, 1978.

Hammond, William M. *Public Affairs: The Military and the Media, 1962–1968.* U.S. Army in Vietnam series. Washington, DC: Center of Military History, 1988.

Hayden, Tom. *Reunion: A Memoir.* New York: Random House, 1988.

Hotze, Henry, and Lonnie A. Burnett. *Henry Hotze, Confederate Propagandist: Selected Writings on Revolution, Recognition, and Race.* Tuscaloosa: University of Alabama Press, 2008.

Hunt, Richard A. *Pacification: The American Struggle for Vietnam's Hearts and Minds.* Boulder, CO: Westview, 1995.

Jacobs, Seth. *Cold War Mandarin: Ngo Dinh Diem and the Origins of America's War in Vietnam, 1950–1963.* Lanham, MD: Rowman & Littlefield, 2006.

Johnson, Timothy D. *Winfield Scott: The Quest for Military Glory.* Lawrence: University Press of Kansas, 1998.

Jowett, Garth S., and Victoria O'Donnell. *Propaganda and Persuasion,* 5th ed. Thousand Oaks, CA: Sage Publications, 2012.

Kellen, Konrad. *A View of the VC: Elements of Cohesion in the Enemy Camp.* Santa Monica, CA: Rand Corporation, October 1968.

Koch, J. A. *The Chieu Hoi Program in South Vietnam, 1963–1971.* Santa Monica, CA: Rand Corporation, January 1973.

Kodosky, Robert J. *Psychological Operations American Style: The Joint United States Public Affairs Office, Vietnam and Beyond.* Plymouth, UK: Lexington Books, 2007.

Kretchik, Walter E. *U.S. Army Doctrine: From the American Revolution to the War on Terror.* Lawrence: University Press of Kansas, 2011.

Lam, Quang Thi. *The Twenty-Five Year Century: A South Vietnamese General Remembers the Indochina War to the Fall of Saigon.* Denton: University of North Texas Press, 2002.

Linebarger, Paul M. A. *Psychological Warfare.* 1948. Reprint, New York: Arno Press, 1972.

Linn, Brian McAllister. *The Philippine War: 1899–1902.* Lawrence: University Press of Kansas, 2000.

Logavall, Frank. *Embers of War: The Fall of an Empire and the Making of America's Vietnam.* New York: Random House, 2012.

Lowell, Edward J. *The Hessians and the Other German Auxiliaries of Great Britain in the Revolutionary War.* New York: Harper & Brothers, 1884.

Mackay, Andrew, Steve Tatham, and Lee Rowland. *The Effectiveness of U.S. Military Information Operations in Afghanistan, 2001–2010: Why RAND Missed the Point.* Central Asia series. Shrivenham, UK: Defense Academy of the United Kingdom, December 2012.

McMaster, H. R. *Dereliction of Duty: Lyndon Johnson, Robert McNamara, the Joint Chiefs of Staff, and the Lies that Led to Vietnam.* New York: HarperCollins, 1997.

McNeill, Ian. *To Long Tan: The Australian Army and the Vietnam War, 1950–1966.* Sidney: Allen & Unwin, 1993.

Military History Institute of Vietnam. *Victory in Vietnam: The Official History of the People's Army of Vietnam, 1954–1975.* Translated by Merle L. Pribbenow. Lawrence: University Press of Kansas, 2002.

Miller, Edward. *Misalliance: Ngo Dinh Diem, the United States, and the Fate of South Vietnam.* Cambridge, MA: Harvard University Press, 2013.

Mintz, Jeanne S., et al. *A Short Guide to Psychological Operations in the Republic of Vietnam, 1965.* Washington, DC: Special Operations Research Office, American University, 1965.

Molnar, Andrew R. *Human Factors Considerations of Undergrounds in Insurgencies.* Washington, DC: Special Operations Research Office, American University, 1966.

Moyar, Mark, *Triumph Forsaken: The Vietnam War, 1954–1965.* New York: Cambridge University Press, 2006.

Moyar, Mark, and Harry G. Summers. *Phoenix and the Birds of Prey: The CIA's Secret Campaign to Destroy the Viet Cong.* Annapolis, MD: Naval Institute Press, 1997.

Nguyen, Cao Ky, and Marvin J. Wolf. *Buddha's Child: My Fight to Save Vietnam.* New York: St. Martin's Press, 2002.

Nguyen, Lien-Hang T. *Hanoi's War: An International History of the War for Peace in Vietnam.* Chapel Hill: University of North Carolina Press, 2012.

Osgood, Kenneth. *Total Cold War: Eisenhower's Secret Propaganda Battle at Home and Abroad.* Lawrence: University Press of Kansas, 2006.

Paddock, Alfred H. *U.S. Army Special Warfare: Its Origins.* Modern War Studies series. Lawrence: University Press of Kansas, 2002.

Page, Caroline. *U.S. Official Propaganda during the Vietnam War, 1965–1973: The Limits of Persuasion.* London: Leicester University Press, 1996.

Panne, Jean-Louis. *The Black Book of Communism: Crimes, Terror, Repression.* Cambridge, MA: Harvard University Press, 1999.

Parker, William D. *U.S. Marine Corps Civil Affairs in I Corps, Republic of Vietnam: April 1966 to April 1967.* Washington, DC: Historical Division, Headquarters, US Marine Corps, 1970.

Pettee, George S. *U.S. PSYWAR Operations in the Korean War.* Washington, DC: Operations Research Office, 1951.

Phillips, Rufus. *Why Vietnam Matters: An Eyewitness Account of Lessons Not Learned.* Annapolis, MD: Naval Institute Press, 2008.

Plaster, John. *SOG: The Secret Wars of America's Commandos in Vietnam.* Boulder, CO: Paladin Press, 2008.

Prados, John. *Vietnam: The History of an Unwinnable War.* Lawrence: University Press of Kansas, 2009.

Quinn-Judge, Sophie. *Ho Chi Minh: The Missing Years, 1919–1941.* Berkeley: University of California Press, 2002.

Race, Jeffrey. *War Comes to Long An: Revolutionary Conflict in a Vietnamese Province.* Berkeley: University of California Press, 2010 [1972].

Rochester, Stuart I. *The Battle behind Bars: Navy and Marine POWs in the Vietnam War.* Washington, DC: Department of the Navy, Naval History and Heritage Command, 2010.

Rostow, Walt W. *The Stages of Economic Growth: A Non-Communist Manifesto.* Reprint. Cambridge, UK: Cambridge University Press, 1990 [1960].

Russell, Bertrand. *Autobiography of Bertrand Russell.* Boston: Little, Brown, 1967.

Sandler, Stanley. *"Cease Resistance: It's Good for You!" A History of U.S. Army Combat Psychological Operations.* Washington, DC: United States Army Special Operations Command, Directorate of History and Museums, 1999.

Schlight, John. *The War in South Vietnam: The Years of the Offensive, 1965–1968.* The United States Air Force in Southeast Asia series. Washington, DC: Office of Air Force History, 1988.

———. *A War Too Long: The History of the USAF in Southeast Asia.* Washington, DC: Air Force History and Museums Program, 1996.

Schwab, Orrin. *Defending the Free World: John F. Kennedy, Lyndon Johnson, and the Vietnam War, 1961–1965.* Westport, CT: Praeger, 1998.

Scotton, Frank. *Uphill Battle: Reflections on Viet Nam Counterinsurgency.* Lubbock: Texas Tech University Press, 2014.

Sheehan, Neil. *The Pentagon Papers. As Published by the* New York Times. New York: Quadrangle Books, 1971.

———. *A Bright Shining Lie: John Paul Vann and America in Vietnam.* New York: Random House, 1988.

Shrader, Charles R. *History of Operations Research in the U.S. Army, Volume 2: 1961–1973.* Washington, DC: Office of the Deputy Under Secretary of the Army for Operations Research, US Army, 2006.

Shultz, Richard H. *The Secret War against Hanoi: Kennedy's and Johnson's Use of Spies, Saboteurs, and Covert Warriors in North Vietnam.* New York: HarperCollins, 1999.

Sorley, Lewis. *The Vietnam War: An Assessment by South Vietnam's Generals.* Modern Southeast Asia series. Lubbock: Texas Tech University Press, 2010.

Spector, Ronald H. *Advice and Support: The Early Years, 1941–1960.* Washington, DC: Center of Military History, 1983.

Stewart, Richard W., ed. *American Military History, Volume 2: The United States Army in a Global Era, 1917–2003.* Washington, DC: Center of Military History, 2005.

Toczek, David. *The Battle of Ap Bac, Vietnam: They Did Everything but Learn from It.* Annapolis, MD: Naval Institute Press, 2007.

Tolson, John J. *Vietnam Studies: Airmobility, 1961–1971.* Washington, DC: Department of the Army, 1989.

Truong, Nhu Tang. *A Vietcong Memoir: An Inside Account of the Vietnam War and Its Aftermath.* New York: Random House, 1986.

Truong, Tran Cong, and Phan Thanh Vinh. *The Failure of the Special War, 1961–65.* Hanoi: Foreign Languages Publishing House, 1967.

———. *The N.F.L.* [*sic*] *Symbol of Independence, Democracy and Peace in South Viet Nam.* Hanoi: Foreign Languages Publishing House, 1967.

Wiest, Andrew, and Michael Doidge. *Triumph Revisited: Historians Battle for the Vietnam War.* New York: Routledge, 2010.

Willbanks, James H. *The Tet Offensive, A Concise History.* New York: Columbia University Press, 2007.

Yates, Lawrence A. *The US Military's Experience in Stability Operations, 1789–2005.* Fort Leavenworth, KS: Combat Studies Institute Press, 2006.

Journal Articles

Ang, Cheng Guan. "Decision-making Leading to the Tet Offensive (1968—The Vietnamese Communist Perspective)." *Journal of Contemporary History* 33, no. 3 (July 1998): 341–353.

———. "Khe Sanh—From the Perspective of the North Vietnamese Communists." *War in History* 8, no. 1 (2008): 87–98.

Birtle, Andrew J. "PROVN, Westmoreland, and the Historians: A Reappraisal." *Journal of Military History* 72, no. 4 (October 2008): 1213–1247.

Briscoe, Charles H. "The 1st L&L in Korea, A Photographer's Record, 1952–53." *Veritas, Journal of Army Special Operations History* 3, no. 4 (2007): 14–25.

———. "A Clearer View of Psywar at Fort Riley & Fort Bragg, 1951–1952." *Veritas, Journal of Army Special Operations History* 5, no. 4 (2009): 60–72.

Burstein, Paul, and William Freudenburg. "Changing Public Policy: The Impact of Public Opinion, Antiwar Demonstrations, and War Costs on Senate Voting on Vietnam War Motions." *American Journal of Sociology* 84, no. 1 (July 1978): 99–12.

Busch, Peter. "The 'Vietnam Legion': West German Psychological Warfare against East German Propaganda in the 1960s." *Journal of Cold War Studies* 16, no. 3 (Summer 2014): 164–189.

Butterfield, Lyman H. "Psychological Warfare in 1776: The Jefferson-Franklin Plan to Cause Hessian Desertions." *Proceedings of the American Philosophical Society* 94 (June 1950): 240.

Grossheim, Martin. "'Revisionism' in the Democratic Republic of Vietnam: New Evidence from the East German Archives." *Cold War History* 5, no. 4 (November 2005), 451–477.

Head, William P. "The March to Oblivion: The Defeat at Ap Bac Hamlet and the Americanization of the Vietnam War." *Journal of Third World Studies* 32 (Fall 2014): 57–81.

Jones Jr., Robert W. "The OSS (Office of Strategic Services)." *Veritas, Journal of Army Special Operations History* 3, no. 4 (2007), 34–51.

Kamps, Charles T. "The JCS 94-Target List: A Vietnam Myth That Still Distorts Military Thought." *Air and Space Power Journal* (Spring 2001): 67–80.

Kinney, D. J. "Selling Greenland: The Big Picture Television Series and the Army's Bid for Relevance during the Early Cold War." *Centaurus* 55, no. 3 (August 2013): 344–357.

Laurie, Clayton D. "The Chanting of Crusaders: Captain Heber Blankenhorn and AEF Combat Propaganda in World War I." *Journal of Military History* 59 (July 1995): 457–481.

Linnett, Richard. "The Curious Case of Private McKinley Nolan." *Vietnam* (June 2011): 46–51.

Lucas, Scott. "Campaigns of Truth: The Psychological Strategy Board and American Ideology, 1951–1953." *International History Review* 18, no. 2 (May 1996): 279–302.

Lunch, William L., and Peter W. Sperlich. "American Public Opinion and the War in Vietnam." *Western Public Opinion Quarterly* 32, no. 1 (March 1979): 21–44.

McAllister, James. "'Only Religions Count in Vietnam': Thich Tri Quang and the Vietnam War." *Modern Asian Studies* 42, no. 4 (July 2008): 751–782.

Mehta, Harish C. "North Vietnam's Informal Diplomacy with Bertrand Russell: Peace

Activism and the International War Crimes Tribunal." *Peace & Change* 37, no. 1 (January 2012): 64–94.

Piasecki, Eugene G. "The A Team Numbering System." *Veritas, Journal of Army Special Operations History* 5, no. 4 (2009): 39–42.

Schreiber, E. M. "Anti-War Demonstrations and American Public Opinion on the War in Vietnam." *British Journal of Sociology* 27, no. 2 (June 1976): 225–236.

Seese, Gregory, and Paul N. Smith. "Measuring PSYOP Effectiveness." *Special Warfare* (November–December 2008): 31–34.

Manuscripts

Barger, Michael G. "Psychological Operations Supporting Counterinsurgency: 4th PSYOP Group in Vietnam." Master's thesis, Army Command and General Staff College, 2007.

Chandler, Robert W. "U.S. Psychological Operations in Vietnam, 1965–72." PhD diss., George Washington University, 1974.

Davis, Ginger. "The Visible Persuaders: A Study of America's Use of Psychological Warfare Using the Chieu Hoi Program as a Case Study." Master's thesis, Texas Tech University, 1998.

De Heer, Derrill. "Psychological Operations Conducted by the Australian Army in Phuoc Tuy Province South Vietnam: 1965–1971." Master of Arts (Research) History, University of New South Wales School of Humanities and Social Science, Australian Defence Force Academy, 2009.

Hoffer, Thomas William. "Broadcasting in an Insurgency Environment: USIA in Vietnam, 1965–1970." PhD diss., University of Wisconsin, 1972.

McDonough, Joseph J. "Analysis of Official U.S. Military Psychological Warfare Efforts in the Vietnam Conflict." MPIA thesis, School of Public Communications, Boston University, 1968.

Peale, James N. "Victory over Viet Cong Terror. " Unpublished manuscript, AHEC.

Rodham [Clinton], Hillary D. "'There Is Only the Fight . . . ' An Analysis of the Alinsky Model." BA thesis, Wellesley College, 1969.

Newspapers and Periodicals

"An American Officer [Lansdale]. 'The Report the President Wanted Published.'" *Saturday Evening Post*, 20 May 1961, 69–71.

Arnett, Peter. "Map Error Makes Saigon Psywar Faces Red." *Washington Post,* 22 April 1967.

"Clergymen Score U.S. Aid to Diem." *New York Times Magazine,* 15 August 1963.

Dallas Morning News (1960–1968).

"Pilot's Despair over Vietnam: Last Letters of a U.S. Flyer Killed in Action." *Life,* 8 May 1964.

Shenon, Philip. "Hanoi Hannah Looks Back, With Few Regrets." *New York Times,* 26 November 1994.

"Steinke on Duty." *Pacific Stars and Stripes,* 30 June 1965.

Vietnam Courier. Foreign Languages Publishing House Weekly, May 1964–1973. Hanoi.

"Viet Nam Defeats Vex U.S." *Winnipeg Free Press,* 9 January 1963, 45.

Online Resources

"1st, 14th PsyOp Battalions and the 7th PsyOp Group USAB&VAPAC and VUNC." timyoho .com/BVApage/BVAHomePage.htm.

"Army Pictorial Center: Catalog of the Big Picture." www.armypictorialcenter.com/Catalog_of_the_Big_Picture.htm.
Chief of Army Field Forces. "Dissemination of Combat Information." 29 August 1952, 17. www.koreanwar-educator.org/topics/reports/after_action/combat_information_bulletins/combat_ information_bulletins_520829_350_05 _56.pdf.
"Democratic Party Platform of 1900." www.presidency.ucsb.edu/ws/index.php?pid=29587.
"Donovan Unofficial." donovan-unofficial.com/home.html.
Fifth Army. "Functions of the 5th Army Combat Propaganda Team, 5th Mobile Radio Broadcasting Company." 5 April 1944. www.psywar.org/psywar/reproductions/5ACPTPWB.pdf.
"George Washington to Continental Congress." 11 May, 1776. *George Washington Papers at the Library of Congress, 1741–1799.* memory.loc.gov/cgi-bin/query/P?mgw:1:./temp/~ammem_KXcS:.
Government Accountability Office. "Use of Agent Orange in Vietnam," 16 August 1978. www.gao.gov/products/CED-78–158.
"History of Project Jenny." www.afvn.tv/ProjectJenny/history.html.
"Library of Congress." memory.loc.org.
Marcuse, Herbert. "Repressive Tolerance." www.marcuse.org/herbert/pubs/60spubs/65repressivetolerance.htm.
Moise, Edwin. "Bibliography of the Vietnam War." www.clemson.edu/caah/history/facultypages/EdMoise/bibliography.html#spe.
Newport, Frank, and Joseph Carroll. "Iraq Versus Vietnam: A Comparison of Public Opinion." www.gallup.com/poll/18097/Iraq-Versus-Vietnam-Comparison-Public-Opinion.aspx.
North, Don. "The Search for Hanoi Hannah." www.psywarrior.com/hannah.html.
Psychological Warfare Division, Supreme Headquarters, Allied Expeditionary Force (PROV). "Report of the 6808th Allied Information Service Field Consolidation Team." N.p., n.d. www.psywar.org/psywar/reproductions/6808AIS.pdf.
"PSYOP and Military Links." www.psywarrior.com/links.html.
"Report of the Executive Committee of the Anti-Imperialist League, February 10, 1899." archive.org/details/ReportOfTheExecutiveCommitteeOfTheAnti-imperialistLeagueFeb.101899.
United Nations Office for Disarmament Affairs. "United States of America Ratification of 1925 Geneva Protocol." disarmament.un.org/treaties/a/1925/unitedstatesofamerica/rat/paris.
United States Air Force. "Operation Ranch Hand: The Air Force and Herbicides in Southeast Asia: 1961–1971." www.afhso.af.mil/shared/media/document/AFD-100928-054.pdf.
US Air Force. "5th Special Operations Squadron." www.919sow.afrc.af.mil/library/factsheets/factsheet.asp?id=16212.
———. "9th Special Operations Squadron." www.hurlburt.af.mil/AboutUs/FactSheets/FactSheets/tabid/140/Article/204527/9th-special-operations-squadron.aspx.
"U.S. Committee to Aid the National Liberation Front of South Vietnam." www.redandgreen.org/CANLF/Index.htm.
"U.S. Navy Beach Jumpers Association." www.beachjumpers.com/History/1960sJBD.htm.
US State Department. "Smith-Mundt Act," also known as "United States Information and Education Exchange Act of 1948," 27 January 1948, www.state.gov/documents/organization/177574.pdf.
"Vietnamese Declaration of Independence, 1945." www.fordham.edu/halsall/mod/1945vietnam.html.
"The Virtual Wall." www.VirtualWall.org/dh/HarrimanAB01a.htm.
"WW1 Propaganda Posters." www.worldwar1gallery.com/propaganda/index2.html.

Oral Histories

Knotts, Joseph B. Interview with United States Marine Corps History and Museums Division Oral History Project, n.d. TTVA.

Shelton, William. Interview with Steve Maxner. Vietnam Archive Oral History Project, 5 June 2000. TTVA.

Shultz, Richard H. (interviewer). *Oral History Interviews with Chiefs, SOG.* Unpublished manuscript. Fletcher School of Law and Diplomacy, July 1997. TTVA.

———. *Oral History Interviews with Military & CIA Officers, Who Served in MACVSOG OP 39 Covert Psychological Operations against North Vietnam.* Unpublished manuscript. Fletcher School of Law and Diplomacy, July 1997. TTVA.

Index